TECHNIQUES
OF SAFETY
MANAGEMENT

TECHNIQUES OF SAFETY MANAGEMENT

A SYSTEMS APPROACH
Third Edition

DAN PETERSEN
Management Consultant
Safety/Organizational Behavior

PROFESSIONAL & ACADEMIC PUBLISHER

215 GREENWICH AVENUE
GOSHEN, NEW YORK

Library of Congress Cataloging-in-Publication Data

Petersen, Dan.
   Techniques of safety management/by Dan Petersen.
      p. cm
   Includes index.
   ISBN 0-913690-14-7 : $39.50 (est.)
   1. Industrial safety. I. Title.
TP55.P37 1989   89-6642
658.4'08-dc20   CIP

To Nadine

Table of Contents

Preface

Since the first edition of this book was published in 1971, a great deal has happened in safety management. The most publicized change was the advent of OSHA, the Occupational Safety and Health Act, which went into effect shortly after the first edition appeared. The most important change, which occurred in spite of OSHA, has been the "growing up" of safety. OSHA emphasized a refocusing of attention on physical conditions, and for a while we in the safety profession followed in this direction. Perhaps our refusal to continue focusing on conditions indicates the maturation of safety management. We know that things do not cause accidents and we have refused to let the passage of a federal law direct us back into nonproductive activities that have proved ineffective in the past.

In recent years we have made tremendous strides toward professionalism. We have defined our scope and functions as professionals and we are now being recognized as certified safety professionals (CSPs) and as registered professional engineers in safety (PEs). Curricula for baccalaureate and graduate degrees in safety have been established, and many people in this country now hold master's degrees and doctorates in the field. Not long ago only a handful of schools taught safety; now many do. We have also developed principles to guide us. This is important, for without principles—without our own body of knowledge— we cannot consider ourselves professionals.

Another indication of our maturation is the development and open discussion of different philosophies of, and approaches to, the practice of safety management. We used to be able to talk only about a traditional approach, based on the "three Es" of education, engineering, and enforcement. We now discuss total loss-control approaches, operational-error approaches, systems safety approaches, psychological approaches, and sociological approaches.

The second edition published in 1978 incorporated many major changes. Those changes, however, were minor compared to the changes in this third edition. The changes in the field of safety that occurred between 1971 and 1978 pale in comparison to the changes since 1978. In almost every facet of safety management, we think and act differently today than we did in 1978. We don't manage the same way today. *In Search of Excellence, The One Minute Manager*, and a host of other books led a management revolution in the early 1980s which will totally change the way safety is managed from this point on.

While the passage of OSHA changed safety in the 1970s (we're not sure for better or for worse), the additional legislation which followed OSHA changed safety even more in the 1980s (we're still not sure whether for better or for worse). In the almost twenty years since OSHA's passage, the agency enforcing the law has changed emphasis more times than years it has existed.

Research since 1978 provides us with new insights and valuable lessons on safety management. The research in accident causation led us from the domino theory to multiple causation to human error reduction concepts. The research in effective safety programming has challenged almost every facet of traditional safety program elements. Behavior research has finally been accepted by safety managers who finally figured out how to use all that "soft" stuff. And using all that psychological "soft" stuff led to breakthroughs–changes in the accident record of 40 and 50 percent reductions almost overnight.

Traditional roles changed. The supervisor began to fade from sight as the "key man" and was replaced by the worker as they key. Middle management became infinitely more important than previously.

The legal world changed markedly. Workers' Compensation concepts began to break down. Executives are being indicted with criminal charges. In the 1960s only serious injuries were looked at by management (and perhaps the insurance carrier). In the 1970s OSHA investigated the serious injury; in the 1980s the state police are often the first on the scene of the serious industrial injury to investigate for criminal negligence and liability.

And of course some real attention has been focused on occupational safety. Bophal, Chernobyl, and Challenger incidents put more focus on safety than ever before. Our corporate world leaders in safety have been fined millions of dollars by OSHA for "inaccurate" record keeping.

These things and many more required a major restructuring of this third edition. It is more limited and also more complete than the first two:

- It is more limited in that it looks at the systems approach to safety programming, with only a light touch on the behavioral.
- It is more complete in that the look at the systems approach is much more comprehensive than in the other editions.

This third edition is limited to the systems, or management approach, because this revision coincided with the revision of *Safety Management–A Human Approach;* so the two have become companion books. The content in each and their interface is shown in Figure I.1 in the introduction.

This book is not a comprehensive book on safety management. I have yet to see any book that is. What this book does do is to look at one facet of safety management–the systems approach . It is broken down into five parts. First, we look at our background, the changes that have happened and are happening, and at our principles. Second, we examine the concepts and techniques needed to manage safety in an organization. Then we look at some of those elements of a safety system; in Part III we discuss the elements that are proactive–what we must and can do before the accident, while Part IV looks at the reactive elements of the safety sytem. Part V reviews the other elements that are also important.

A number of sections of this third edition are totally new–in fact most of the book is new. For instance:

- A new historical perspective to safety is included in Chapter 1.
- Five new principles have been added to the five principles of safety management in the first two editions.
- A new section on management trends is in Chapter 2.
- Clearer role identification is discussed in Chapter 4.
- A close look at what drives performance in an organization is in Chapter 5.
- A summary of our behavioral options is in Chapter 7.
- A complete look at our hazard control options is in Chapter 8.
- There is a new section on using ergonomics.
- A more detailed look at auditing and other more valid performance measures is in Chapter 10.
- More on the analysis of records is included in Chapter 11.

— There is a totally new chapter on the use of Statistical Process Control concepts in safety.
— In Part IV there is information on those important programs adjunct to the safety system that we must not forget—OSHA compliance, fleet programs, product safety, wellness, substance abuse, stress and others.

This edition of the book reflects the fundamental changes in our knowledge and beliefs that have taken place in recent years. The first edition preached line accountability as the key to safety performance, and this edition does the same. However, we know more now about measurement, and since measurement is the key to accountability, it must be discussed in more detail. Effective measurement of the line organization's safety performance is the most important item in the safety program. With effective measurement you get motivation, performance, and results. Without effective measurement, you get none of these.

The ideas put forth in this book are not original. Ideas rarely are—they are transmitted from person to person and they grow in the process. This book attempts merely to record current ideas and assess their value. The thoughts expressed come from many sources and are the product of years of collecting, developing, polishing, teaching, and managing. Certain individuals, however, have contributed heavily to the material presented in the book.

D. A. Weaver helped develop many—perhaps most—of the ideas in this book. He and I worked closely for many years.

Dr. Phil Brereton (Illinois State University) also helped develop some of the ideas in the book. Ray Campbell, former safety analyst for Industrial Indemnity Company, did most of the work on the special-emphasis program described in Appendix C. Ray, with the help of Norm Leaper (Industrial Indemnity), worked on the concepts and wrote the manual for the campaign described in the Appendix and for my discussion of computer techniques, I am indebted to Bill Pope, president of Safety Management Information Systems, Inc., one of the true pioneers in safety management. For the ideas on profiling, I wish to thank Jim Tye, director general of the British Safety Council, and Bunny Matthysen, managing director of the National Occupational Safety Association of South Africa.

And there are many others who are quoted or cited throughout the book. They are mentioned in the applicable chapters. My thanks to all of them.

DAN PETERSEN

Introduction

The intent of this book is to fill some voids left by other safety textbooks and references, to challenge outdated theories on which we sometimes rely, to spotlight new techniques that are available but not often used, and to offer outlines for work in areas we often evade. This text is not a comprehensive reference book for the student or professional; rather, it is intended to challenge thinking. It seeks to add to (and in some cases question) the basic existing textbooks in accident prevention. The reader who needs basic technical safety information will find that hundreds of sources supply it. Once a need has been defined, information can be found. This book has not attempted to repeat all this excellent information.

Techniques of Safety Management is intended for the practicing safety professional in industry, in government, and in insurance. It also is slanted toward the line manager in business who has a particular interest in loss control. Its purpose is not to teach basic safety, but rather to describe areas of weakness in the safety profession and to point to some directions for tomorrow. The secondary audience is the student of industrial safety.

THE SCOPE OF THIS BOOK

The book consists of five parts and the appendices.

Part I. Concepts

This part deals with beliefs concerning safety and is included in the book because of my firm conviction that safety needs a better conceptual framework than it now seems to have.

Part II. Managing Safety

It is the intent of Part II to show how the theory of Part I can be applied in practice. The practical application is discussed in terms of what management and the safety professional can do to accomplish loss control—what procedures and policies management can set up to accomplish its defined safety goals.

Part III. Proactive System Elements

The most effective safety system elements are those that occur before the fact of the accident; what people do every day to try to prevent that accident from happening. Part III looks at the elements that need to be in place.

Part IV. Reactive System Elements

After the accident has happened little good can be achieved unless we have some systems in place to learn from the event. Part IV examples what systems can do this.

Part V. Other System Elements

There are a number of programs that can be put into place that have a major impact on the safety record. These are often overlooked in many organizations. Part V describes some of them.

The Appendices

The appendices present examples of the principles discussed in Parts I to V.

THE STUDENT

I would suggest that the student—whether studying on the undergraduate or graduate level or just starting in industry but new to the safety profession—use this book in addition to other texts, not as a basic text. The basic texts offer the technical knowledge that any safety professional needs; this book offers an additional challenge to thought.

THE SAFETY PROFESSIONAL

In most cases this person did not choose the profession but was chosen by it. In industry it is common for management to select its safety specialist not for abilities and knowledge in the field of safety, but rather because of demonstrated abilities in some other field. One day the future safety professional is tapped on the shoulder by the boss and the job is assigned, either as a full-time position in management or in addition to other duties. The new specialist is usually then in the position of having to learn an entire profession overnight without outside help.

THE INDUSTRY SPECIALIST

It is extremely difficult for the industry specialist starting out in this field. There are few good orientation "schools," although some have been established in recent years. Information about them is available from the National Safety Council, the American Management Association, insurance carriers, and governmental agencies. Some courses are far better than others, and the new safety specialist should look at course content critically and should also check with safety professionals who might provide valuable information before selecting a course.

Even with such assistance as this, the position is largely self-learned and self-defined. This book, directed to the experienced safety professional as well as the neophyte, attempts to provide some help. It has been said over and over that "there is nothing new in safety—we just keep saying the same old things in new ways." This is foolishness—there is more new in safety right now than any one of us can know or comprehend—we have only to look to find it.

In recent years exciting things have happened in safety research and theory. Most of these new developments have not yet been effectively used. This is the job of the experienced safety professional in industry today, a job not performed well in present—to know and comprehend all the new developments in safety.

THE PART-TIME SAFETY SPECIALIST IN INDUSTRY

Most part-time safety people make safety the smaller part of their job. They keep up with new developments in the portion of their work that they consider to be the main part, but they do not keep up in safety.

This book is directed to part-time safety specialists who need to modernize their thinking even more than full-time professionals.

THE LINE MANAGER IN INDUSTRY

Although this book is not aimed at line managers, I hope many of them read it, as well as people in top-management positions. The responsibility—the actual carrying out of all safety work—is a line job. Parts II, III, and IV ought to be of particular interest to line and top management.

THE PRIVATE CONSULTANT

The consultant has a well-defined role, that of adviser to industry management. The consultant analyzes the customer's problems and suggests controls and therefore must be even more up-to-date than the customer. This book is intended for the consultant also.

THE INSURANCE SAFETY ENGINEER

The role of the insurance safety engineer or consultant (or whatever the title may be) is sometimes not as clear as the private consultant's role. Insurance safety engineers should perform the same role as private consultants. Some do; some do not. Some insurance engineers only inspect—some of their customers prefer it this way. Most insurance engineers are competent and well prepared to advise and assist industrial management in its job of loss control. Today insurance engineers themselves choose the level of safety service that they provide—as consultants, engineers, or inspectors. Each engineer ought to determine on what level he or she presently is operating by honestly answering each of the following questions:

1. How am I measured?
   a. By number of recommendations
   b. By number of calls
   c. By results achieved with the customer
2. What do I leave with my clients?
   a. Lists of things that are wrong
   b. Lists of solutions
   c. The results of my analysis
3. How do I schedule my contacts with customers?
   a. Regularly—routinely
   b. On the basis of need
   c. According to current projects
4. What is my means of communication?
   a. A recommendation pad
   b. A letter to the customer
   c. A formal report to management
5. Who is my customer contact?
   a. The plant engineer
   b. The plant superintendent
   c. The manufacturing vice president

6. What do my customers expect from me?
   a. Trouble
   b. Help
   c. Advice
7. If my customer has a poor loss record, what do I do?
   a. Inspect more frequently
   b. Call more often
   c. Define the problems
8. If my customer has an excellent loss record, what do I do?
   a. Inspect less frequently
   b. Call less often
   c. Analyze why
9. What do I believe causes accidents?
   a. Conditions
   b. People
   c. Systems failures
10. What is the key to safety?
    a. Inspection
    b. Employee participation
    c. Management direction
11. Who is the "key person?"
    a. The employee
    b. The supervisor
    c. Management
12. What is the purpose of inspection?
    a. To unearth hazards
    b. To remove hazards
    c. To find symptoms of problems
13. Which is best?
    a. Safety first
    b. Production with safety
    c. Safe production
14. If we reduce frequency, will severity also be reduced?
    a. Always
    b. Usually
    c. The causes may be different
15. What is the single greatest factor we must work on?
    a. Improvement of physical conditions
    b. Line acceptance of responsibility for safety
    c. Better procedures to fix accountability for safety

Each of the above 15 questions has three possible answers. In each instance, answer *a* is typical of the inspector, answer *b* is typical of the engineer, and answer *c* is typical of the consultant. Scoring one, two or three points for each answer as indicated, if you scored 15 to 30, you are behind the times by 30 years; if you scored 30 to 40, you are behind 10 years. A score of 45 says you are up-to-date.

The questions give some insight into the subject matter presented in the following pages. This book is slanted toward the insurance engineer as much as it is toward staff safety specialists in industry.

	COVERED IN	
	TSMIII	**SMHA II**
Historical Overview	✔	✔
Changes that require change in safety	✔	✔
Principles of Safety Management	✔	✔
Managing Safety		
— Role Definition	✔	
— Accountability Systems	✔	
— Measurement	✔	
— Rewards	✔	
Proactive System Elements		
— Climate		✔
— Conflict		✔
— Participation		✔
— Communications		✔
— Selection		✔
— Training		✔
— Positive Reinforcement		✔
— Ergonomics	✔	✔
— Finding Hazards	✔	
— Prioritizing Hazards	✔	
— Removing Hazards	✔	
— Auditing	✔	
— Perception Surveys	✔	
Reactive System Elements		
— Records	✔	
— Analysis of Losses	✔	
— Computers	✔	
— Statistical Process Control	✔	
— Cost Containment Strategies	✔	
Other System Elements		
— OSHA Compliance	✔	
— Fleet Safety	✔	
— Product Safety	✔	
— Subjective Injury		✔
— Stress Management		✔
— Substance Abuse		✔
— Wellness Programs		✔

Exhibit 1.1

THE GOVERNMENT INSPECTOR

One of OSHA's primary problems concerns the function of compliance officers: they have been pure inspectors in the past. This will change as their role enlarges to include helping organizations as well as enforcing the law. As the inspector evolves into an engineer and a consultant, the job will be subject to the criteria that apply to the insurance engineer. I would urge the government safety person to also answer the 15 questions above in grading his or her service to assigned companies.

As this revision is being written, several government enforcement agencies are discussing other ways to help American business achieve safety standards. Some excellent approaches and ideas are being considered. Often the laws, as written, do not allow such help. Our laws, written for the worker's safety, often hinder achievement of their avowed goals. It is hoped that this will change. If government *can* and *will* help companies—through better methods, better approaches, and better people—the worker will benefit, as will the company.

The purpose of this book, then, is to fill some voids in other safety references and to challenge outdated concepts. It will aid the safety professional—whether in industry, insurance, or government—and the student, as an addition to other safety textbooks.

As indicated in the preface, this revision of Techniques in Safety Management is intended to be a companion book to *Safety Management—A Human Approach, Second Edition.* I revised the two together, this concentrating on the management approach and the other on the behavioral approach. Neither approaches can nor should stand alone; they go together. Exhibit I.1 shows how the two books interface, and what each covers to assist the reader.

Part I

Safety Concepts

Chapter One

Historical Framework

The safety movement of today has been a long time hatching. Probably the best way to understand today's safety management direction is to trace its roots briefly. There are a number of excellent writings that do this.

The bicentennial issue (September, 1975) of *Occupational Hazards* magazine contained one of the best descriptions of the history of safety in the United States. Following are excerpts from that publication:

> In 1850 America was a country divided against itself. Sectional strife grew increasingly bitter, then climaxed in the Civil War. The Union survived intact and was soon caught up in a rampaging industrial revolution that by the 1890's made the United States the world's premier manufacturing nation.
>
> Amid the dizzying industrial growth and political turbulence of these crucial fifty years, the safety and health of the growing army of industrial workers were often ignored. But throughout the period and especially toward the end of it, concern for safety was growing—among workers, employers, and a few government officials who were the trailblazers leading the industrial society into the safety movement of the 20th Century. The late 19th Century hatched new industries, and safety techniques were primitive. It's doubtful the early capitalists would have implemented safety techniques even if any had been available. This was the age of exploitation, of profit, whatever the cost in wasted resources or in wasted men. Industrial safety did not appear to be much of a problem in a country with only 1.5 million industrial workers.
>
> In 1850 the factory system was common only to textile making. The nation had just one big steel mill and less than 10,000 miles of railroad track.
>
> By the 1870's industry was a way of life in America. Coal mining and railroading had replaced textiles as the leading industries, and both went through bloody labor troubles in the 1870's. Safety-related issues played a part, though not a central one, in both these confrontations. During the late 1860's miners and railroaders were intensely concerned about the hazards of their jobs and had begun to agitate for better conditions.
>
> Anthracite mining in Eastern Pennsylvania was a notoriously dangerous occupation. Toxic or flammable gases ("stink damp," "fire damp," "rotten gas"), rock falls, and fire were constant threats. In one 7-year period, unmarred by major disasters, 566 miners were killed and 1,655 injured in Schuylkill County, Pennsylvania.
>
> Miners in Schuylkill County, heart of the anthracite country, pushed hard for mine safety legislation throughout the mid 1860's. Finally in 1869, the Pennsylvania legislature passed a mine safety act for "better regulation and ventilation of the mines." It authorized one inspector whose authoirty was limited to Schuylkill County. That same year, 109 miners died in an explosion in nearby Luzerne

3

County, and the lawmakers were shaken by the public outcry into extending inspection to all anthracite producing counties.

Early railroad unions organized not to fight for wages or hours, but for safety and accident compensation. The engineers who organized the Brotherhood of the Footboard in 1863 (later called the Grand International Brotherhood of Locomotive Engineers) wanted accident and death benefits for themselves and their families. The Order of Railway Conductors (1868) and the Brotherhood of Firemen (1873) organized for the same purposes.

Carroll D. Wright was director of the pioneer Massachusetts Bureau of the Statistics of Labor from 1873 until he became first United States Commissioner of Labor in 1886. One of his many innovations came in 1878 when he surveyed on-the-job working conditions.

Notable among his findings: The average worker missed 13½ days a year because of illness; many regarded their jobs as hazardous.

On the Federal front, the Interstate Commerce Act, which formed the Interstate Commerce Commission, passed in 1887.

Thereafter Congress asserted the Federal government's right to regulate safety among firms involved in interstate commerce. Laws passed in 1890 and 1891 ordered coal mines inspected for defective conditions, ventilation, wetting dust, shot fire, escape shafts, furnace shafts, and safety cages. In 1893 came a law requiring safety appliances, among them the automatic coupler on railroads. In 1897 Congress took on the safety of boats propelled by gas, of tugs, and of freighters. Also in 1897 the United States Commissioner of Labor began a study of European workmen's compensation laws. In 1892 an in-plant safety department was organized at the Joliet Works of the Illinois Steel Co.

America had no clear idea of how many workers were being killed and disabled annually in industrial accidents. The Russell Sage Foundation sought to find out by means of its Pittsburgh Survey, so-called because it was confined to greater Pittsburgh. The results were shocking. Sage investigators counted 526 on-the-job fatalities and 500 seriously and permanently disabled workers in one year's time in a single county.

The Pittsburgh Survey is credited with impelling the formation of a safety establishment and the advent of organized safety programs in industry. They point out that survey findings were published in 1907 and set off this chain reaction:

1908.	Congress enacted the nation's first workmen's compensation law covering Federal employees.
1910.	The U.S. Bureau of Mines was established.
1910–15.	State safety and industrial hygiene departments were established in many jurisdictions, and plant inspections by the States were either undertaken or stepped up.
1911.	The American Society of Safety Engineers was established.
1911–15.	New York was the first State to pass a workmen's compensation law (1911). Thirty states enacted workmen's compensation laws.
1913.	The National Council for Industrial Safety, the forerunner of the National Safety Council, was organized.

In 1911, the United States held its first national conference on industrial diseases. In 1912, the U.S. Public Health Service, which had been established in 1902, was expanded to include an occupational health division. In 1916, the American Occupational Medical Association was chartered. Industrial safety programs, modelled along today's lines, had started to be developed. The Safety Establishment was formed. For the first time in our history, occupational health was being systematically addressed. And perhaps, most significant, workmen's compensation systems had been adopted by most States thus providing dollar incentives for industrial accident programs.

Safety knowledge gained in wartime was now incorporated into industrial safety programs. Using an old trench helmet from the war as a model. E. W. Bullard designed the first protective safety cap for industry in 1919. The first industrial respirators appeared shortly after the war when someone had an idea to salvage gas masks left over from the battle and use them for protection in industry. The 1920's saw the first industrial first aid kit.

Organized industrial safety programs became more prevalent during the twenties. Companies began competing for safety awards for working without a lost-time accident. In 1926, Carnegie Steel Co. boasted 2,600,000 man-hours without a lost-time accident; Illinois Steel reported 3,000,000 man-hours; and Clark Thread Co. over 10,000,000 man-hours.

National Safety Council figures for 1926 showed 24,000 industrial fatalities and over 3,000,000 non-fatal personal injuries in industry, at an estimated cost to business of over one billion dollars! The all-industry IFR rate was 31.87.

For those injured on the job, workmen's compensation now provided some relief. In 1917, the Supreme Court reaffirmed the employer's inherent responsibility under workmen's compensation law for an employee's safety and health—regardless of who was at fault. The court pointed out that under workmen's compensation the worker forfeited chances to sue for higher damages (including pain and suffering) in return for a fixed schedule of payments—usually a percentage of his weekly income.

By 1921, 46 states had enacted workmen's compensation laws. Two years later, the National Council on Compensation Insurance was founded and functioned as a rate-making and rate-administering body for all classes of carriers writing workmen's compensation insurance. The cost of work accidents was recognized as a cost of doing business.

While the safety movement moved forward, occupational health lagged. For the most part, occupational diseases were not clearly understood and properly defined. Many occupational diseases were not specified under disease schedules nor could they meet the court's definition of a "traumatic" injury.

The National Safety Council published over 80 pamphlets on safety practices in industry before 1927. Included were tips on protective clothing, goggles, ventilation, housekeeping, materials handling, fire extinguishment, and safety inspections. More were being developed by the American Engineering Standards Committee (AESC) founded in 1918.

In the late 1920's, the AESC shed its committee status and became a full-fledged association: the American Standards Association, predecessor of ANSI. At the time of reorganization, 193 standards had been approved and 160 were in the mill. Among those approved were:

- Safety Code for the Protection of Industrial Workers in Foundries.
- Safety Code for Building Exits.
- Safety Code for the Use, Care, and Protection of Abrasive Wheels.
- Safety Code for the Prevention of Dust Explosions.
- Safety Code for Power Presses and Hand Foot Presses.
- Safety Code for the Protection of Heads and Eyes of Industrial Workers.
- Safety Code for Industrial Sanitation (underway).

The 1930's began with the shock of the stock market crash and ended with the shock of World War II. Between them came the pain of the Great Depression.

In the midst of these hard times, the industrial safety movement made slow, steady gains. In 1930 National Safety Council members reported an accident frequency rate of 17.50. In 1939 the rate was 11.04.

Safety equipment had reached a fair level of sophistication by the 1930's. Versions, though sometimes crude, of most devices used today were available. During the 1930's some of safety's leading theoreticians began their careers, and institutions of lasting importance to the industrial safety movement were founded.

In 1936 came the Walsh-Healey Public Contracts Act. This law established rules under which Federal contracts of $10,000 or more would be granted and retained. The Division of Labor Standards enforced the act.

A key provision of the Walsh-Healey Act stated: "No part of such contract will be performed. . .in any plants, factories, buildings, or surroundings, or under working conditions which are hazardous, unsanitary, or dangerous to the health and safety of employees engaged in the performance of said contract. Compliance with safety, sanitary, and factory inspection laws of the State. . .shall be prima facie evidence of compliance. . . ."

Repeated, willful violation of safety or other standards established in the law could result in blacklisting and loss of all Federal contracts. With the proliferation of government contracts during World War II and the Cold War, Walsh-Healey requirements grew more important to more employers.

The Dominoes

During the early 1930's many safety directors relied on Sidned J. Williams' *The Manual of Industrial Safety,* published in 1927, for professional guidance. Williams' approach was pragmatic, aimed at dispelling the fatalistic idea that accidents couldn't be avoided.

In 1931 H. W. Heinrich published *Industrial Accident Prevention*. This work presented Heinrich's "domino theory of accident prevention." Heinrich pictured five factors in the accident sequence as dominoes. Knock over the first and a chain reaction begins which ends in an injury.

Heinrich also held that 88 percent of all accidents were caused by unsafe acts, while only 10 percent were caused by unsafe conditions. (The remaining 2 percent couldn't be categorized.) The safety director could reduce accidents by controlling unsafe acts with enforcement of safety rules, discipline, and safety education.

In 1940 the world reeled as nations toppled before Hitler's war machine: Poland, Denmark, Norway, the Netherlands, Belgium, and France.

With the war on, the manpower scene shifted almost at once from the labor surplus of the depression to a frantic need for workers. Even as early as 1940, the shift had begun as war orders rolled in. By December 1941 the number of employed had shot from 35 to 41 million. After Pearl Harbor the 7 million still-unemployed found jobs. In addition some 6½ million new wage earners were recruited, including 4 million women and 2 million teenagers. Men near retirement were encouraged to stay on. The work week grew longer, plants ran three shifts, 7 days a week. The U.S. Department of Labor went on in 1942 to publish safety and health standards in what was then known as the "Green Book." It contained what were considered to be the minimum standards. Enforcement was left primarily to the States and compliance with State regulations was deemed prima facie evidence of compliance with Walsh-Healey.

Since nearly everyone in the war economy was now operating under government contract, many companies had to take a hard look at safety and charge a person or committee with direct responsibility for the first time. Programs were now established along formal lines.

Factories were now flooded with women, teenagers, and inexperienced men—a condition that gave rise to large numbers of accidents. In 1939, just before the war boost, the all-industry injury frequency rate was 11.83 according to the National Safety Council. In 1940 the figure rose to 12.52 and rocketed to 15.39 in 1941. From there the figures edged down slightly for the rest of the war as safety programs took the situation in hand. In 1945 the figure stood at 13.39.

One of the more prominent books used as a guide to safety in the forties was *Industrial Safety* by Roland P. Blake. It was first printed in 1943 and continued in wide use through the decade. Blake set out four fundamentals of safety: 1) discover accident causes, 2) control environmental causes, 3) control behavioristic causes, and 4) provide supplementary activities, by which he meant rule books, posters, booklets, movies, contests, meetings, and employee magazines.

Blake recommended such practices as investigating all accidents, keeping records and tabulating all facts concerning accidents, analysis of those facts, inspections of property and equipment, job analysis, and job training.

The National Safety Council was reporting that the all-industry, injury frequency rate had dropped dramatically from 15.39 in 1941 to 9.30 in 1950. There had been a comparable drop in severity for that period. And the dramatic descent continued through the first five years of the 1950's. In 1955, IFR was at 6.96.

Safety directors found their status in industry spiraling. The popular wisdom among them was that virtually all the safety engineering had been done. Enforcement was soft-pedaled, and there was great enthusiasm over enlisting the production supervisor in the day-to-day of safety supervision within his department. Educational courses for supervisors, tailgate talks with employees, and in some companies, joint-management-union safety committees were all part of the safety educational and motivational process.

The three E's—Engineering, Education, and Enforcement—were securely in place, and the emphasis was on Education. The dictum that 85 percent of accidents were caused by unsafe acts and only 15 percent by unsafe conditions had near-universal adherence in safety management circles, and so the emphasis appeared appropriate. Consensus safety standards were rarely challenged.

In the early 1950's psychological testing for accident proneness was enjoying a vogue. Accident proneness as a concept had been bouncing around at least since the 1920's. Safety managers in the late 1950's began putting some of their more favored managerial theorems under the microscope.

A few began asking if, after all, safety managers hadn't handed over too much safety managerial responsibility to the foremen, and wondered about his capacity to deliver in view of his other duties. Some others questioned whether the time-honored 85-15 percent (unsafe acts to unsafe conditions) theory might not owe more to shoddy accident investigations than to scientific observation. A few brave safety directors ventured the belief that safety management was too statistics-conscious and wondered how valid safety statistics really were. Briefs were held for safety directors having a line rather than a staff function.

Meanwhile, the theory on accident proneness was being revised. In 1956, M. S. Schulzinger, in his book, *The Accident Syndrome*, established that the accident prone are a shifting, rather than a fixed, group. All of us at some time in our lives are temporarily accident prone during periods when distress adversely affects our concentration to the task at hand. The search by safety directors for the accident prone tailed off and a treasured old theory was discarded.

For the industrial safety movement the 1960's were tumultuous. Accident frequency rates started to climb. Reappraisal of safety management theorems intensified. And finally, the Federal government began inserting its presence in the fields of industrial safety, occupational health, fire protection, and workmen's compensation.

During the 1960's, accident frequency rates, which had declined steadily since 1926 and had leveled off in the late 1950's, suddenly started an upward trend. Between 1961 and 1969, the National Safety Council's all-industry accident frequency rate climbed from 5.99 to 8.87.

Industrial safety managers disagreed over what was causing the trend. Some blamed larger, less experienced workforces during economic boom conditions, others blamed new machines and production methods, a few even blamed the hippies' outlook on life. Time-honored safety management theories were being widely examined and reappraised.

Industrial safety managers were affected by several other pieces of Federal legislation during the 1960's. In 1968, Congress enacted the Fire Research and Safety Act, which would set up a national fire data collection system, establish a clearinghouse for fire information, and provide funds for fire education courses.

In 1969, Congress passed the Construction Safety Act, which enabled the Secretary of Labor to set safety standards for the construction industry. A year later, these would be incorporated among the standards of the Occupational Safety and Health Act.

Congress also enacted several pieces of environmental legislation in response to a growing environmental movement. The Water Quality Control Act of 1965 established the Federal Water Pollution Control Administration to help the States establish and enforce clean water standards. The Air Quality Act of 1967 established the National Center for Air Pollution Control within the Department of HEW. In December 1970, both agencies became part of the new Environmental Protection Agency.

On an unprecedented scale, more and more safety directors were becoming loss control managers, whose duties encompassed safety, health, fire protection, plant security, workmen's compensation, and insurance.

In this climate, the Federal government began expressing increasing interest in industrial safety—an interest which would culminate in enactment of the Occupational Safety and Health Act of 1970.

In early 1970, the Nixon Administration had set passage of an omnibus safety bill as a major domestic priority, compromises had been struck in the Congress, and momentum was building for enactment. Congressman William A. Steiger (R-Wis.) was shepherding a compromise bill through the House and Senator Harrison A. Williams, Jr. (D-N.J.) was floor managing a bill, dissimilar from Steiger's in some respects, in the Senate. Both bills passed their respective chambers, were reconciled in conference, and on December 29, 1970, President Nixon signed the Occupational Safety & Health Act of 1970 into law to become effective April 28, 1971. The essence of the bill was the promulgation of safety and health standards by the Secretary of Labor that have the force of law, with the establishments to be covered subject to inspection by OSHA compliance officers empowered to cite for violations, propose penalities, and set abatement deadlines.

The OSH Act authorized the establishment of an Occupational Safety and Health Administration under the direction of an Assistant Secretary of Labor—OSHA. A body of ANSI and NFPA consensus standards were recast in legal language and promulgated within 60 days of the OSH Act's effective date.

REFERENCES

Occupational Hazards, "History of the Safety Movement," *Occupational Hazards,* September, 1976.

Petersen, D. *Safety Management—A Human Approach, Second Edition,* Goshen, NY: Aloray, 1988.

Chapter Two

The Changing Scene

It is obvious from the last chapter that we have been living with almost constant change since the middle of the last century. It is also obvious that changes seem to be coming at us more rapidly each year. In all of the changes that have been thrust upon us, some are far more important than others. Certain changes actually require us to adjust how we work and what we do. In today's world it is simply unacceptable for corporate executives to ignore safety as they could in 1900. Today it is dangerous for the executive or the safety professional to take the chances that might have been taken ten years ago. The safety manager of today faces the possibility of malpractice suits, civil liability, even criminal liability. So do the executives.

It is the purpose of this chapter to focus on some of the changes that require us to change. Organizations are managed differently today than they were twenty years ago. Workers respond differently.

There are also changes that take our eye off the real goal of building systems that get results. Exhibit 2-1 suggests some of these. As Chapter 1 indicated, since the first safety programs safety management has taken a number of different directions. While they occupied most of our time then, in retrospect many were less important in terms of results achieved.

If you were a Safety Director in	You would be paying attention to:	And you would overlooked:
1912	New Workers' Compensation Laws	Everything Else
1922	New Standards	Worker Behavior, Management
1932	Occupational Diseases	" " "
1942	Walsh-Healey Compliance	" " "
1947	Training	" " "
1952	Noise Abatement	" " "
1957	Management Policies and Manuals	" " "
1962	Product Liability	" " "
1967	Off Job Safety Programs	" " "
1972	OSHA Compliance	" " "
1977	Systems Safety & Audits	" " "
1982	Hazard Communication Standard	" " "
1983	Wellness Programs	" " "
1984	Asbestos Standard	" " "
1987	Recordkeeping	" " "

Exhibit 2.1 The safety manager's major concerns.

We have historically reacted to the environment of the day and in many cases have not had particularly impressive results. For instance, after the passage of Workers' Compensation laws, after the passage of OSHA or Walsh-Healey, or many other pieces of legislation, we would spend most (or all) of our time reacting to the content of the law. Sometimes our reaction prevented some loss, sometimes it didn't, sometimes it was counter productive. For instance:

— In the mid-1950s, while working for an insurance company, I spent perhaps 80 percent of my time attempting to abate noise problems while the accident record continued unabated.
— In the late 1950s, we believed management had to have a safety policy statement, and we spent great effort getting these written, signed and placed in the manual. Result? No change in anything.
— In the early 1960s, product liability was hot, so we ignored the rest of our safety problems.
— In 1971, all efforts were spent on OSHA compliance, ignoring what we had learned during the previous sixty years about what works and what doesn't.
— In the 1980s, it was hazard communication, asbestos and recordkeeping, as that is what OSHA pushed. The result? Unknown, but probably not much.

Perhaps this is a learning process or perhaps we knew all along that accidents can be prevented with behaviorally and managerially sound approaches. Management research told us what would work in the famous 1920's Hawthorne Studies. Behavioral research has outlined similar things for us throughout the years since. Somehow, we listened to the government and to short term financial fears, rather than solid in-depth research which could have guided us.

How do we figure out what to pay attention to now? And how do we know what not to pay attention to? This chapter focusses on those changes we ought to look at now that should direct our future efforts. Exhibit 2.2 lists those discussed in this chapter and suggests other changes, more publicized, but less important in terms of results to be achieved. In Exhibit 2.2 the left column (What we ought pay attention to) is what this book (and its companion, *Safety Management—A Human Approach, Second Edition*) is all about. The changes in the second and third columns must be recognized, must be dealt with, but will not materially help control losses. OSHA's current thrust (whatever it might be) typically is a good example. Whether it is recordkeeping honesty, hazard communication, or whatever, you must be aware of the direction and do what must be done, but complying typically won't do much for your bottom line or safety record. Similarly, writing safety manuals, buying packaged audits, writing more rules and regulations, will cost a bunch with little pay off. Hiring outsiders to do these things fits into the same category.

CHANGES TO KEEP AN EYE ON

Here are some crucials items to be concerned about:

— *Workers' Compensation Legislative Changes.* As the Workers' Compensation system breaks down, it means more hazard to you personally, to your executive personally, and to your company's financial shape. As indicated in the last chapter, Workers' Compensation originally was the sole recourse for the injured employee. This is no longer true. The likelihood of law suit against you, your executive and the organization is more likely each year.
— *The Federalization of Workers' Compensation.* There are continuing signs that the federal government could take this over. If so, look for liberalization and tremendously escalating costs.

Pay Close Attention To:	Pay Some Attention To:	Just do what you have to on these:
— Assessing the Climate and the Culture of your Company — Defining the Drivers in your Company — Your Performance Appraisal System — The Daily Numbers Game in the Company — Your Legal Liability (civil & criminal) — Your Executive's Legal Liability (civil and criminal) — The Breakdown of the Workers' Compensation System — The Possible Federalization of the Workers' Compensation System — The Changing Span of Control of the Supervisor — The Trend to Employee-run Safety Programs	— This year's OSHA thrust	— Recordkeeping — Hazard Communication — Safety Manuals — Packaged Audit Concepts* — Rules, Regulations * Forget Entirely

Exhibit 2.2 Trends to watch.

- *Continued Civil and Criminal Liability*. It appears we have just started.
- *Stress Claims*. In five years stress claims have grown from a negligible dollar figure under Workers' Compensation to our single largest Workers' Compensation health pay out. It has just begun.
- *Management Style Trends*. How we manage has changed; and continues to change. In many cases our safety programs have become irrelevant.
- *Employee Trends*. Over time the entire workforce has changed and we often haven't recognize the changes. It will continue to change.
- *New Ways to Manage*. Some companies call it high performance systems; some call it autonomous work grouping, some participative management, some survival. It totally changes our safety management concepts.

It is beyond the purview of this book to deal with all of the above crucial issues. Each is extremely important and must be tracked carefully by both the safety professional and the executives. The legislative and legal issues change daily and must be tracked daily. To attempt to deal with these in this book would be impossible, as they change constantly, and anything said on these issues should be said by legal writers, not safety writers. The current trend in stress claims will be mentioned only lightly later; they are, however, dealt with more completely in *Safety Management—A Human Approach, Second Edition*. In addition there are many other books dealing with the subject.

This book does include: management style changes, employee trends, and new ways to manage.

MANAGEMENT STYLE CHANGES

"Safety programs" exist in an organizational environment. This organization has a structure, a culture, a climate, and a personality. One of the fundamental tenets of safety management (see Chapter 3) is that we ought to structure our "safety programs" so that they fit the organization. This has and is posing some major problems for safety people today since the way organizations are managed is a constantly changing thing. The changes in management thinking have been fast and furious in recent years; as each new book is published management changes its direction and philosophy on how to manage.

The transition in management thinking started a number of years ago when managers began to question the dictates of *Classical Management,* or *Scientific Management,* as described in the early years by Frederick W. Taylor. Actually Taylor was reemphasizing the approach to management spelled out even earlier by a German author, Max Weber, who originally structured out the rules of a bureaucracy—those elements that should be in place for an effective organization. Outgrowths of those 19th century German military rules (for that was Weber's orientation) are still evident today in most organizations:

- *Job Descriptions*—the concept that a job can be described on a piece of paper and a belief that anyone going into that job will do what that paper says; that organizations do not change people do.
- *Standards of Performance*—the concept that minimum levels of acceptable performance can be defined and used in appraisals.
- *Organization Charts*—the concept that says we can describe an organization with boxes and lines and that these lines actually describe something—authority networks, communication flows, power, etc.

Following the Classical School of management thinking came the *Human Relations* school of thought. To a degree the Human Relations approach, popular in management until the 1970s and still taught today, was an outgrowth of the misinterpretation of the research of the 1920s, '30s, '40s, and '50s. Many interpreted the research to say "happy" workers are productive workers, thus the role of management was to make workers happy. Actually the research did not say it quite that way, as it is considerably more complicated than that.

The underlying assumptions of the above two schools of management thought were similar, but slightly different. Classical Management is based on the assumption that "everybody is alike—we can get the behaviors we want through manipulation." Thus we used wage incentive plans, piece rates, and other schemes to get more productivity. Human Relations Management has an underlying assumption which says "Everybody's alike—we can get the behaviors we want by making them all happy." Neither assumption had any basis in fact.

In the 1970s the *Contingency School of Management Thinking* emerged, and for the first time, management's underlying assumptions about people were in tune with psychological reality. The assumption behind the contingency theory of management is that "everybody's different," therefore our management style, and how we deal with a worker, must be contingent upon the situation, upon the workers, and upon their needs. How a manager manages must be appropriate to the situation.

For decades, people in the field of management have been involved in a search for a "best" style of leadership. Yet, the evidence from research clearly indicates that there is no single all-purpose leadership style. Successful leaders are those who can adapt their behavior to meet the demands of their own unique situation.

The Situational Leadership theory is based on the amount of direction (task behavior) and the amount of socio-emotional support (relationship behavior) a leader must provide given the situation and "the level of maturity" of the follower or group.

The developers of the *Situational Leadership* theory were Paul Hersey and Ken Blanchard. They explain it like this:

> The recognition of task and relationship as two critical dimensions of a leader's behavior has been an important part of management research over the last several decades. These two dimensions have been labeled various things ranging from "autocratic" and "democratic," to "employee-oriented" and "production-oriented."
>
> For some time, it was believed that task and relationship were either/or styles of leadership and, therefore, could be shown as a continuum, moving from very authoritarian leader behavior (task) at one end to very democratic leader behavior (relationship) at the other.
>
> In most recent years, the idea that task and relationship were either/or leadership styles has been dispelled. In particular, extensive leadership studies at Ohio State University questioned this assumption and proved it wrong.
>
> By spending time actually observing the behavior of leaders in a wide variety of situations, the Ohio State staff found that they could classify most of the activities of leaders into two distinct and different behavioral categories or dimensions. They named these two dimensions "Initiating Structure" (task behavior) and "Consideration" (relationship behavior). Definitions of these two dimensions follow:
>
> *Task behavior* is the extent to which a leader engages in one-way communication by explaining what each follower is to do as well as when, where, and how tasks are to be accomplished. *Relationship behavior* is the extent to which a leader engages in two-way communication by providing socio-emotional support, "psychological strokes," and facilitating behaviors.
>
> In the leadership studies mentioned, the Ohio State staff found that leadership styles tended to vary considerably. The behavior of some leaders was characterized mainly by directing activities for their followers in terms of task accomplishment, while other leaders concentrated on providing socio-emotional support in terms of personal relationships between themselves and their followers. Still other leaders had styles characterized by both task and relationship behavior. There were even some leaders whose behavior tended to provide little task or relationship for their followers. No dominant style appeared. Instead, various combinations were evident. Thus, it was determined that task and relationship are not either/or leadership styles. Instead, these patterns of leader behavior can be plotted on two separate and distinct axes as shown in Exhibit 2.3.
>
> Since research in the past several decades has clearly supported the contention that there is no "best" style of leadership, any of the four basic styles shown in Exhibit 2.5 may be effective or ineffective depending on the situation.
>
> Situational Leadership Theory is based upon an interplay among (1) the amount of direction (task behavior) a leader gives, (2) the amount of socio-emotional support (relationship behavior) a leader provides, and (3) the "maturity" level that followers exhibit on a specific task, function, or objective that the leader is attempting to accomplish through the individual or group (follower(s)).

This Situational Leadership approach of Hersey and Blanchard became very popular in the late 1970s and early 1980s when one of the original authors, Blanchard, restated the basic thesis in different terms in the book, *The One Minute Manager*, which became a best seller. The

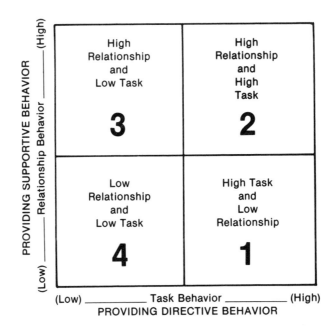

Exhibit 2.3 Four basic leader behavior styles.

one minute approach initiated a wave of instant management solutions in this period. But others cautioned against buying into one simple concept as the way to go. Jay Lorsch cautioned:

> In spite of their potential for wide application, however, these ideas have been only sparingly used. Surely, General Foods, Volvo, and Procter & Gamble have introduced innovations in some factory organizations, and some management organizations have done so as well, but how many other company managements have failed to use the available knowledge? Further, why have the companies that have claimed success in one location or division been so reluctant to apply the ideas in other appropriate places?
>
> One obvious reason seems to be the confusion, skepticism, and controversy about the relevance of these ideas in the minds of many managers. For example: Is participative management a suitable style for all managers? Can job enrichment be applied in a unionized factory? Will managers set realistic goals with a management by objectives program?
>
> One major reason for the difficulties in applying behavior science knowledge has been the interpretation that such ideas are applicable to all situations.
>
> Along with this search for the universal went a tendency to invent specific techniques for applying the theories, which it was argued would lead to improved results in all situations. Examples are management by objectives, autonomous work groups, laboratory training, job enrichment, and participative leadership.
>
> In spite of the rush to simple popular solutions in the last decade, some behavioral scientists have become aware that the universal theories and the techniques they spawned have failed in many situations where they were inappropriate. These scholars are trying to understand situational complexity and to provide managers with tools to analyze the complex issues in each specific situation and to decide on appropriate action. Examples of these efforts are listed in Figure 2.4.

For the reasons indicated above, not all organizations have rushed to implement the situational management concepts, even though for the first time this management theory has assumptions which make psychological sense.

The *Cultural Assessment* approach to management became popular with the publishing of *In Search of Excellence.* This became one of the most popular management books in years. The authors quite simply have looked at the companies in this country that consistently seem to be the best performers, the most effective in terms of bottom line performance. From this in-depth look at the best run companies, they have distilled some keys to effective management—some rather interesting keys. For instance:

— The best companies have "A Bias for Action." They are ready to innovate; they try new things; they even encourage failure.
— The best foster autonomy and enterpreneurship down through the organization.

Author	Publication	Major focus
Fred E. Fiedler	*A Theory of Leadership Effectiveness* (New York: McGraw-Hill, 1967).	Leadership of a work unit
John P. Kotter	*Organizational Dynamics* (Reading, Mass.: Addison-Wesley, 1978).	Organizational change
Edward E. Lawler	*Pay and Organizational Effectiveness: A Psychological View* (New York: McGraw-Hill, 1971).	Employee motivation
Paul R. Lawrence and Jay W. Lorsch	*Organization and Environment* (Division of Research, Harvard Business School, Harvard University, 1967).	Organizational arrangements to fit environmental requirements
Harry Levinson	*Men, Management and Mental Health* (Harvard University Press, 1962).	Employee motivation
Jay W. Lorsch and John Morse	*Organizations and Their Members* (New York: Harper and Row, 1975).	Organizational arrangements and leadership in functional units
Edgar H. Schein	*Career Dynamics: Matching Individual and Organizational Needs* (Reading, Mass., Addison-Wesley, 1978).	Life stage careers, and organizational requirements
Robert Tannenbaum and Warren H. Schmidt	*"How To Choose A Leadership Pattern"* (HBR May-June 1973).	Leadership
Victor H. Vroom and Phillip W. Yellon	*Leadership and Decision-Making* (University of Pittsburgh Press, 1973).	Leadership behavior, for different types of decisions
Joan Woodward	*Industrial Organization: Theory and Practice* (Oxford University Press, 1965).	Organizational design

Exhibit 2.4 Examples of situational frameworks.

— The best have hands-on, value driven executives; they are out on the shop floor. Management is not seen as the supervisor's exclusive task. The supervisor is clearly not the key person.
— The best utilize a simple organizational form, a lean staff indicating fewer safety staff people will get more line management action.
— The best have simultaneous loose-tight properties. They allow great autonomy while adhering to firm tight values.

As our individual organizations begin to shift from one school of thought to another, we must make sure that our safety programs shift in accordance with the current management approaches and styles. Safety programs that "fit" the culture and climate of the organization must shift with the times. Too often our safety programs have remained firmly rooted in the classical school of thought, depending upon standards of managerial performance, corporate safety manuals, rules and regulations to get the job done. While this is perfectly appropriate to 1950 style management, it is most appropriate to 1980 style management thinking.

Management beliefs and approaches have changed tremendously in the last forty years. These changes in and of themselves suggest the absolute need for safety programs to be in tune with the times and in harmony with the way the organization is being managed.

EMPLOYEE TRENDS

Another major reason for reassessing safety system approaches is to ensure the approaches are in tune with the attitudes and values of the constant change of work force. Some of these are discussed in detail in *Safety Management—A Human Approach, Second Edition.* A few thoughts from that are excerpted here:

Individuals are not born with values or ideas about what is right and wrong, good and bad, desirable and undesirable. Rather, values are learned as the result of inter-actions with certain institutions—primarily the family, church, and school—and with lifetime experiences. The reason why the youth of today are different is that they have experienced different institutional influences and lifetime experiences.

These three institutions (family, church and school) have undergone a series of profound changes that have altered their influence on people's values. For instance children used to begin school at the age of five or six. This has changed. Many children now attend pre-school at ages two to four. Thus, children are out of the home at an earlier age and are exposed to people of different social classes. Since about 80 percent of a child's personality is formed before the age of six, this change becomes significant. Also, with the great amount of moving in today's families, children are less exposed to grandparents and other relatives. Most children leave home for college today so the influence of the family on them is removed at an earlier age. Also, divorce rates rose at an alarming rate from the past.

Another factor in the changed family influence is the role of women in the family. In 1970, 41 percent of women were employed compared to 20 percent employed twenty-five years ago. The majority of these women work full-time, and a large percentage of them have children—most commonly between the ages of five and seventeen. Changes have also occurred in the man's job which tends to make changes in his influence in the development of values in his children. Men travel more in business and are away from home more than ever before. For all of these reasons, and others, the family is not as influential in instilling values in its children as it used to be. Therefore, a substantial percentage of youth believe in different things than their parents did. The influence of the church and the school is also different than it used to be. The church has always had much to say about what people should and should not do, and what they should and should not believe in. This influence has declined.

Schools, like the other institutions mentioned above, have also changed in recent times. Of the changes, the major one has been the emergence of different teaching techniques. It used to be that schools taught by means of memorization: memorize dates, states, the Presidents, the multiplication tables. Students would keep at it until they could get it right. Teaching was authoritarian in nature. Now, the trend has been toward different approaches to teaching which encourages students to think through problems for themselves, not to accept anything that they cannot work out themselves, and not to accept things just because they always have been that way. This teaching encourages analysis and thought and the questioning attitudes that the youth of today have about life, the world, and all else.

The institutional influences that have shaped the young people of today are quite different from the institutional influences that shaped the older generations. The other major area that shapes a person's value system is experience. Here too there are major differences between the young people and the older generation.

For instance, growing up during World War II leaves a person with different values than growing up during Vietnam. Being a child during the depression of the 1930s leaves an imprint that is different from the child growing up in the affluent 1960s. The older generation's value system was shaped to a great extent by two major events:

The first critical experience was the Great Depression, particularly the 1929–1933 period. During this unfortunate episode in our history. Gross National Product dropped by 49.6 percent, personal income dropped by 49.8 percent, and unemployment rose from 3.2 to 23.6 percent. Behavioral scientists maintain that this event left an indelible psychological scar on nearly everyone who experienced it. It tended to foster very strong attitudes about risk and speculation and saving money. It also tended to produce a strong sense of company loyalty becuase an individual was sincerely grateful to a company for employing him while one out of every four of his friends were without work.

The second critical event was the Second World War because it touched the lives of nearly everyone in the country in one way or another. Over 16 million Americans served an average of 33 months. Nearly 292,000 Americans died, and over 671,000 were wounded. Despite the initial controversy about the propriety of the United States involvement in the war, it became a unifying device in our country, tending to weld people together in a common cause and promoting and intensifying a feeling of patriotism.

These are the critical events that have shaped the values of the younger generations:

The first thing that must be remembered about this generation is that none of them were alive during the Great Depression, nor the Second World War—nor were they old enough to experience even the Korean War in an adult way. Consequently, the critical experiences of the pre-war generation have about as much impact on the values of the post-war generation as the Boston Tea Party had on the pre-war generation. At best, it is something that they listen to politely.

Instead, most studies indicate that the critical experiences of the post-war generation are as follows:

- The nuclear age
- The civil rights movement
- Poverty
- Space exploration
- The Vietnam War
- Ecology
- University experiences
- The communications revolution

As a result of the critical experiences of the post-war generation, certain key perspectives have emerged. They are:

- Belief that national priorities are inverted
- Distaste for social rigidity
- Concern about institutional rigidity
- Lack of political influence and power
- Loss of self to technology and institutions
- Intolerance of hypocrisy
- Absence of meaningful relationships

These changed values have resulted in changed life styles. Consider some of these changes:

1. *Instant Everything.* It used to be that a person in debt was viewed as a social outcast; debt was considered something to be avoided. We used to save before buying. Our attitude about credit has completely changed. This postponed gratification value has been replaced by an emphasis on instant satisfaction.
2. *Pleasure.* There is a trend toward living in the present rather than the future, toward having fun now rather than later. The call for "doing our own thing" has enjoyed remarkable success.
3. *The Changed Work Ethic.* The Puritan ethic states that work is good. It should be maximized and fun minimized. People used to feel guilty about having fun. Executives justify taking vacations on the grounds that it will allow them to achieve more after getting away. According to Margaret Mead, "People used to live to work, now they work in order to live." The younger people often reject the old work ethic.
4. *Simplification.* Another change is the trend toward simplication. People are interested in products and services that take the work out of life, that allow them to do things quicker and easier, leaving them more time and energy to do what they want.
5. *Safety and Health.* People are becoming increasingly interested in safety and health products, and in their appearance. Interest in safety is at an all time high in this country.
6. *Naturalism.* The young tend to reject artificial behavior. They don't dig "phonies." They like to be themselves and expect others to also be natural.
7. *Personal Creativity.* This trend is apparent on the job (being recognized with job enrichment, participation) and off the job. Everyone wants to "do his own thing."
8. *Reliance on Others.* Problems seem of such a magnitude (pollution, energy, traffic) that people feel they personally can do little. We must rely on "them,"

on government, on business, or others, to solve our problems. The trend is away from self-reliance.

9. *Loss of Confidence in Government, Business, Management,* People were pessimistic about their own personal futures, but were optimistic about the future of the country as a whole. In recent times, people are optimistic about their individual futures, but are very pessimistic about the future of society. There is an important linkage between people's confidence in society and their attitudes toward business. The loss of confidence in institutions becomes, in reality, the loss of confidence in business. This is one of the underlying reasons for the growth of the consumerism movement.

10. *Consumerism.* As we know, this major change is upon us totally in the safety profession.

How do today's workers react to a bureaucratic safety program? It seems obvious that they react rather strongly and negatively to most programs. Where does this leave us? With the need to change.

REFERENCES

Blanchard, K. & S. Johnson. *The One-Minute Manager,* Morrow & Co., New York, 1982.

Business Week, Cover Story. "Changes in Management," *Business Week,* January 20, 1986.

Hersey, P. & K. Blanchard. *Situational Leadership,* Center for Leadership Studies, 1976.

Kollatt, D., & R. Blackwell. "Direction 1980, Changing Life Styles," Columbus, *Management Horizons,* 1972.

Lorsch, J., "Making Behavioral Sciences More Useful," *Harvard Business Review,* March-April, 1979.

Peters, T. & R. Waterman, *In Search of Excellence,* Harper & Row, New York, 1982.

Chapter Three

Safety Management's Principles

As indicated in Chapters 1 and 2, safety and safety management have slowly but certainly been changing over the last century, with the changes becoming more rapid recently. A part of that change has been in what we believe about how accidents are caused and about how they can be controlled.

Our first real causation model was Heinrich's "domino" theory.

DOMINOES: UNSAFE ACTS AND CONDITIONS

From the very beginning we have recognized that certain conditions are involved in accident causation. Thus as soon as it became economically feasible to try to control accidents, we immediately started working on physcial conditions. Then in 1931 Henirich informed us of what now is a painfully obvious and simple truth—that people, not things, cause accidents.

Heinrich stated it this way:

> The occurrence of an injury invariably results from a completed sequence of factors, the last one of these being the injury itself. The accident which caused the injury is in turn invariably caused or permitted directly by the unsafe act of a person and/or a mechanical or physical hazard.

He likened this sequence to a series of five dominoes standing on edge. These dominoes are labeled:

1. Ancestry or social environment.
2. Fault of a person.
3. Unsafe act or condition.
4. Accident.
5. Injury

Most safety people have preached this theory many times. Many of us have actually used dominoes to demonstrate it; as the first one tips, it knocks down the other four unless at some point a domino has been removed to stop the sequence. Obviously, it is easiest and most effective to remove the center domino—the one labeled "unsafe act or condition." This theory is easy to understand; it is also a practical approach to loss control. Simply stated, it says, "If you are to prevent loss, remove the unsafe act or the unsafe condition."

We use this theory in two fundamental areas today: accident investigation and inspection. In accident investigation, almost invariably the forms that we use, or that we give to our supervisors to use, ask that one unsafe act and/or unsafe condition be identified and removed. This, of course, seems very logical, considering the principles expressed by the "domino theory." It is in fact a very practical and pragmatic approach. At least something constructive comes out of an accident. In inspection we are looking for the unsafe act or the unsafe condition so that we can remove it. Here again our approach has been simple, practical, pragmatic, and successful for many years.

The domino theory was intended to be a very practical system for removing the things that cause accidents. Perhaps, however, our interpretation of the domino theory has been too narrow. For instance, using the investigation procedures of today, when we identify an act and/or condition that "caused" an accident, how many other causes are we leaving unmentioned? When we remove the unsafe condition that we identify in our inspection, have we really dealt with *the* cause of a potential accident?

Today we know that behind every accident there lie many contributing factors, causes, and subcauses. The theory of multiple causation states that these factors combine together in random fashion, causing accidents. If this is true, our investigation of accidents ought to identify as many as these factors as possible—certainly more than one act and/or condition.

Let us look briefly at the contrast between the multiple-causation theory and our too narrow interpretation of the domino theory. If we investigate a person falling off a stepladder using our present investigation forms, we identify one act and/or one condition:

The unsafe act: Climbing a defective ladder.
The unsafe condition: A defective ladder.
The correction: Getting rid of the defective ladder.

This would be typical of a supervisor's investigation of this accident under the domino theory.

Let us look at the same accident in terms of multiple causation. Under the multiple-causation theory we would ask what some of the contributing factors surrounding this incident were:

1. Why was the defective ladder not found during normal inspections?
2. Why did the supervisor allow its use?
3. Didn't the injured employee know it should not be used?
4. Was the employee properly trained?
5. Was the employee reminded not to use the ladder?
6. Did the supervisor examine the job first?

The answers to these and other questions would lead to the following corrections:

1. An improved inspection procedure.
2. Improved training.
3. A better definition of responsibilities.
4. Prejob planning by supervisors.

Our narrow interpretation of the domino theory has put blinders on us and has severely limited us in finding and dealing with root causes of accidents.

Symptoms versus Causes

With the above accident, as with any accident, we must find fundamental root causes and remove them if we hope to prevent a recurrence. Defining an unsafe act of "climbing a defective ladder" and an unsafe condition of "defective ladder" has not led us very far toward any meaningful safety accomplishments. When we look at the act and the condition, we are looking only at symptoms, not at causes. Too often our narrow interpretation of the domino theory has led us only to accident symptoms. If we deal only at the symptomatic level, we end up removing symptoms and allowing root causes to remain and thus lead to another accident or some other type of opertional error.

Over the years we have tended to become confused concerning our definition of accident causes. We have considered unsafe acts and conditions to be causes of accidents, and the things that allowed the acts or produced the conditions we have thought of as personal factors or sub-causes. Many times we have looked for the "proximate cause" of an accident, as some laws and codes use this terminology.

One fact seems clear: When we look only deep enough to find the act or condition, we deal only at the symptomatic level. This act or condition may be the "proximate cause," but invariably it is not the "root cause." To effect permanent improvement, we must deal with root causes of accidents.

THE MANAGEMENT SYSTEM AND ACCIDENTS

Root causes often relate to the management system. They may be due to management's policies and procedures, supervision and its effectiveness, or training. In our example of the defective ladder, some root causes could be a lack of inspection procedures, a lack of management policy, poor definition of responsibilities (supervisors did not know they were responsible for removing the defective ladder), and a lack of supervisory or employee training.

Root causes are those whose correction would effect permanent results. They are those weaknesses which not only affect the single accident being investigated but also might affect many other future accidents and operational problems.

The root causes of accidents (weaknesses in the management system) are also the causes of other operational problems. Though this fact is not immediately obvious, the more we consider it, the more obvious it becomes. Consider, for instance, how often our safety problems stem from lack of training—and how often our quality problems also stem from this same lack of training. Or consider how poor selection of employees creates safety problems and other management problems. The fundamental root causes of accidents are also fundamental root causes of many other management and operational problems.

This is not a new thought. Heinrich expressed it in a slightly different way:

> Methods of most value in accident prevention are analogous with the methods for
> the control of quality, cost, and quantity of production.

We believe today that this statement expresses some of Heinrich's best thinking, but for some reason this is the one principle presented by Heinrich that safety people have *not* lived by in the past.

Consider the difference between the way we handle safety and the way we handle quality, cost, and quantity of production. How does management accomplish other things that it wants—a certain level of production, for instance? When management officials decide they want a certain level of production, they first tell somebody what level they want, or they set a policy and definite goals. Then they say to someone, "You do it." They define responsibility. They say, "You have my permission to do whatever is necessary to get this job done." They grant authority. And finally they say, "I'll measure you to see whether you are doing it." They fix accountability. This is the way management motivates its employees to do what it wants in production, in cost control, in quality control, and in all other areas except safety.

In safety, industry has taken a quite different tack. Management officials have not effectively used the above tools of communication, responsibility, authority, and accountability. Rather, they have chosen committees, safety posters, literature, contests, gimmicks, and a raft of other things that they would not consider in quality, cost, or production. In those other areas management has not worried too much about motivating people. There, management has decided what it wants and then has made sure that it gets exactly that. In safety, we are in the ludicrous position of pleading for management support, instead of advising management on how to better direct the safety effort to attain its specified goals.

The concept that the root causes of accidents are weaknesses in the management system and thus are causes of other operational problems is discussed further throughout this book.

Severity versus Frequency

Most of us in safety work have believed in a predictable relationship between the frequency of accidents and their severity. Many studies have been made over the years to determine this relationship, with varying results. Originally, studies seemed to show that for every serious accident we can expect to suffer 29 minor accidents and 300 no-injury accidents or "near misses." This relationship has been discussed and taught for many years. Most recent studies show a similar relationship but with completely different numbers.

Common sense dictates totally different relationships in different types of work. For instance, the steel erector would no doubt have a different ratio from that of the office worker. This very difference might lead us to a new conclusion. Perhaps circumstances which produce the severe accident are different from those which produce the minor accident.

Safety professionals for years have been attacking frequency in the belief that severity would be reduced as a by-product. As a result, our frequency rates nationwide have been reduced much more than our severity rates have. One state reported a 33 percent reduction in all accidents between 1965 and 1975, while during the same period the number of permanent partial disability injuries actually increased. This state is typical of others, and its figures are typical of our national figures. In the period 1926 to 1967, the national frequency rate improved 80 percent, while the permanent partial disability rate improved only 63 percent. (This could, of course, be due partly to the changing definition of a partial disability, as laws vary.)

If we study mass data, we can readily see that the types of accidents resulting in temporary total disabilities are different from the types of accidents resulting in permanent partial disabilities or in permanent total disabilities or fatalities. For instance, the National Safety Council (Exhibit 3.1) shows that handling materials accounts for 25 percent of all temporary total disabilities and for 21 percent of all permanent partial injuries but for only 6 percent of all permanent total injuries and fatalities. Electricity accounts for 13 percent of all permanent totals and fatalities but for a negligible percentage of temporary totals and permanent partials. These percentages would not differ if the causes of frequency and severity were the same. They are not the same. There are different sets of circumstances surrounding severity. Thus if we want to control serious injuries, we should try to predict where they will happen. Today we can often do just that.

The Key Person

Another fundamental tenet of safety has stated that the supervisor is the key in accident prevention. This seems axiomatic in our thinking. The supervisor is the person between manage-

Type of accident	Temp. Total	Perm. Partial	Perm. Total
Handling Materials	24.3%	20.9%	5.6%
Falls	18.1	16.2	15.9
Falling Objects	10.4	8.4	18.1
Machines	11.9	25.0	9.1
Vehicles	8.5	8.4	23.0
Hand Tools	8.1	7.8	1.1
Electricity	3.5	2.5	13.4
Other	15.2	10.8	13.8

Exhibit 3.1 Accident types.

ment and the workers who translates management's policy into action. The supervisor has eyeball contact with the workers.

Is this the key person? In a way, yes. However, although the supervisor is the key to safety, management has a firm hold on the keychain. It is only when management takes the key in hand and does something with it that the key becomes useful. Safety professionals have sometimes used the keyperson principle to focus their efforts on frontline supervision, forgetting that the supervisor will do what the boss wants, not what the safety specialist preaches.

THE FUNCTION OF SAFETY

It is only in recent years that most safety professionals have been able to define their role in the safety work that is being accomplished. What they do has changed and will continue to change as our concepts and principles continue to evolve. If permanent results can be effected by dealing with root causes, safety professionals must learn to work well below the sympotomatic level.

If accidents are caused by management system weaknesses, safety professionals must learn to locate and define these weaknesses. They must evolve methods for doing this. This may or may not lead them to do the things they did in the past. Inspection may remain one of their tools—or it may not. Investigation may be one of their tools—or it may not. Certainly safety professionals must use new tools and modernize old tools, for their direction is different today—their duties must also be different.

In the safety profession, we started with certain principles that were well explained in Heinrich's early works. We have built a profession around them, and we have succeeded in progressing tremendously with them. And yet in recent years we find that we have come almost to a standstill. Some believe that this is because the principles on which our profession is built no longer offer us a solid foundation. Others believe that they remain solid but that some additions may be needed. Anyone in safety today ought to at least look at that foundation—and question it. Perhaps the principles discussed here can lead to further improvements in our approach and further reductions in our record.

THE TEN BASIC PRINCIPLES OF SAFETY

Principle 1

An unsafe act, an unsafe condition, and an accident are all symptoms of something wrong in the management system. We know that many factors contribute to any accident. Our thinking, however, has always suggested that we select one of these as the "proximate" cause of the accident or that we select one unsafe act and/or one unsafe condition. Then we remove that condition or act.

The theory of multiple causation suggests, however, that we trace all the contributing factors to determine their underlying causes. For instance, the amputation of a finger in a power press might start with an act (putting the hand under the die) and a condition (an unguarded point of operation). Tracing back from this point might lead, however, into an inquiry concerning why the operator was selected for the job, why the operator was poorly trained, why the supervisor allowed the act, why the supervisor was poorly selected and trained, why the maintenance on the press was poor, or why the policy of management allowed an unguarded press.

This principle suggests not that we boil down our findings to a singe factor but rather that we widen our findings to include as many factors as seem applicable. Hence, every accident opens a window through which we can observe the system and the procedures. Different accidents would unearth similar things that might be wrong in the same management system. Also, the theory suggests that besides accidents, other kinds of operational problems result from the same causes. Production tie-ups, problems in quality control, excessive costs, customer complaints, and product failures, have the same causes as accidents. Eliminating the causes of one organizational problem will eliminate the causes of others.

If we were actually to utilize this theory, we would redesign our accident investigation procedures in a way that would enable us to identify as many contributing factors as possible in any single incident. Most of the changes would be directed at improving the organizational system—not at finding fault.

Identifying the act, the condition, and the accident as a starting point to learn why the act and the accident were allowed to happen and why the condition was permitted to exist will lead to effective loss control. We view the accident, the act, and the condition as symptoms of something wrong in the system. Then we try to identify what is wrong in the organizational system that allows an unsafe act to be performed and an unsafe condition to exist.

If in the instance of the finger amputation, we said merely that the cause was the unguarded press, the correction would consist of putting a guard on the press. However, we would have treated only the symptom, not the cause, for tomorrow the press would again be unguarded. Only when we diagnose causes and treat them do we effect permanent control. The function of the safety professional, then, is similar to that of a physician who diagnoses symptoms to determine causes and then treats those causes or suggests appropriate treatment.

Principle 2

We can predict that certain sets of circumstances will produce severe injuries. These circumstances can be identified and controlled. This principle states that we can predict severity of accidents under certain conditions and thus turn our attention to severity per se instead of merely hoping to reduce it by attacking frequency.

Statistics show that we have been only partially successful in reducing severity by trying to control frequency. National Safety Council figures show an 80 percent reduction in the frequency rate over the last 40 years. The same source shows that during this period there has been only a 72 percent reduction in the severity rate, a 67 percent reduction in the fatal and permanent total rate, and a 63 percent reduction in the permanent partial disability rate.

A number of recent studies suggest that severe injuries are fairly predictable in certain situations. Some of these situations involve:

- *Unusual, nonroutine work.* This includes the job that pops up only occasionally and the one-of-a-kind solution. Nonroutine work may arise in both production and nonproduction departments. The normal controls that apply to routine work have little effect in the non-routine situation.
- *Nonproduction activities.* Much of our safety effort has been directed to production work. However, there is a tremendous potential exposure to loss associated with nonproduction activities such as maintenance and research and development. In these types of activities most work tends to be nonroutine. Since it is nonproduction work, it often does not get much attention regarding safety, and usually the work is not carried out according to standardized procedures. Severity is predictable here.
- *Sources of high energy.* High energy sources can usually be associated with severity. Electricity, steam, compressed gases, and flammable liquids are examples.
- *Certain construction situations.* Included are high-rise erection, tunneling, and working over water. (Actually, construction severity is an amalgam of the previously described high-severity situations.)
- *Many lifting situations.* Since the 1960s the back strain has been our largest problem in both frequency and cost.
- *Repetitive motion situations.* In the early 1980s industry has faced numerous tendonitis and karpol tunnel syndrome claims. Many result in surgery and considerable lost time.
- *Psychological stress situations.* The mid 1980s noted the emergence of a whole new safety problem—injuries, illnesses and claims resulting from employees exposed to stressful environments. Since these often result in much lost time, and in long term, even permanent disability, they tend to be very costly.
- *Exposure to toxic materials.* Since Bophal toxic materials are recognized as a contributing factor to serious long term liabilities.

In recent years, severity has taken on a whole new meaning. Incidents like the chemical fumes escaping at Bophal, the nuclear disaster at Chernobyl, the shuttle disaster of 1986, and chemical plant explosions have focussed unprecedented attention on the severe incident.

Principle 3

Safety should be managed like any other company function. Management should direct the safety effort by setting achievable goals and by planning, organizing, and controlling to achieve them. Perhaps this principle is more important than all the rest. It restates the thought that safety is analogous with quality, cost, and quantity of production. It also goes further and brings the management function into safety (or, rather, safety into the management function). The management function by definition should include safety, but in practice it has not done so. Management has too often shirked its responsibility here, has not led the way, and at best has given "support."

We in the safety profession are often partly at fault. We have not made management lead the way—only asked for (or hoped for) management support. We have not demonstrated that safety is a management responsibility requiring goal setting, proper planning, good organization, and effective management-oriented controls. At times, we have not even spoken management's language. It is only when management manages safety as it does other functions that we will see results to reverse our present trends.

Inherent in this principle is the fact that safety is and must be a line function. As management directs the effort by goal setting, planning, organizing, and controlling, it assigns responsibility to line managers and grants them authority to accomplish results. The word "line" here refers not only to first-level supervisors but also to all management-level supervisors above those on the first level, up to the top. Part II of this book is devoted to management's role in safety.

Principle 4

The key to effective line safety performance is management procedures that fix accountability. Any line manager will achieve results in those areas in which he or she is being measured by management. The concept of "accountability" is important for this measurement, and the lack of procedures for fixing accountability is safety's greatest failing. We have preached line responsibility for many years. If we had spent this time devising measurements for fixing accountability of line management, we would still be achieving a reduction in our accident record.

A person who is held accountable will accept the given responsibility. In most cases, someone who is not held accountable will not accept responsibility—he or she will devote the most attention to the things that management is measuring: production, quality, cost, or any other area in which management is currently exerting pressure.

This principle is extremely important for the implementation of principle 3. Principle 4, in effect, makes principle 3 work. Several chapters in Part II are devoted to accountability. There we discuss its importance in more detail and present some specific techniques for fixing accountability to the line.

Principle 5

The function of safety is to locate and define the operational errors that allow accidents to occur. This function can be carried out in two ways: (1) by asking why accidents happen— searching for their root causes—and (2) by asking whether certain known effective controls are being utilized. The first part of this principle is borrowed from the ideas of W. C. Pope and Thomas J. Cresswell as put forth in their article entitled "Safety Programs Management," in the August 1965 issue of the *Journal of the American Society of Safety Engineers.* This article defines safety's function as locating and defining operational errors involving incomplete decision making, faulty judgments, administrative miscalculations, and just plain poor management practices.

Pope and Cresswell suggest that to accomplish our purposes, we in safety would do well to search out not what is wrong with people but what is wrong with the management system that allows accidents to occur. This thinking is borne out in the ASSE publication entitled "Scope and Functions of the Professional Safety Position," where the position of safety is diagrammed as in Exhibit 3.2.

In further describing the four major areas under identification and in appraising the problem, the publication states that it is the function of safety to:

> Review the entire system in detail to define likely modes of failure, including human error, and their effects on the safety of the system. . . .

> Identify errors involving incomplete decision making, faulty judgment, administrative miscalculation and poor practices. Designate potential weaknesses found in existing policies, directives, objectives, or practices.

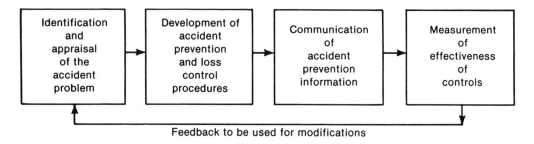

Exhibit 3.2 The professional task.

This new concept directs the safety professional to look at the management system, not at acts and conditions.

The second part of principle 5 suggests a two-pronged attack: (1) tracing the symptom (the act, the condition, the accident) back to see why it was allowed to occur and (2) looking at the company's system (procedures) and asking whether certain things are being done in a predetermined manner that is known to be successful.

We have discussed item 1 above: tracing the symptom back to its underlying root causes. The accident described earlier illustrated the difference between removing the symptom (the act and the condition) and asking why the act was allowed to be performed and the condition was permitted to exist (tracing back to root causes).

We have not yet discussed item 2 above: looking at the system to determine what controls are in effect or asking whether the company has the controls that are needed for the exposures it faces. Much of the rest of this book pertains to this second approach to safety. It suggests that you analyze your company by asking such questions as:

— What is management's safety policy?
— How is your company organized?
— What is the function of your safety department?
— How is safety niched into the organization?
— What are the relationships between staff people?
— How are line safety responsibilities defined?
— How does management hold people accountable?
— How are supervisors measured in safety?
— How are employees selected?
— How are employees trained?
— How are supervisors motivated?
— How is management motivated?

In discussing some of these topics, there are outlines for thinking so that the readers may take an objective look at their own companies—and analyze their own needs.

But before we begin our discussion of managing the safety function and motivating for safety, let us define once more the role of safety—as it fits in an industrial enterprise.

For many years the slogan of the National Safety Council was "safety first," and many professionals believed and preached this. Today we realize that we really do not want "safety first" any more than we want "safety last." In other words, we do not want to think of safety as being separate from the other aspects of production. Obviously, we want effective production first, but we want it to be accomplished in such a manner that no one is hurt and losses are minimized.

In the past, safety professionals were oriented to safety programs for their companies—the aim was to superimpose a safety program on the organization. Today, safety professionals realize that what is really needed is "built-in" safety, "integrated" safety, not an artificially introduced program. Safety must be an integral part of a company's procedures. That is:

> We do not want production and a safety *program,* or production *and* safety, or production *with* safety—but, rather, we want *safe production.*

We can better understand the concept of integrated safety if we discuss how safety fits into production. The goal of management is efficient production—production which maximizes profit. To obtain this goal, it has two basic resources: (1) employees and (2) facilities, equipment, and materials (Exhibit 3.3). To the company's personnel, management applies such influences as training, selection and placement processes, employee health programs, and employee relations practices. The resource of facilities, equipment, and materials is influenced by maintenance, research, and engineering.

These influences and basic resources are brought together through various procedures (Exhibit 3.4). The function of safety is to:

— Build safety into these procedures.
— Continually audit the carrying out of these procedures to ensure that the controls are adequate.

Safety accomplishes these tasks by asking *why* certain acts and conditions are allowed and asking *whether* certain known controls exist.

Safety is not a resource; it is not an influence; it is not a procedure; and it certainly is not a "program." Rather, safety is a state of mind, an atmosphere that must become an integral part of each and every procedure that the company has. This, then, is what we mean by "built-in" or "integrated" safety. It is the only brand of safety that is permanently effective.

Since any accident, unsafe act, or unsafe condition indicates a system failure, the safety professional must become a systems evaluator. The following chapters are intended to aid in this task.

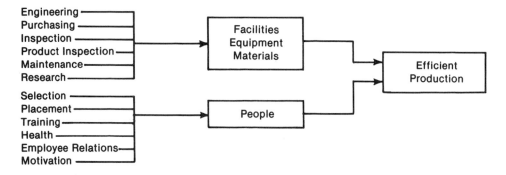

Exhibit 3.3 Management's process for attaining efficient production.

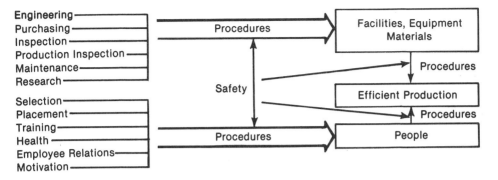

Exhibit 3.4 The role of safety in efficient production.

Principle 6

The causes of unsafe behavior can be identified and classified. Some of the classifications are Overload (the improper matching of a person's capacity with the load); Traps, and the Worker's Decision to Err. Each cause is one which can be controlled.

Principle 6 summarizes the causation model in Exhibit 3.5. It suggests that management's task with respect to safety is to identify and deal with the causes of unsafe behavior, not the behavior itself. The model was developed by the author and explained in the book, *Human Error Reduction,* published in 1981. The model suggests that human error (unsafe behavior) is involved in every accident (incident) and that there are many reasons behind this behavior, not just four. It further suggests that these reasons can be identified and classified and are caused by specific things. The model suggests that there are many actions that managers can take to reduce the likelihood of the unsafe act beyond the two (or three) suggested originally by Heinrich. We simply are not limited to education and enforcement. Actually, we know today that education and enforcement are the two *least* effective things that we can do to improve behavior of the worker.

The model states that an injury or other type of financial loss to the company is the result of an accident or incident. The incident is the result of a systems failure and a human error. Systems failure includes most subjects traditional safety management might include, such as

— Management's statement of policy on safety.
— Who is designated as responsible and to what degree?
— Who has what authority, and authority to do what?
— Who is held accountable? How?
— How are those responsible measured for performance?
— What systems are used for inspections to find out what went wrong?
— What systems are used to correct things found wrong?
— How are new people oriented?
— Is sufficient training given?
— How are people selected?
— What are the standard operating procedures? What standards are used?
— How are hazards recognized?
— What records are kept, and how are they used?
— What is the medical program?

These items are discussed in the following chapters.

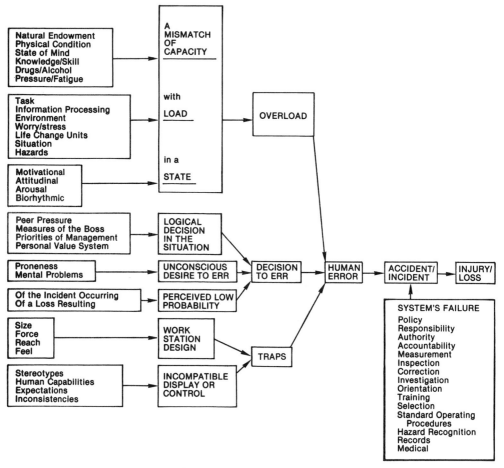

Exhibit 3.5 The Causation Model.

The second and always present aspect and cause of an incident or accident is human error. Human error results from one or a combination of three things: (1) overload, which is defined as a mismatch between worker capacity and the load we place on them in given state; (2) a decision to err; and (3) traps that are left for the worker in the workplace.

Overload. The human being cannot help but err when given a heavier work load than he or she has the capacity to handle. This overload can be physical, physiological, or psychological. To deal with overload as an accident cause, we have to look at an individual's capacity, work load, and current state. To deal with overload as an organizational cause, we have to identify the safety controls available for dealing with capacity, work load, and state.

A human being's capacity refers to physical, physiological, and psychological endowments (what the person is naturally capable of); current physical condition (and physiological and psychological condition); current state of mind; current level of knowledge and skill relevant to the task at hand; and temporarily reduced capacity owing to drugs or alcohol use, pressure or fatigue.

Load refers to the task and what it takes physically, physiologically, and psychologically to perform it. Load also refers to the amount of information processing the person must do; the working environment; the amount of worry, stress, and other psychological pressure; and the person's home life and total life situation. Load refers to a person's work situation per se and to hazards he or she faces daily at work. State refers to a person's level of motivation, attitude, arousal, and to his or her biorhythmic state.

Decision to Err. In some situations it seems logical to the worker to choose the unsafe act. Reasons for this might be the following:

— Because of the worker's current motivational field it makes a lot more sense to operate unsafely than safely. Peer pressure, the boss's pressure to produce, and many other reasons might make unsafe behavior seem preferable.
— Because of the worker's mental condition, it serves him or her to have an accident. (This is called *proneness.*)
— Because the worker just does not believe he or she will have an accident. (This is called *low perceived probability.*)

Traps. The third cause of human error is the traps that are left for the worker. Here we are talking primarily about human factor concepts. One trap is incompatibility. The worker errs because the situation he or she works in is incompatible with the worker's physique or with what he or she is used to. The second trap is the design of the workplace–it is conducive to human error.

Traps are examined in some depth in Chapter 9 on Using Ergonomics. Overload and Decision to Err are looked at briefly in Chapter 7 on Changing Behavior, and examined in depth in *Safety Management–A Human Approach, Second Edition.*

Principle 7

In most cases, unsafe behavior is normal human behavior; it is the result of normal people reacting to their environment. Management's job is to change the environment that leads to the unsafe behavior.

Principle 7 is an extension of Principle 6. It suggests that when people act unsafely they are not dumb, are not careless, are not children that need to be corrected and changed to make them "right." Rather, it suggests that unsafe behavior is the result of an environment that has been constructed by management. In that environment, it is completely logical and normal to act unsafely.

Principle 8

There are three major subsystems that must be dealt with in building an effective safety system: (1) the physical, (2) the managerial, and (3) the behavioral.

Principle 7 reemphasizes that our task is to change the physical and psychological environment that leads people to unsafe behavior.

The role of safety was defined in four elements: (1) analysis, (2) developing systems of control, (3) communicating those systems to the line organization implementing them, and (4) monitoring the results achieved. Principle 8 suggests these four must include systems in all three areas: (1) physical condition control, (2) the management system, and (2) the behavioral environment.

Traditionally "safety programs" dealt with the physical environment. Later we looked at management and attempted to build management principles into our safety programs. Today we recognize the need also to look at the behavioral environment—the climate and culture in which the safety system must live.

Principle 9

The safety system should fit the culture of the organization.

As indicated earlier, times have changed. The way we manage has changed markedly. And the way we manage safety must also change to be consistent with other functions (Principle 9). To strive for an open and participative culture in an organization and then use a safety program that is directive and authoritarian simply does not work.

NEW PRINCIPLES OF SAFETY MANAGEMENT

1. An unsafe act, an unsafe condition, an accident: all these are symptoms of something wrong in the management system.
2. Certain sets of circumstances can be predicted to produce severe injuries. These circumstances can be identified and controlled:

   Unusual, nonroutine   High energy sources
   Nonproductive activities   Certain construction situations

3. Safety should be managed like any other company function. Management should direct the safety effect by setting achievable goals and by planning, organizing, and controlling to achieve them.
4. The key to effective line safety performance is management procedures that fix accountability.
5. The function of safety is to locate and define the operational errors that allow accidents to occur. This function can be carried out in two ways: (1) by asking why — searching for root causes of accidents, and (2) by asking whether or not certain known effective controls are being utilized.
6. The causes of unsafe behavior can be identified and classified. Some of the classifications are overload (improper matching of a person's capacity with the load), traps, and the worker's decision to err. Each cause is one which can be controlled.
7. In most cases, unsafe behavior is normal human behavior; it is the result of normal people reacting to their environment. Management's job is to change the environment that leads to the unsafe behavior.
8. There are three major subsystems that must be dealt with in building an effective safety system: (1) the physical, (2) the managerial, and (3) the behavioral.
9. The safety system should fit the culture of the organization.
10. There is no one right way to achieve safety in an organization; however, for a safety system to be effective, it must meet certain criteria. The system must:
    1. Force supervisory performance.
    2. Involve middle management.
    3. Have top management visibly showing their commitment.
    4. Have employee participation.
    5. Be flexible.
    6. Be perceived as positive.

Exhibit 3.6 New principles of safety management.

Principle 10

There is no one right way to achieve safety in an organization; however, for a safety system to be effective, it must meet certain criteria. The system must:

— Force supervisory performance.
— Involve middle management.
— Have top management visibly showing their commitment.
— Have employee participation.
— Be flexible.
— Be perceived as positive.

These ten principles, restated in Exhibit 3.6, are the underlying theme of this entire book.

These are the new principles that our reexamination has developed. Many of these are as yet only principles without tested techniques available for implementation while others have well-tested methods of implementation available.

Many people in safety management today believe that we are no longer progressing because those principles which Heinrich enumerated and which we have relied on are suspect. Many people believe that some of the other principles which Heinrich stated (such as his belief in unsafe acts as a primary problem) are very true, but that we have ignored them. And even though we may believe that accidents are caused by people, we still concentrate a great deal of our time and energy on the environment—the physical surroundings and conditions.

While this book concentrates on the systems approach to safety, which includes the physical and the managerial, the whole thrust is that all systems must be constructed with behavioral soundness.

REFERENCES

American Society of Safety Engineers. "Scope and Function of the Professional Safety Position," 1966.

Blake, R. P. *Industrial Safety,* Prentice-Hall, Inc., Englewood, NJ, 1943.

Heinrich, H., D. Petersen, & N. Roos. *Industrial Accident Prevention, 5th Edition,* McGraw-Hill, New York, 1980.

National Safety Council. "Accident Facts," 1976.

Petersen, D. *Human Error Reduction and Safety Management,* Aloray, Goshen, NY, 1980.

Petersen, D. *The OSHA Compliance Manual,* McGraw-Hill, New York, 1975.

Pope, W. C., and T. J. Cresswell. "Safety Programs Management," *Journal of the ASSE,* August 1965.

Part II

Managing Safety

Chapter Four

Roles

In Chapter 3 we looked at ten principles of safety management. Principle 3 suggested that we should manage safety as we manage anything else, through defining roles, defining tasks, measuring performance of those tasks, and rewarding appropriately.

In this chapter we will define roles. In the next chapter we'll look at the systems that drive performance, and in Chapter 6 we will look at measurement in safety.

ROLE DEFINITION

The roles of several layers of the line hierarchy are as follows:

1. The role of the first line supervisor is to carry out some agreed upon tasks to an acceptable level of performance.
2. The roles of middle and upper management are to:
   a. ensure subordinate performance.
   b. ensure the quality of that performance.
   c. personally engage in some agreed upon tasks.
3. The role of the executive is to visibly demonstrate the priority of safety.
4. The role of the safety staff is to advise and assist each of the above.

THE SUPERVISORY ROLE

The supervisor's role as defined above is relatively singular and simple—to carry out the tasks agreed upon. What are those tasks? While it may depend upon the organization, the tasks might fall into these categories:

TRADITIONAL TASKS	NON-TRADITIONAL TASKS
— Inspect	— Give Positive Strokes
— Hold Meetings	— Ensure Employee Participation
— Perform One-on-ones	— Do Worker Safety Analyses
— Investigate Accidents	— Do Force-Field Analyses
— Do Job Safety Analyses	— Assess Climate and Priorities
— Make Observations	— Crises Intervention
— Enforce Rules	
— Keep Records	

In addition to the above they no doubt will be responsible for certain day-to-day actions not easily spelled out or measured (following standard operating procedures (SOPs)).

How well do they understand their responsibilities? You may wish to find out by having them fill out the checklist in Exhibit 4.1. See what agreement your supervisors reach on their responsibilities, then on their perception of their authority (Exhibit 4.2), and finally on their perception of how they are now being measured for their safety performance (Exhibit 4.3).

Usually this exercise is well worth the time. It tells you just exactly how well your supervisors understand their role and their tasks, how much authority they think they have, and how closely they think they are being measured. In this exercise we invariably find much confusion in the perception of supervisor responsibilities. They often simply do not know what is expected of them, particularly in safety. Almost always they have no idea of the extent of their authority, and usually they are unclear as to how their performance is being measured, again particularly in safety.

As mentioned earlier, the role of the supervisor is to engage regularly (daily) in some pre-defined tasks. What are those tasks? There are many that have been traditionally perceived as crucial and others thought today to be very meaningful while somewhat non-traditional. For the most part all of this is based mostly on tradition and opinion, not on research. As the various tasks are discussed in this chapter, we'll attempt to describe the tasks and where possible share whatever research there is on that task.

What are the crucial tasks for a supervisor? What *must* a supervisor do regularly to achieve safety success? The answer to this is that we don't know. The research seems to say that there are no crucial "must-do" tasks. The research seems to say that it doesn't make any difference what a supervisor does, as long as something is done regularly, daily, to emphasize the importance of safety.

We will look at both traditional and non-traditional tasks.

THE TRADITIONAL SUPERVISORY TASKS

Investigating

Accident investigation is a device for preventing additional accidents. According to the National Safety Council's *Accident Prevention Manual for Industrial Operations,* the principal purposes of an accident investigation are:

- To learn accident causes so that similar accidents may be prevented by mechanical improvement, better supervision, or employee training.
- To determine the "change" or deviation that produced an "error" that, in turn, resulted in an accident (systems safety analysis).
  - To publicize the particular hazard among employees and their supervisors and to direct attention to accident prevention in general.
  - To determine facts bearing on legal liability.

What Should Be Investigated?

An accident that causes death or serious injury obviously should be thoroughly investigated. The "near-accident" that might have caused death or serious injury is equally important from

Listed below are 48 tasks and responsibilities. (Some of them may not be a part of your job.) Put a check mark in the column under the head that best indicates your responsibility for the tasks listed at the left.

Usually you will have responses in all three columns. This is normal. Do not be dismayed to find you have a number of check marks in the third column. Everyone does.

	Yes	No	I Don't Know
New Employees			
1. Hire			
2. Accept or reject applicants			
3. Report on probationary employees			
Training Employees			
4. Orient new employees			
5. Explain safe operation/rules			
6. Hold regular production meetings			
7. Hold safety (tool box) meetings			
8. Coach employees on the job			
Production			
9. Control quantity			
10. Control quality			
11. Stop a job in progress			
12. Authorize changes in setup			
13. Requisition supplies			
14. Control scrap			
15. Establish housekeeping standards			
Safety			
16. Take unsafe tools out of production			
17. Investigate accidents			
18. Establish inspection committees			
19. Inspect your own department			
20. Correct unsafe conditions			
21. Correct unsafe acts			
22. Send employees to the doctor			
Discipline			
23. Recommend promotions or demotions			
24. Transfer employees out of your department			
25. Change an employee to a less desirable job			
26. Grant pay raises			
27. Issue warnings			
28. Suspend			
29. Discharge			
Assigning Work			
30. Prepare work schedules			
31. Assign specific work			
32. Delegate authority to leaders			
33. Authorize overtime			
Employee Affairs			
34. Prepare vacation schedules			
35. Grant leaves of absence			
36. Lay off for lack of work			
37. Process grievances			
Coordination			
38. Authorize maintenance and repairs			
39. Make suggestions for improvement			
40. Discuss problems with management			
41. Recommend changes in policy			
42. Improve work methods			
Cost Control			
43. Reduce waste			
44. Keep production records			
45. Budget			
46. Approve expenditures			
Other			
47. Keep up employee morale			
48. Reduce turnover			

Exhibit 4-1 Your responsibilities as a supervisor.

Listed below are the same tasks and responsibilities presented in the exercise just completed. For each you may have some degree of authority to get things done, whether over people, over expenditures, or both. Put a check mark in the column under the head that best describes the degree of authority you have over the situations listed at the left.

Again you will usually have responses in all columns. If you have difficulty in deciding, enter a question mark under the column which seems right.

	Complete Authority	Decide and Implement but inform	Decide but Check	Boss Decides	Influence Only
New Employees					
1. Hire					
2. Accept or reject applicants					
3. Report on probationary employees					
Training Employees					
4. Orient new employees					
5. Explain safe operation/rules					
6. Hold regular production meetings					
7. Hold safety (tool box) meetings					
8. Coach employees on the job					
Production					
9. Control quantity					
10. Control quality					
11. Stop a job in progress					
12. Authorize changes in setup					
13. Requisition supplies					
14. Control scrap					
15. Establish housekeeping standards					
Safety					
16. Take unsafe tools out of production					
17. Investigate accidents					
18. Establish inspection committees					
19. Inspect your own department					
20. Correct unsafe conditions					
21. Correct unsafe acts					

Exhibit 4-2 Supervisor authority.

	Complete Authority	Decide and Implement but inform	Decide but Check	Boss Decides	Influence Only
22. Send employees to the doctor					
Discipline					
23. Recommend promotions or demotions					
24. Transfer employees out of your department					
25. Change an employee to a less desirable job					
26. Grant pay raises					
27. Issue warnings					
28. Suspend					
29. Discharge					
Assigning Work					
30. Prepare work schedules					
31. Assign specific work					
32. Delegate authority to leaders					
33. Authorize overtime					
Employee Affairs					
34. Prepare vacation schedules					
35. Grant leaves of absence					
36. Lay off for lack of work					
37. Process grievances					
Coordination					
38. Authorize maintenance and repairs					
39. Make suggestions for improvement					
40. Discuss problems with management					
41. Recommend changes in policy					
42. Improve work methods					
Cost Control					
43. Reduce waste					
44. Keep production records					
45. Budget					

Exhibit 4.2 Continued.

	Complete Authority	Decide and Implement but Inform	Decide but Check	Boss Decides	Influence Only
46. Approve expenditures **Other** 47. Keep up employee morale 48. Reduce turnover					

Degrees of Authority
   To help you clarify how much authority you have on your job, we have defined the five degrees of authority.
   1. *Almost complete authority.* You can make any decision related to an area of responsibility and can implement that decision without having to check with anybody. You don't even have to inform anyone else.
   2. *Decide and implement but keep them informed.* You can make a decision and you can implement your decision, but you must tell the boss what you've done.
   3. *Decide, but check.* The decision is yours, but you cannot implement your decision until you've cleared it with the boss.
   4. *Let the boss decide.* You can gather the facts but are required to have the boss make the decision.
   5. *Influence only.* You have no authority except your personal influence.

Exhibit 4.2 Continued.

the safety standpoint and should be investigated, for example, the breaking of a crane hook or a scaffold rope or an explosion associated with a pressure vessel.

Each investigation should be made as soon after the accident as possible. A delay for only a few hours may permit important evidence to be destroyed or removed, intentionally or unintentionally. Also, the results of the investigation should be made known quickly, as their publicity value is greatly increased by promptness.

Any epidemic of minor injuries demands study. A particle of emery in the eye or a scratch from handling sheet metal may be a very simple case. The immediate cause may be obvious and the loss of time may not exceed a few minutes. However, if cases of this or any other type occur frequently in the plant or in your department, an investigation might be made to determine the underlying causes.

With any accident, we must find the fundamental root causes and remove them if we hope to prevent a recurrence. Root causes often relate to the management system. They may be due to management's policies and procedures, supervision and its effectiveness, or training. Root causes are those which would effect permanent results when corrected.

The Key Facts in Accidents

We have concentrated up to this point on root causes. It could be that in your company additional information will be needed in an investigation. There is a national standard in investigations with which you should at least be familiar.

For each task that is part of your job, identify whether or not your boss is, in fact, measuring your performance (yes or no). Briefly state how it is measured (numerical or other—state specifically the measure used). Then indicate whether the measure is a result or an activities measure. Finally, check what degree of accountability is used: strong, medium, or light.

	Measured?		Type			Degree		
	Yes	No		R	A	S	M	L
New Employees								
1. Hire								
2. Accept or reject applicants								
3. Report on probationary employees								
Training Employees								
4. Orient new employees								
5. Explain safe operation/rules								
6. Hold regular production meetings								
7. Hold safety (tool box) meetings								
8. Coach employees on the job								
Production								
9. Control quantity								
10. Control quality								
11. Stop a job in progress								
12. Authorize changes in setup								
13. Requisition supplies								
14. Control scrap								
15. Establish housekeeping standards								
Safety								
16. Take unsafe tools out of production								
17. Investigate accidents								
18. Establish inspection committees								
19. Inspect your own department								
20. Correct unsafe conditions								
21. Correct unsafe acts								
22. Send employees to the doctor								
Discipline								
23. Recommend promotions or demotions								
24. Transfer employees out of your department								
25. Change an employee to a less desirable job								
26. Grant pay raises								
27. Issue warnings								
28. Suspend								
29. Discharge								
Assigning Work								
30. Prepare work schedules								
31. Assign specific work								
32. Delegate authority to leaders								
33. Authorize overtime								
Employee Affairs								
34. Prepare vacation schedules								
35. Grant leaves of absence								
36. Lay off for lack of work								
37. Process grievances								

Exhibit 4-3 Supervisor accountability.

	Measured?		Type			Degree		
	Yes	No		R	A	S	M	L
Coordination								
38. Authorize maintenance and repairs								
39. Make suggestions for improvement								
40. Discuss problems with management								
41. Recommend changes in policy								
42. Improve work methods								
Cost Control								
43. Reduce waste								
44. Keep production records								
45. Budget								
46. Approve expenditures								
Other								
47. Keep up employee morale								
48. Reduce turnover								

Results Versus Activity Measures
   There are two ways that management uses to measure your performance in a single area. It can measure the end result of your effort or it can measure whether or not you are carrying out your defined tasks. An example is in safety. *Result measurement* asks how many accidents have happened to your people, how many days were lost by them, how much it cost the company. *Activity measurement* asks how many times you have talked to your people about safety or how many times you've inspected your department.

Degrees of Accountability
   No one escapes being measured in some way on the job. How you are assessed depends on what your tasks are and how much authority you have. Just as there are degrees of authority, there are degrees of accountability. For instance, one company describes three levels of accountability.
   1. *Strong accountability* must accept heavy contribution for the achievement and effectiveness of end results; requires active participation and the initiation of action.
   2. *Medium accountability* must accept a secondary contribution to the achievement and effectiveness of end results; may be through an interrelated function or may be heavy in one facet of a many-faceted accountability. Also requires active participation and the initiation of action within the area of the persons concerned.
   3. *Light accountability* must accept accountability for contributing but not for initiating any action.

Exhibit 4.3 Continued.

   As explained in American National Standards Institute (ANSI) Standard Z16.2, *Method of Recording Basic Facts Relating to Nature and Occurrence of Work Injuries,* the purpose of the standard is to identify certain key facts about each injury and the accident that produced it. These facts are to be recorded on a form which will permit summarization to show general patterns of injury and accident occurrence in as great analytical detail as possible. These patterns are intended to serve as guides to the areas, conditions, and circumstances to which accident prevention efforts may be directed most profitably. Such facts as the following are recorded:

   — Nature of injury—the type of physical injury incurred.

— Part of body—the part of the injured person's body directly affected by the injury.
— Source of injury—the object, substance, exposure, or bodily motion that directly produced or inflicted the injury.
— Accident type—the event that directly resulted in the injury.
— Hazardous condition—the physical condition or circumstance that permitted or occasioned the occurrence of the accident type.
— Agency of accident—the object, substance, or part of the premises in which the hazardous condition existed.
— Agency of accident part—the specific part of the agency of accident that was hazardous.
— Unsafe act—the violation of a commonly accepted safe procedure that directly permitted or occasioned the occurrence of the accident event.

While much information is needed for legal and company purposes, keep in mind the real reason for the investigation is to unearth problems that can lead to other incidents. In addition to the above information, the form used for investigation should lead the investigator into looking for underlying accident causes (see Exhibit 4.4). This form can be graded for quality by the supervisor's boss (see Exhibit 4.5).

Inspecting for Hazards

Inspection is one of the primary tools of safety. At one time, it was virtually the only tool and it still is the one most used.

Name of Injured _____ Date of Accident _____ Time _____
Seriousness: Lost Time   Doctor   First Aid Only   Near Miss
Nature of Injury _____
What Happened? _____
_____

What acts and conditions* were involved? What caused them? How were they corrected?   *At least five (use back also)

Unsafe Act/Condition/ Symptom	Possible/Probable Cause	Correction/ Suggested Correction
1.		
2.		
3.		
4.		
5.		

Supervisor _____ Department _____

Exhibit 4.4 Supervisor's report of accident investigation.

	Circle One	
1. Was it on time?	Yes-5 pts.	No-0 pts.
2. Was seriousness indicated?	Yes-5 pts.	No-0 pts.
3. Does it say where it happened?	Yes-5 pts.	No-0 pts.
4. Can you tell exactly what the injury is?	Yes-5 pts.	No-0 pts.

	Circle One
5. How many acts and conditions are listed?	5 4 3 2 1 0
6. How many causes are identified?	5 4 3 2 1 0
7. How many corrections were made or suggested?	5 4 3 2 1 0
8. How many of the listed corrections would have prevented this accident?	5 4 3 2 1 0
9. How many corrections are permanent in nature?	5 4 3 2 1 0
10. In how many of the corrections listed is the supervisor **now** doing something differently?	5 4 3 2 1 0

Total of Circled Points
Multiply X 2

Reviewed by _____ SCORE _____
General Foreman

Exhibit 4.5 Investigation rating sheet.

Many articles have been written about safety inspections and many have asked the question: "Why inspect?" Some typical answers have been:

— To check the results against the plan.
— To reawaken interest in safety.
— To reevaluate safety standards.
— To teach safety by example.
— To display the supervisor's sincerity about safety.
— To detect and reactivate unfinished business.
— To collect data for meetings.
— To note and act upon unsafe behavior trends.
— To reach first hand agreement with the responsible parties.
— To improve safety standards.
— To check new facilities.
— To spot conditions.

According to the National Safety Council, inspections have two objectives: (1) maintaining a safe work environment and controlling the unsafe actions of people, and (2) maintaining product quality and operational profitability.

Inspection Checklists

Systematic inspection is the basic tool for maintaining safe conditions and checking unsafe practices. Each company, plant, or department should develop its own checklist. Sample check-lists, stressing work areas or work practices or both, are shown in Exhibits 4.5 to 4.8.

SAFETY INSPECTION CHECK LIST

Plant or Department _____ Date _____

This list is intended only as a reminder. Look for other unsafe acts and conditions, and then report them so that corrective action can be taken. Note particularly whether unsafe acts or conditions that have caused accidents have been corrected. Note also whether potential accident causes, marked "X" on previous inspection, have been corrected.

(✓) indicates *Satisfactory*   (X) indicates *Unsatisfactory*

---

1. FIRE PROTECTION
Extinguishing equipment .... ☐
Standpipes, hoses, sprinkler
   heads and valves .... ☐
Exits, stairs and signs .... ☐
Storage of flammable material ... ☐

2. HOUSEKEEPING
Aisles, stairs and floors .... ☐
Storage and piling of material .... ☐
Wash and locker rooms .... ☐
Light and ventilation .... ☐
Disposal of waste .... ☐
Yards and parking lots .... ☐

3. TOOLS
Power tools, wiring .... ☐
Hand tools .... ☐
Use and storage of tools .... ☐

4. PERSONAL PROTECTIVE EQUIPMENT
Goggles or face shields.......... ☐
Safety shoes................... ☐
Gloves .... ☐
Respirators or gas masks .... ☐
Protective clothing .... ☐

5. MATERIAL HANDLING EQUIPMENT
Power trucks, hand trucks
Elevators .... ☐
Cranes and hoists.............. ☐
Conveyors .... ☐
Cables, ropes, chains, slings..... ☐

6. BULLETIN BOARDS
Neat and attractive .... ☐
Display changed regularly .... ☐
Well illuminated .... ☐

7. MACHINERY
Point of operation guards .... ☐
Belts, pulleys, gears, shafts, etc... ☐
Oiling, cleaning and adjusting .... ☐
Maintenance and oil leakage ..... ☐

8. PRESSURE EQUIPMENT
Steam equipment .... ☐
Air receivers and compressors... ☐
Gas cylinders and hose .... ☐

9. UNSAFE PRACTICES
Excessive speed of vehicles...... ☐
Improper lifting .... ☐
Smoking in danger areas .... ☐
Horseplay .... ☐
Running in aisles or on stairs ..... ☐
Improper use of air hoses........ ☐
Removing machine or other
   guards .... ☐
Work on unguarded moving
   machinery .... ☐

10. FIRST AID
First aid kits and rooms.......... ☐
Stretchers and fire blankets ...... ☐
Emergency showers............ ☐
All injuries reported............. ☐

11. MISCELLANEOUS
Acids and caustics .... ☐
New processes, chemicals
   and solvents .... ☐
Dusts, vapors, or fumes.......... ☐
Ladders and scaffolds........... ☐

Signed _____

Exhibit 4.6 Safety inspection check list.

	SUMMARY OF UNSAFE PRACTICES	Area or Division _____ Force _____ District _____ Period Covered _____					
	GROUP (FORCE, DISTRICT, DIVISION OR AREA)						TOTAL
1	SUPERVISORS REPORTING UNSAFE PRACTICES						
2	SUPERVISORS NOT REPORTING UNSAFE PRACTICES						
3	TOTAL NO. OF SUPERVISORS						
4	TOTAL NO. OF EMPLOYEES IN GROUP						
5	NUMBER OF UNSAFE PRACTICES REPORTED						
	CAUSE OF UNSAFE PRACTICES						
6	LACK OF ANALYZING, OR PLANNING THE WORK						
7	INADEQUATE BASIC TRAINING						
8	LACK OF DEFINITE OR SPECIFIC INSTRUCTIONS						
9	IMPROPER ASSIGNMENT OF EMPLOYEE						
10	FAILURE TO SEE THAT INSTRUCTIONS WERE FOLLOWED						
11	OTHER						
12	LACK OF ANALYZING, OR PLANNING THE WORK						
13	DISREGARD FOR KNOWN SAFE PRACTICES						
14	LACK OF EXPERIENCE						
15	ABSTRACTION OR FORGETFULNESS						
16	HASTE						
17	OTHER						
18	TOTAL						
	UNSAFE PRACTICES						
19	MOTOR VEHICLES—OPERATION AND MAINTENANCE						
20	POLES—WORKING ALOFT						
21	ACTION, BOTH IN AND OUT OF BUILDINGS THAT MIGHT RESULT IN SLIPS OR FALLS						
22	LADDERS—EXTENSION C.O. AND STEPLADDERS						
23	BODY BELTS AND SAFETY STRAPS						
24	CLIMBERS, PADS, AND STRAPS						
25	GUARDING EMPLOYEES AND PUBLIC						
26	TOOLS AND MATERIALS						
27	GOGGLES						
28	USE OF RUBBER GLOVES AND OTHER PROTECTIVE DEVICES AND PRECAUTIONS TAKEN AROUND LIVE WIRES						
29	MANHOLES, CONDUIT AND EXCAVATIONS						
30	FIRST AID FOR AND CARE OF INJURIES						
31	MISCELLANEOUS						
32	TOTAL						

(Note: Each organization should substitute in this section the types of unsafe practices which are most likely to be encountered in its own operation.)

Left margin labels: Supervisor (rows 6–11), Employee (rows 12–17)

RECOMMENDATIONS _____

Observed by _____

Title _____

Exhibit 4.7 Your key safety tasks.

UNSAFE PRACTICES OBSERVED

	Reference	Number		Reference	Number
19. Motor Vehicles-operation and maintenance			26. TOOLS AND MATERIALS		
Improper parking			Improper use of tools and materials		
Failure to conform to traffic laws			Use of defective or unauthorized tools		
Stepped off or on vehicle in motion			Tossing tools and materials		
Backing without taking proper precautions			Dropping tools and materials from aloft		
Poor housekeeping on truck or car			Unsafe lifting or handling		
Lights, brakes, horn, etc. not tested			Unsafe carrying of tools, nails, tacks, etc.		
Unsafe handling of Derrick			Improperly stored or placed		
Under Derrick in operation			Unsafe use of cable car		
Hands on or near winch line sheave or drum			Handline dangling		
Unsafe handling and use of trailer			Tools and materials left lying around		
Unsafe performance of maintenance operations			Failure to use tree sling		
			Unsafe cutting of wire-flying ends or pieces		
20. POLES-WORKING ALOFT			Not using flashlight where required		
Failure to test or make safe before climbing					
Failure to compensate for unbalanced load			27. GOGGLES-FAILURE TO USE WHEN:		
Working in dangerous position			Drilling concrete brick or other masonry		
Lack of care in climbing			Grinding, chipping, handling brush, etc.		
Standing under workman aloft					
Lack of care-approaching or leaving pole			28. USE OF RUBBER GLOVES ETC. AND PRECAUTIONS TAKEN AROUND LIVE WIRES		
21. ACTION BOTH IN AND OUT OF BUILDINGS, THAT MIGHT RESULT IN SLIPS OR FALLS			Failure to use rubber gloves when required		
			Failure to test and maintain properly		
Unnecessary running, stairs and across floors			Working too close, live wires when aloft		
Lack of ordinary care outside buildings			Lack of precautions to prevent wire from		
Sitting on tilted chairs			Flipping up into live wires		
Standing on chairs, boxes, cans, etc.			Sagging or falling onto live wires		
22. LADDERS-EXTENSION, C.O. AND STEP LADDERS			Reel tender not protected		
			Unsafe use or chain hoist, tent, steel tape, etc.		
Failure to make secure (footing lashing, holding)			Incomplete survey after suspected E.L. Contact		
Failure to place leg thru ladder or use safety			Carelessness around electric circuits and equipment		
Pull rope not tied			Failure to use adequate protective devices		
Defective ladder, spurs not turned, etc.			29. MANHOLES, CONDUIT AND EXCAVATIONS		
Overreaching too high up on ladder, etc.			Failure, proper tests for gas or oxygen deficiency		
Improper angle for climbing			Insufficient ventilation (with sail or blower)		
Failure to inspect			Entered manhole without manhole guard		
Unguarded at hazardous location			Failure to use ladder in manhole or excavation		
Failure to use where required or used wrong kind			Insufficient guarding, manholes or excavations		
Not lashed to motor vehicle or carried properly			Unsafe removal and handling of cover		
Failure to extend spreaders or stepladder			Smoking or open flames near or in manholes		
Tools or material left on steps or top					
Shoved C.O. ladder endangering person on ladder			30. FIRST AID FOR AND CARE OF INJURIES		
			Failure to give proper first aid		
23. BODY BELTS AND SAFETY STRAPS			Did not continue proper care of injury		
Failure to use while aloft (On pole, platform, etc.)			Failure to report injury		
Failure to look, feel and know snaphook is secure			First aid kit not properly maintained		
			31. MISCELLANEOUS		
Not inspected or properly maintained			Horseplay on-the-job		
Pushing keeper against objects			Debris left lying around		
24. CLIMBERS, PADS AND STRAPS			Unsafe position on st., highway, rd., r.r. track, etc.		
Wearing in trees, vehicles, on ladder, ground, etc.			Handling of hot solder, paraffin, etc.		
Failure to inspect and maintain properly			Poison ivy or oak-lack of protection or care		
Unsafe climbing habits			Clothing and shoes-poor, insufficient		
25. GUARDING EMPLOYEES AND PUBLIC			Trees or brush-cutting and handling		
Failure to use adequate warning signs or flags					
Keeping children and others away from operation					
Guarding public at dangerous location					

Exhibit 4.8 Summary of unsafe practices.

The format of Exhibit 4.6 is typical of many companies' approach. Exhibit 4.7 and 4.8 concentrate on unsafe practices, while Exhibit 4.9 is designed for the concept of multiple causation and symptomatic safety discussed in the last chapter.

Coaching

We use the term coaching instead of training because coaching connotes a wider range of activities than does training. We are interested in other activities that should be done to improve people's job performance and not in a dusty, classroom approach that the term training brings to mind.

To coach means to help a person do better. Some organizations have extensive training programs and others have very limited training. In either case, there is coaching on the job. Any supervisor will readily admit having a responsibility to help people do better.

For our purposes, we'll say that coaching consists of explaining, demonstrating, correcting, encouraging, and reprimanding. Putting the coaching process in its simplest form, we can say that it has three steps: (1) finding out where the employee is now; (2) finding out where we want the person to be; and (3) providing the difference.

Unfortunately in safety we usually spend almost all our time determining content and method. Theorists tell us to spend the bulk of our effort and concentration on (1) and (2) above. If we do a good job in analyzing these, the third step usually falls into place naturally. Theorists tell us that content is no more than the difference between the first two steps and that the method of presentation is almost immaterial to learning.

Every supervisor has two different coaching tasks: (1) to coach each person individually and (2) to coach the group collectively.

The Individuals

The supervisor must look at each person individually and go through the process described above of finding out the person's needs—where the person is now in comparison to where you want him or her to be. Then with explanations, demonstrations, corrections, encouragements, and reprimands, each individual must be coached to achieve the performance level you want.

When should the individual be coached? Obviously, coaching should occur whenever you feel the person needs to do better in any area. Coaching seems to be particularly appropriate in the following kinds of situations:

When new persons are assigned. The supervisor needs to show them around, introduce them to people, explain the purpose of the department, instruct them on safety rules and regulations, point out special hazards, and in other ways get them oriented to the job to be done. By giving it conscious attention, the supervisor will do it better and improve the person's performance.

When an unsafe act is seen. Often a word or comment or brief explanation is sufficient to correct it. Sometimes there is a need for greater explanation or demonstration. An unsafe act continued long enough will finally produce an accident and safety is not a goal which can be measured in daily or hourly figures. People know safety is important by what the supervisor does in daily and hourly contacts with them. When the supervisor's back is turned on unsafe acts, when coaching is lax, they quite readily accept the fact that safety is not an important goal or important in their work.

When we see anything well done. It is easier to improve performance by praise of the job well done than it is by criticism. Individuals react positively to compliments and want to continue to receive them.

```
┌─────────────────────────────────────────────────────────────────────────┐
│                    SUPERVISOR'S INSPECTION FORM                           │
│   Name _____ Date _____          │
├──────────────────────────┬──────────────────────┬────────────────────────┤
│    Symptom Noted         │      Causes          │   Corrections Made or   │
│  Act/Condition/Problem    │   Why—What's Wrong   │      Suggested          │
│                          │                      │    By you — By others   │
├──────────────────────────┼──────────────────────┼────────────────────────┤
│                          │                      │                         │
│                          │                      │                         │
│                          │                      │                         │
│                          │                      │                         │
│                          │                      │                         │
│                          │                      │                         │
└──────────────────────────┴──────────────────────┴────────────────────────┘
```

Exhibit 4.9 Supervisor's inspection form.

When giving orders or work assignments. Orders should clearly express what is wanted.

When we see anything being done incorrectly. If we coach when we see things well done, people are more ready to accept coaching when something is done incorrectly. When praising establish what is wanted. When something is done incorrectly, establish what is *not* wanted and use the opportunity to establish what is wanted.

Whenever assigning an unusual job. Much work falls into rather usual patterns but in every organization there are some situations that are unusual. Every unusual situation or unusual job suggests a need to coach, to explain, to demonstrate, to make sure that people know what is wanted, for it is in these circumstances that a severe injury can be expected.

The Group

The supervisor must also look at the department as a whole and assess its needs and look for occasions or situations when coaching can be most effective. Some of these are:

When there is a better way to do things.

When new products, methods, or machines are introduced. Even though only some people will operate the machines, it is well to acquaint the whole group with the new product, method, or machine. Let all understand how it fits into the total group effort.

When unsafe conditions crop up. An unsafe act is performed by an individual but unsafe conditions require a group effort. A supervisor cannot see or be present every time an unsafe condition crops up. They must be seen and corrected or reported as a group effort.

When establishing goals. Supervisors often do not clearly establish and explain their goals to their people. Usually there is a general idea of an end result desired by the work of the whole group and a general idea of the performance required and desired by the whole group. This needs to be explained and discussed with the group. The supervisor's job is to get the group to want to achieve the goals of the supervisor.

When any unusual condition affects the whole group. This may come about as the result of an accident, a slow-down in work with necessary layoffs, an unusual or special job, or any occasion in which the understanding of the whole group and the cooperation of the whole group is important to accomplish the end desired.

How to Coach

While specific methods are not discussed, here are a few ideas on human learning. There has been more research on this topic than on almost any other in psychology. Here are some of the "knowns":

Motivation and learning. Learning theorists agree that individuals will learn best if they are motivated toward some goal which is attainable by learning the subject matter presented. The behavior of people is oriented toward relevant goals, whether these goals are safety, increased recognition, production, or simply socializing. People attempt to achieve those goals which are important to them at the moment, regardless of what's important to the supervisor.

Reinforcement and learning. Positive rewards for certain behavior increase the probability that the particular behavior will occur again. Negative rewards decrease that probability. Reinforcement is most important to learning.

Practice and learning. An individual learns best through practice and involvement.

Feedback and learning. Telling workers how they are doing is essential for learning. It is difficult for workers to improve performance unless given some knowledge of that performance. What are the errors? How can they be corrected? This is essential.

Meaningfulness and learning. In general, meaningful material is learned better than material which is not meaningful. In order to simplify the workers' learning task, make the material as meaningful as possible to them personally.

The task of coaching will be discussed in two separate categories, individual coaching and group coaching. Group coaching (safety meetings) have been much more traditional and used much more than individual coaching (one-on-ones). Here is one of the many places where safety tradition flies in the face of research. There is little evidence that safety meetings (group coaching) results in improved accident records.

There is much research that suggests individual coaching (one-on-ones) would be better. In the one-to-one relationship between two people, a boss and a subordinate, the closer it is to a levelled relationship (that is, a relationship of equality rather than superior-subordinate), the better off both parties will be. You can have that levelled relationship one to one; you cannot in a group or meeting format. Motivation is defined as a person attempting to satisfy his or her current needs. A boss can only find out what those needs are in a one-to-one format—not in a group or meeting format. In addition there is the power of the informal group, it being the largest single determinant of individual worker behavior. Management can only break through that power one-on-one; in a meeting or group format, that power is only strengthened.

One-on-one communication by a supervisor is absolutely crucial to safety success.

Motivating

Another responsibility is to motivate people—to turn them on. Chapter 7 discusses in some detail what turns people on (and off) today.

Job Safety Analyses

Job Safety Analysis (JSA) is a procedure that identifies the hazards associated with each step of a job and develops solutions for each hazard that either eliminate it or control it. A job safety analysis worksheet is illustrated in Exhibit 4.10. In the left column, the basic steps of the job are listed in the order in which they are performed. The middle column describes how to

JOB: Using a Pressurized Water Fire Extinguisher		
WHAT TO DO (Steps in sequence)	**HOW TO DO IT** (Instructions) (Reverse hands for left-handed operator.)	**KEY POINTS** (Items to be emphasized. Safety is always a key point)
1. Remove extinguisher from wall bracket.	1. Left hand on bottom lip, fingers curled around lip, palm up. Right hand on carrying handle palm down, fingers around carrying handle only.	1. Check air pressure to make certain extinguisher is charged. Stand close to extinguisher, pull straight out. Have firm grip, to prevent dropping on feet. Lower, and as you do remove left hand from lip.
2. Carry to fire.	2. Carry in right hand, upright position.	2. Extinguisher should hang down alongside leg. (This makes it easy to carry and reduces possibility of strain.)
3. Remove pin.	3. Set extinguisher down in upright position. Place left hand on top of extinguisher, pull out pin with right hand.	3. Hold extinguisher steady with left hand. Do not exert pressure on discharge lever as you remove pin.
4. Squeeze discharge lever.	4. Place right hand over carrying handle with fingers curled around operating lever handle while grasping discharge hose near nozzle with left hand.	4. Have firm grip on handle to steady extinguisher.
5. Apply water stream to fire.	5. Direct water stream at base of fire.	5. Work from side to side or around fire. After extinguishing flames, play water on smouldering or glowing surfaces.
6. Return extinguisher Report use.		

Exhibit 4.10 Job analysis worksheet.

perform each job step. The right column gives the safety procedures that should be followed to guard against hazards. The basic steps in making a JSA are:

— Select the job to be analyzed.
— Break the job down into successive steps.
— Identify the hazards and potential accidents.
— Develop ways to eliminate the hazards listed.

A blank worksheet used for OSHA compliance is shown in Exhibit 4.11.

Job operation _____

Presently required personal protective equipment _____

Sequence of job steps   Hazards or OSHA violations

1.   1.
2.   2.
3.   3.
4.   4.

Recommended safe procedure

1.
2.
3.
4.

Exhibit 4.11 OSHA job safety analysis worksheet.

Selecting the Job

In selecting jobs to be analyzed and in establishing the order of job priorities, the National Safety Council suggests the following factors:

— *Frequency of accidents.* A job that has a repeated number of accidents is a good candidate for an early JSA.
— *Severity of accidents or injuries.* Any job that has produced disabling injuries might be considered for an early JSA.
— *A high potential for severity.* If the potential for a serious accident or injury is present, a JSA might well be warranted.
— *New jobs or changed jobs.*

Breaking the Job Down

First, the job is broken down into its basic steps. These steps should describe what is being done in order of occurrence. The National Safety Council gives this example:

JOB: PLANTING A TREE

1. Select the site.
2. Bring tools, equipment, and the tree to the site.
3. Dig the hole.
4. Prepare the hole.
5. Put the tree into the hole.
6. Backfill, tamp, and water.
7. Brace the tree.
8. Clean up and return equipment.

And the council suggests these key points in breaking down a job:

— Select the right person to observe. The person selected must be experienced, capable, cooperative, and willing to share thoughts.

— Brief the person on the purpose. If the worker selected has never worked on a JSA before, thoroughly explain it. Ask cooperation.
— Observe the job for the breakdown.
— Record each step on the worksheet.
— Check the breakdown with the worker when finished to get input.

Identifying Hazards

After the breakdown, each step should be analyzed in detail to identify hazards and potential accidents. Each should be recorded on the worksheet in the center column. Keep hazards parallel with the steps recorded. Check with the employee for ideas. Check also with other employees who have knowledge of that job.

Developing Solutions

When the hazards have been identified, the next step is to begin to develop solutions to the problems identified. Solutions might incorporate:

— An entirely different way to do the job.
— A change in physical conditions, layout, or environment.
— A changed job procedure.
— A change in frequency of how often the job is performed.

For each hazard on the sheet, ask "What can be done differently and how should it be done?" Answers and solutions should be very specific and very concrete to be of value. Solutions which merely state "Be more alert" or "Use More Caution" or something similar are worthless. Solutions must state exactly what to do and how to do it.

Benefits

While performing the JSA, you'll learn more about the job observed than ever before. You also will have involved an employee and demonstrated that you care about employee safety on that job. And you'll be creating safer conditions for the job observed.

Provide at least 25 of the worksheets in Exhibit 4.12 for each supervisors who will be involved in the JSA program. Then set a schedule as follows:

— List all jobs in the department.
— Schedule them for analysis. Take no more than perhaps three per week, and no less than one per week.
— Carry out each analysis as outlined in this section.
— Upon completion of each analysis, review it in detail with the worker involved, who may have valuable additions.
— Provide the worker with a copy of the final analysis.
— Keep several copies yourself and use these in future orientation and training sessions.

JOB SAFETY ANALYSIS		
Job:		**Date:**
Employee:	**Supervisor:**	
Job Steps	Hazards/Potential Accidents	Controls

Exhibit 4.12 Worksheet for analyzing job safety.

Job Safety Observations

A Job Safety Observation (JSO) provides a device to learn more about the work habits of people. Following the procedures described below, you can use this opportunity to check on the results of past training; make immediate, on-the-spot corrections and improvements in work practices; and compliment and reinforce safe behavior. Through your comments you can encourage proper attitudes toward safety. Follow these steps when implementing a JSO program:

— Select the worker and job to be observed.
— Make the observation.
— Record (see Exhibit 4.13).
— Review the results with the employee observed.
— Follow up.

Worker Selection

Consider the following possibilities when determining which person to observe first:

— The new person on the job.

— People who have recently completed training programs.
— Below-average performers.
— Accident repeaters.
— Risk takers.
— Workers with special problems.

Making the Observation

In most cases tell the person what you are doing prior to the JSO. Then simply observe normal operation, making any applicable notes on the worksheet about the normal work practices and procedures.

The Review

When the JSO is completed sit down with the worker and give the conclusions. Express appreciation for the cooperation and lay out your honest feelings about work habits and

JOB SAFETY OBSERVATION		
Employee:	Supervisor:	
Job:	Date:	Time:
Notes on Any Job Practices that Are Unsafe:		
Notes on Any Practices that Need Change or Improvement:		
Notes on Any Practices that Deserve Complimenting:		
Notes on the Review and Discussion:		

Exhibit 4.13 Worksheet for job safety observation.

practices. The first time you go through this with a worker both of you will be nervous, and the worker may be concerned or apprehensive. Keep the discussion informal and friendly. Do not let the discussion be a one-way communication. Encourage the worker to talk and discuss any problems or barriers to working safely.

The Follow-up

Follow up the observation as needed. In some cases follow-up will be often. In others, it will be seldom. How often depends on the person and on the job. Follow-up JSO's are usually a good idea after a job change.

Benefits of JSOs

The JSO is a feedback device. It provides excellent information on the effectiveness of training and on the adequacy of job procedures. Through the JSO any substandard practices can usually be identified before an accident happens. JSO's also provide the opportunity to discuss performance with people individually and to compliment or correct their work habits. In addition, you get to know each worker better and thus can spot any physical or psychological problems more readily.

One-on-Ones

While slightly less traditional than some procedures mentioned, one-on-ones are becoming more common and popular every year. As indicated under coaching, there is good rationale behind the one-on-one contact. It makes considerably more sense than the traditional safety meeting; it allows levelling, it allows the supervisor the opportunity to understand the worker's needs (which is essential to motivation) and it provides a way to break through the power of the informal group.

Most organizations utilizing one-on-ones set up some structure to ensure they do in fact happen. This may be a quota system or some other mechanism or report.

NON-TRADITIONAL TASKS

There are many non-traditional tools being used today. We'll mention only a few to give a flavor of the possibilities.

Positive Reinforcement

One of the best approaches is to force the utilization of positive reinforcement. Following the contacts and observations mentioned above, many organizations are requiring supervisors to provide some positive strokes when safe behavior has been observed. There have been many experiments on the use of positive reinforcement and all to date have shown excellent results. A recent study of some 35 of these experiments showed an average improvement in safe behavior of 60 percent in a relatively short time period. It appears to be a powerful tool. This is discussed in some detail in Chapter 7 and in more detail in *Safety Management-A Human Approach, Second Edition* and in *Safe Behavior Reinforcement* by the author.

Safety Sampling

Safety Sampling (SS) is a well-tested technique in accident prevention. It is a little different from the other techniques described in this section in that it normally is implemented on a company-wide basis by management or the safety director rather than within one department by the supervisor. Nevertheless, because it has been so effective in safety programs, we believe a supervisor can use this tool advantageously.

What Is Safety Sampling

Safety Sampling measures the effectiveness of the line manager's safety activities but not in terms of how many or how few accidents occur. It measures effectiveness before the fact of the accident by taking a periodic reading of how safely the employees are working.

Like all good accountability systems or measurement tools, Safety Sampling is also an excellent motivational tool. Each employee finds it important to be working as safely as possible when the sample is taken. Many organizations that have conducted Safety Samplings report a good improvement in their safety record as a result of the increased interest in safety on the part of employees.

Safety Sampling is based on the quality control principle of random sampling inspection, which is widely used by inspection departments to determine quality of production output without making 100 percent inspections. The degree of accuracy desired dictates the number of random items selected which must be carefully inspected. The greater the number inspected, the greater the accuracy.

Safety Sampling Procedure

There are four steps in a Safety Sampling:

1. *Prepare a code.* The element code list of unsafe practices is the key to Safety Sampling and supervisory training. This list contains specific unsafe acts which occur in your department—the "accidents about to happen." The element code list is developed from the past accident record and from known possible causes. The code is then placed on an observation form (see Exhibit 4.14)
2. *Take the sample.* With the code list attached to a clipboard and with a counter, start sampling. Proceed through your area, observe every employee who is engaged in some form of activity, and record whether the employee is working in a safe or an unsafe manner. Each employee is observed only long enough to make a determination. Once the observation is recorded, it should not be changed. If the observation of the employee indicates the task is being performed safely, it is recorded on the counter. If the employee is observed performing an unsafe practice, a check is made on the form in the column which indicates the type of unsafe practice.
3. *Validate the sample.* The number of observations required to validate the sample is based on the degree of accuracy desired. Count the total number of observations made, and

SAFETY SAMPLING

A. Number of Safe Observations:

B. Number of Unsafe Observations:

Unsafe Act Noted:						Sample Date:				
No Safety Glasses Worn										
Improper Use of Tools										
Working on Unguarded Machine										
Not Using Pushsticks										
Working Near Tripping Hazard										
Improper Use of Air Nozzles										
Using Machine Improperly										
Wearing Loose Clothing										
Wearing Rings										
Improper Lifting										
Improper Positioning for Lift										
Climbing on Racks										
Unsafe Loading/Piling										
Using Defective Equipment										
Other: (Specify)										

C. Percentage

Sample Date	%	Sample Date	%	Sample Date	%

Supervisor: _____ Department: _____

Exhibit 4.14 Worksheet for sampling.

Percentage of Unsafe Observations	Observations Needed	Percentage of Unsafe Observations	Observations Needed
10	3,600	30	935
11	3,240	31	890
12	2,930	32	850
13	2,680	33	810
14	2,460	34	775
15	2,270	35	745
16	2,100	36	710
17	1,950	37	680
18	1,830	38	655
19	1,710	39	625
20	1,600	40	600
21	1,510	41	575
22	1,420	42	550
23	1,340	43	530
24	1,270	44	510
25	1,200	45	490
26	1,140	46	470
27	1,080	47	450
28	1,030	48	425
29	980	50	400

Exhibit 4.15 Number of observations needed for 90 percent accuracy.

determine how many unsafe practices seen. The percentage of unsafe observations is then calculated. Using this percentage and the desired accuracy, the number of observations required can be calculated by using the data in Exhibit 4.15.

4. *Prepare the report.* The results of the observations can be presented in many different forms. However, the report should include the total percentage of unsafe activities and the number and type of unsafe practices observed.

File the reports and periodically compare the findings to spot trends in the types of unsafe practices occurring. Such information will assist in planning effective safety training programs.

Benefits

Research has proven that Safety Sampling shows the same trends as claim costs, number of accidents, and accident cost per man-hour, although it correlates better with the all-accident rate (all reported accidents per 1,000 man-hours).

This seems to mean that Safety Sampling provides an excellent indicator of accident problem areas before the accidents occur. Of course, by far the best value of sampling is motivational. Sampling arouses extreme interest in safety where there was little interest before.

Incident Recall Technique (IRT)

Confidential attitude surveys conducted by consulting firms in a number of companies have revealed that it is fairly common for supervisors to hide accidents. It is even more common for employees to hide them. The reasons usually given for not reporting accidents are:

- Fear of discipline.
- Concern about the accident record.
- Concern about reputation.
- Fear of medical treatment.
- Dislike of medical personnel.
- Desire not to interrupt work.
- Avoidance of red tape involved.
- Desire to keep personal record clear.
- Concern about what others might think.
- Poor understanding of the need to report.

Most of these barriers to reporting accidents are eliminated by the IRT.

The IRT Procedure

The basic objective of the IRT is to gain the cooperation of the employee, so that he or she can and will freely relate all incidents from the past that can be recalled. The success of the IRT, in terms of number of incidents revealed, depends primarily on the skill used in the interview. Here are the steps of the IRT:

1. Put the employee at ease.
2. Explain the purpose of the interview and of the IRT.
3. Give assurance that the IRT is totally confidential. Initial success depends to a great degree on this confidentiality.
4. Point out the benefits of the IRT to everyone—to the employee, the family, the department, the company.
5. Show and explain the report form to the employee. (See Exhibit 4.16).
6. Conduct the interview. Simply ask the employee to recall each near-miss accident seen or heard about on the job. With each incident recalled, be sure to determine how many times he or she has seen it or heard of its happening. Jot down all pertinent information.
7. Ask lots of questions to fill in the gaps. Avoid interrupting but get full information on each incident.;
8. Review your understanding of the incident with the employee. Repeat it to make sure you have it right.
9. Discuss the causes of the incident with the employee and possible remedies. Make it clear you want and need the employee's help.
10. Thank the employee at the conclusion of the interview.

Benefits

The benefits are obvious. First, and foremost, it may prevent accidents from occurring. Second, employee interest in safety is heightened by asking for and receiving employee involvement and participation. More information has been collected on current accident causes than would have been through a more conventional method. Employees sincerely interested in their safety. Furthermore, the IRT can help you check Safety Sampling findings and provide additional input for safety training programs.

INCIDENT RECALL TECHNIQUE		
Person Interviewed:		Date:
Supervisor:	Department:	
Incidents Recalled:		
Analysis of Causes of Recalled Incidents:		
Action Taken on Causes:		

Exhibit 4.16 Worksheet for incident recall.

Techinque of Operations Review (TOR)

The Technique of Operations Review (TOR) was devised to assist in finding some of the multiple, interrelated causes behind the accidents requiring investigation. It is basically a tracing system, but it also can be used as a training technique in safety. It was devised in 1967 by D.A. Weaver, while director of education at Employees Insurance of Wausau. The TOR centers around the cause code shown in Exhibit 4.17. The supervisor's incident investigation report in Exhibit 4.18 and technique of operations review in Exhibit 4.19 were developed for use with the TOR by Paul Mueller (Green Giant Company).

The TOR begins with an incident. Its purpose is to expose the real problems behind this incident, which are viewed as symptoms of more serious trouble. The TOR steps are:

1. Describe the incident. State clearly what happened.
2. Select one number from the cause code (Exhibit 4.17) which seems to be the immediate cause of the incident.
3. Trace and eliminate. The initially chosen number is jotted on the form and the trace step begins. Following this initial number and its description will be other numbers. Jot them down and read their descriptions. Then decide whether or not they could also be contributing causes to the incident. These numbers lead to additional numbers. List them. Read the descriptions of these and decide again. Keep tracing and eliminating numbers until they run out. When the "outs" overtake the "ins" or the final number on the list repeats the number at the top, you have come full circle. The numbers and descriptions that remain are those you have decided are contributing factors.
4. List the contributing factors.
5. Select solutions.

1 TRAINING

10 Training not formulated or need
not forseen 23, 48, 64

11 Instruction was given but results
show it didn't take 44, 47, 56

12 Training available but the employee
was not assigned or did not
attend 26, 35, 87

13 Performance not in accord with
policy or procedure 47, 55, 62

14 Failure to provide training where
need had been specified . . 34, 83, 88

15 Error blamed on faulty training
when in fact the error stemmed
from deficiencies in management
systems 26, 36, 52, 81

3 DECISION & DIRECTION

30 By-passing, conflicting orders,
too many bosses 33, 48, 80

31 Decision too far above the
problem 34, 83

32 Authority inadequate to cope with
the situation 22, 82

33 Decision exceeded author-
ity 13, 47, 86

34 Decision evaded; power to decide
not exercised 25, 85

35 Orders or directives failed to produce desired action. Not clear, not
understood, or not
followed 41, 50, 52

36 Failure to investigate, and to apply
the lessons of similar
mishaps 26, 43, 61

37 Hazard or problem — controls not
developed 26, 64, 66, 86

| TRACE GUIDE | 10 11 12 13 14 15 | | 30 31 32 33 34 35 36 37 |
| | 20 21 22 23 24 25 26 | | 40 41 42 43 44 45 46 47 48 |

2 RESPONSIBILITY

20 Duties and tasks not clear, or
not accepted 22, 25, 40

21 Conflicting goals 30, 48, 83

22 Dual or overlapping responsi-
bility 25, 30, 48, 80

23 Pressure of immediate tasks
obscures full scope of
responsibilities 10, 32, 34, 87

24 Buck passing, responsibility not
tied down 25, 48, 82

25 Job descriptions
inadequate 48, 80, 84

26 Hazard or problem —
not recognized 34, 37, 48, 81

4 SUPERVISION

40 Failure to orient or coach — new
worker, unusual situation,
unfamiliar equipment or
process, etc. 23, 24

41 Supervisor failed to tell
why 23, 35, 48

42 Supervisor failed to
listen 15, 36, 82

43 Unsafe Act. Failure to correct
before accident
occurred 26, 36, 51, 61

44 Failure to supervise closely until
proficiency was
assured 23, 36, 48, 64

45 Honest error. Failure to act, or action
turned out to be
wrong 14, 15, 20, 56

46 Disorder or confusion in work
area 34, 54, 56, 87

47 Job practice out of step with
job training 15, 42, 63

48 Initiative. Failure to see problems
and exert an influence
on them 23, 42, 50, 85

Exhibit 4.17 TOR cause code.

5 WORK GROUPS

50 Morale. Conflict, insecurity. Lack of faith in the boss or the future of the job 56, 31, 66, 86

51 Conduct. Supervisor sets a poor example 50, 42, 31, 86

52 Team spirit. Failure to pull together, uncooperative 15, 21, 81, 84

53 Rules. Not publicized, not clear. Unfair enforcement or weak discipline 40, 41, 80

54 Clutter. Anything not needed in the work area 20, 32, 60

55 Lack of things needed — tools, space, protective equipment, storage bins, etc. 24, 36, 65

56 Voluntary compliance. Work group sees little advantage to themselves 21, 82, 84

7 PERSONAL TRAITS

70 Work assignment — unsuited for this particular individual 11, 33, 45, 63

71 Poor work habits; careless of rules, tools, equipment, procedures, etc. 26, 30, 53, 85

72 Health problem 26, 34, 80, 81

73 Inappropriate behavior or judgement 33, 45, 72, 83

74 Undesirable peer pressures influence work performance and risk taking 20, 56

75 Behavior not adjusted to the workplace 32, 53, 85

TRACE GUIDE	50 51 52 53 54 55 56 60 61 62 63 64 65 66	70 71 72 73 74 75 80 81 82 83 84 85 86 87 88

6 CONTROL

60 Work flow. Inefficient or hazardous. Layout, scheduling, stacking, piling, routing, storing . 66, 32, 46, 48

61 Unsafe condition 65, 36, 43, 86

62 Equipment. Insufficient, unavailable, deficient design, inoperative 26, 31, 84

63 Procedure out of step with available technology; inadequate review and revision 24, 81, 85

64 Procedure not available or not followed 34, 80, 88

65 Deficient inspection, reporting, or maintenance 34, 47, 88

66 Hazard or problem — controls not maintained 26, 34, 88

8 MANAGEMENT

80 Policy. Failure to assert a management will before the mishap at hand 15

81 Goals. Not clear, or not converted into decisions and directions 21, 35, 85

82 Span of attention. Too many irons in the fire. Inadequate development of subordinates 30, 36

83 Conflicting priorities not resolved. Excessive emphasis on short range accomplishments 12, 21

84 Coordination. Departments inadvertently create problems for each other 34, 36

85 Failure to encourage subordinates to exercise their power to decide 31, 50

86 Accountability. Failure to develop appraisal and measurement of key goals and objectives 24, 37, 50

87 Staffing. Inadequate organization to cover necessary functions, or to use available human resources, or to cope with turnover and absenteeism 23, 46, 62

88 Hazard or problem — not properly evaluated 37, 66, 81

Exhibit 4.17 Continued.

INCIDENT INVESTIGATION

Employee (if involved)   Dept.   Clock No.

Incident date   Reported

1. Describe the incident, include location, witnesses, and circumstances surrounding
   incident. Try to identify the causal factors involved.

2. Subject causes to TOR analysis: state, trace, eliminate.

3. List factors for which you will initiate corrective action.

   Factor: Action:

4. List factors which require feasible corrective action by others. Circle routing
   to their attention.

Exhibit 4.18 Supervisor's incident investigation report.

Hazard Hunt

One other method that has been used successfully to spot possible accident causes is the Hazard Hunt. It is also good for involving people. To implement the procedure follow these steps:

1. Make copies of the hazard hunt form (Exhibit 4.20.)
2. Hold a short session with employees to explain the form and the reasons for using it.
3. Have employees jot down anything they feel is a hazard and return the form.
4. Review the forms, correct those hazards that can be corrected, and initiate action on the others.
5. Always inform employees of the actions, even if the hazards they mention are not really problems. Then hash over any disagreements with individuals to clear the air.
6. If something is a hazard that must be corrected, assign a priority to it and schedule it for rectification.

Some Behavioral Tasks

Most of the above supervisory tools have been used in one or more organizations. There are other strategies that could be used for the purpose of getting information and for better handling the worker. Following are a few:

Climate Analysis

A quick, superficial climate reading of an organization can be obtained by filling out the form in Exhibit 4.21 after reading and reviewing the climate factors explained in Chapter 8. If the form is filled out by people of various organizational levels, the evaluation will increase in validity.

Barrier Removal Analysis

Closely allied with climate analysis is the concept of barrier removal analysis. McGregor suggested that the primary role of the Theory Y manager was to remove the barriers that exist

TECHNIQUE OF OPERATIONS REVIEW	
Employee Involved:	Department:
Incident Date:	
1. Description of the Incident:	
2. Your TOR Analysis: Trace and Eliminate	
3. Causes Decided Upon:	
4. Corrections Decided Upon:	

Exhibit 4.19 Worksheet for reviewing technique of operations.

To: _____

From: _____

HAZARD HUNT

I think the following is a hazard: _____

_____

_____

_____

DO NOT WRITE BELOW HERE — TO BE FILLED IN BY SUPERVISOR

Supervisor:

    Agree this is a hazard.

        Corrected by Supervisor on _____ Discussed on _____

        If cannot correct, sent to Personnel on _____

        Job order on _____ Scheduled _____ Discussed on _____

    Do not agree this is a hazard.

        Discussed _____ Conclusion _____

        To Personnel _____ Conclusion _____

DO NOT WRITE BELOW HERE — FOR PERSONNEL USE

Supervisor _____ HH No. _____

Matrix No. _____ (Seriousness)

Exhibit 4.20 Hazard hunt.

between a worker and his or her work. Such barriers might be lack of supplies, equipment problems, management goofs, red tape, or problems coordinating with other departments.

In barrier-removal analysis, one upper-management activity might be to see how lower-level managers are carrying out barrier-removal. This might be done through observation or by asking lower-level managers to keep an activity log for a period of time and then categorizing those activities, for example, supervising, fire fighting, barrier removing, and so on.

Inverse Performance Standards

The traditional way to make performance appraisals is to have the boss evaluate the subordinate. When inverse performance standards are used, the subordinate evaluates the boss. Exhibit 4.22 shows one possible way of doing this. Subordinates fill out the form and submit it anonymously to a third person, who keeps the responses confidential and provides the boss with a summary of all responses.

Organization _____ Date _____
Assessment by _____

Safety-Program Climate

Lively ├──┼──┼──┼──┼──┼──┼──┼──┼──┤ Negligent

Corporate Climate Factors   Good   Poor

Confidence and Trust _____
Subordinate Interest _____
Understanding of Problems _____
Training and Helping _____
Teach How to Solve Problems _____
Giving of Support _____
Information Dispersal _____
Opinions Sought _____
Approachability _____
Recognition _____
Summary _____

Corporate Climate Requirements

Goals _____
Communications Down _____
Departmental Goals _____
Interdependency _____
Participation _____
Freedom to Work _____

Expansion _____
Delegation _____
Innovation _____
Fluid Communication _____
Stability _____

Corporate Personality

The Pacesetters Who _____
            Type _____
        Self-concept _____
          Motives _____
Economic Challenges _____
Our History _____
The Common Denominators are _____
_____
_____
_____

Exhibit 4.21 Climate assessment form.

Leveling Analysis

The rating form in Exhibit 4.22 asks some questions that are pertinent to the leveling analysis. Since the interpersonal relationship between boss and employee is an important factor in the employee's performance, it seems worthwhile to attempt to analyze this relationship. Given the proper training in interpersonal relationships, a supervisor might be expected to maintain a relationship that is motivational. Training in transactional analysis (TA) is particularly helpful for this. Once trained in TA, supervisors might be expected to maintain an adult-to-adult relationship with each employee.

The last questions on the form in Exhibit 4.22 will help to define the relationship. Periodic contacts with lower-level employees by representatives of upper management are informative also. The important thing is that upper management care enough to find out what kinds of relationships exist.

Department

Note: Do not sign your name. Your boss will not see this sheet. He or she will receive a summary of all responses from this department.

Consider your boss and how he or she performs compared to your expectations of him or her.

Does your boss: Better than I would expect   Worse than I
                                                                                would expect
                                            10   1

Know you? _____
Understand you? _____
Know what your needs are? _____

Write any comments here you wish

Back you? _____
Listen to you? _____
Talk to you? _____

Write any comments here you wish

Allow your input? _____
Ask for your ideas? _____
Use your ideas? _____

Write any comments here you wish

Remove any barriers in your way? _____
Have enough influence with his or her boss? _____
Have enough influence with other departments? _____

Write any comments here you wish

Talk down to you? _____
Treat you as a child? _____
Treat you as a subordinate? _____

Write any comments here you wish

Exhibit 4.22 Inverse performance standards rating form.

Worker Safety Analysis

Safety professionals and supervisors are generally aware of the concept of job safety analysis; the systematic analysis of specific jobs to spot situations with accident potential. Worker safety analysis is the systematic analysis of a worker. One tool that we can use in worker safety analysis is shown in Exhibit 4.23. Management can devise a form to assist the supervisor to look systematically at each worker to determine whether he or she is highly likely to make human errors. Are there logical reasons why any particular employee is likely to make such errors? Worker safety analysis can uncover these reasons.

Some of the subjects covered on the form in Exhibit 4.23 are optional, for example, information about biorhythms and life change units (LCUs). These items can be left off a form if the information is not available, although LCU information might be available to a supervisor who knows his or her people. Personality type and accident risk (discussed in Chapter 7) are also optional and might not be filled in by many. These items may be less valuable, from a safety standpoint, than the other items on the form. The value analysis is quite arbitrarily decided by the supervisor, and might or might not be useful.

Current motivational analysis is the most valuable item on this form. The supervisor should look at each item listed and determine what motivational pull it will have on the employee. Included are most of the important determinants of employee performance we have discussed in this book.

The supervisor should consider current job assignment in terms of the load it places on the person. The last item on the form is force field analysis; the supervisor may choose to perform a small force field analysis to determine the current pulls on the employee. The entire worker safety analysis might lead to some disposition, as shown at the bottom of the form.

Obviously, the supervisor who fills out this kind of form will need considerable training in the concepts involved. The purpose of worker safety analysis is to help the worker; its intent is to spot the causes of human error before an accident can occur.

THE MIDDLE MANAGEMENT ROLE

The traditional picture of middle manager behavior is one which is abstract and static, where the manager is seen as having a nicely bounded, carefully defined, and compartmentalized role. Authority is supposedly commensurate with responsibility. The middle manager receives instructions from the boss and transmits them to subordinates; by this process making sure, by exhorting, directing, or cajoling, that they know what they are supposed to do and that they will be held responsible for doing it. The distinguishing marks of this system of organization are that every individual has a job which is clearly defined (it has a clear beginning and end), and that every member of the organization is required to make discrete decisions within the framework of "delegated" authority. This approach is characterized by organization charts and by terms such as *delegation, line,* and *staff.*

Empirical studies clearly show that successful managers in modern organizations operate in a manner very different from that suggested by the traditional view of management. The facts of organizational behavior make it perfectly plain that the modern manager operates in a dynamic context where things are not as simple as the *principles of management* might suggest. The manager's job today is *not* a job with nearly defined authority and responsibility. The traditional idea that organizations can be characterized as networks of roles connected by single

Worker-Safety Analysis

Name _____ Date _____

Long-term analysis

    Biorhythmic information: dates to watch: _____
    LCU information: approximate units accumulated now: _____

Personality and value analysis
    Personality type _____
    Accident risk _____

	Key importance	No importance
Value of work		
Value of safety		

Current motivational analysis	Turn ons	Turn offs
Peer group		
Me (boss relations)		
Company policy		
Self (personality)		
Climate		
Job-motivation factors		
Achievement		
Responsibility		
Advancement		
Growth		
Promotion		
Job		
Participation		
Involvement		

Current job assignment	High	Low
Pressure involved		
Worry or stress		
Information processing need		
Hazards faced		
Other		

Force-field analysis

Pulls to safety

  ↑
_____
   ↓

Pulls away from safety

_____

Current assessment:

☐ OK   ☐ Discuss with worker   ☐ Training   ☐ Crisis intervention
               ☐ Contract   ☐ Behavior modification
☐ Crisis intervention

Exhibit 4.23 Worker safety analysis form.

lines of authority is not only far from being the case, it is native. There is a multiplicity of relationships between managers in an organization.

Leonard Sayles, in *Managerial Behavior,* discusses some of the "old wives' tales" of management theory and includes the following theories:

- *A manager should take orders from only one person, the boss.* Most managers, in fact, work for, or respond to, many people (customers or those in a position to make demands upon them).
- *The manager does no work himself; but gets things done through the activities of subordinates.* Actually the manager must carry on many relationships with others and participates in all kinds of activities to get things dones.
- *The manager devotes most of his or her time and energy to supervising subordinates.* Actually, the manager is away from subordinates a significant portion of the time.
- *The good manager manages by looking at results.* Actually, methods of continuous feedback are required.
- *To be effective, the manager must have authority equal to responsibility.* Actually, a manager almost never has authority equal to responsibility; but must depend on the actions of many people over whom he or she has not the slightest control.
- *Staff people have no real authority since they are subsidiary to the line organization.* Actually, staff groups have very real power.

Since the middle manager's job is not neatly defined, it would be naive to spell out the safety responsibilities in a neatly defined manner.

The danger, however, is that the middle manager might be left out of the system, which would destroy the entire system. If there is one problem with traditional safety programs that towers over all others, it is ignoring middle management in the program. Today we believe the middle manager is much more the keyperson in safety than the supervisor, and must have a clearly defined role that consists of:

- Ensuring supervisory performance.
- Ensuring the quality of that performance.
- Doing some tasks personally to show safety's importance.

These might be defined further for the middle manager as the following:

- To restate policy.
- To participate in meetings.
- To review employee performance.
- To establish checks to ensure adherence to the program.
- To set an example.
- To use safety performance as a measure of management capability.
- To serve on investigating committees.

These responsibilities can be quantified, observed, and measured. Incentives (rewards) can be structured for each type of behavior wanted and a system devised to link the reward consistently to the behavior.

Middle managers (anyone between the executive on top and the supervisor) must again become an integral part of the safety program for program success. Subordinates (the supervisors) will respond and react primarily to their wishes, so their wish list must include safety performance. If they do not perceive results in safety as their problem, it is unlikely they will put the pressure on beneath them. If they are not required to carry out some performances related to safety, it is unlikely they will see subordinate safety performance as important. If they do not engage in some regular visible safety related performance, it is unlikely their subordinate workers will see safety as a high priority item.

Since middle managers become the driving force in the entire safety system, we will look in more depth at their role in the next chapter when we discuss drivers.

THE EXECUTIVE ROLE

Top management participation is highly important in functioning safety programs. This is not exactly a new or exciting thought. Safety people have spent considerable time thinking about and talking about how to get management's backing, how to enlist its support, or how to get its interest. For the most part they have little to show for their efforts.

Perhaps the real key to our past problems of getting management's "backing" is that the few times we have gotten top management attention we have not been sufficiently clear about exactly what we wanted them to do. We have shown them our figures which either spell out that we are doing a good job, or that "they" (the line) are not doing their job. But we have not stated exactly what we want the executives to do differently from what has been done in the past. About the clearest thing we do ask is that they sign and issue a safety policy.

Our thinking on "managing the safety function" starts with a written safety policy—a definition, if you will, of management's desires concerning safety. Most executives agree that a policy is a fine thing to have, but few agree on what a policy is.

"Policy" is often confused with "rules," "established practices," "procedures," and "precedents"—not only in speech but also in action. But "policy" has certain unique implications. It implies scope for discretion, initiative, and judgment in deciding what ought to be done in specific situations.

Too often in discussions regarding policy, we become bogged down trying to sort out policy from procedures, from SOPs and from rules. Perhaps it is not all that necessary for policy to be pure. The most important thing is probably whether management's interest is accurately communicated.

Defintion of Policy

One dictionary says that policy is "a settled course adopted and followed by a body." When top executives determine and announce a "settled course," they are affirming a shared purpose or one in which they want to enlist the voluntary cooperation of every member of the organization. They are giving out a guide for thinking, and this guide is to be used by other members who have been delegated authority to make decisions in keeping the company on a given course while realizing company purposes.

A policy should do three things:

1. It should affirm long-range purpose.
2. It should commit management at all levels to reaffirm and reinforce this purpose in daily decisions.
3. It should indicate the scope left for discretion and decision by lower-level management.

Why Safety Policy?

If the principle "*Safety is a line responsibility*" is true, it is important not only that we as safety people believe it but also that the line organization believe it. The line organization will believe that safety is its responsibility *only* when safety is definitely assigned to it by management.

Another principle is that "*Management should direct the safety effort by setting safety goals and by planning, organizing, and controlling to achieve them.*" A safety policy is management's expression of direction to be followed. It is management's first step in organizing to accomplish its desires. In almost all cases it is important that management's safety policy be in writing to ensure that there will be no confusion concerning direction and assignment of responsibility.

Safety policy, more than most other policies that come from management, requires some action from each individual in the organization, from the president to the lowest-rated worker. Safe performance of an organization requires that a decision be made by each person in it. The most important factor that each will consider in making the decision for or against safety is, "What does the big boss want from me?"

What is included in the safety policy may vary from company to company. No doubt most organizations will not write "pure" policy. They will include, either intentionally or inadvertently, some procedures, some philosophy, and perhaps even some rules. This is perfectly all right— whatever serves the company best is what should be included.

No one policy is right or wrong—we might best assume that each is right for the organization it serves. We can, however, outline some of the things that should be included in most management policies on safety. As a minimum, the following areas ought to be touched on in a safety policy:

- *Management's intent.* What does management want?
- *The scope of activities covered.* Does the policy pertain only to on-the-job safety? Does it cover off-the-job safety also? Fleet safety? Public safety? Property damage? Fire? Product safety?
- *Responsibilities.* Who is to be responsible for what?
- *Accountability.* Where and how is it fixed?
- *Safety staff assistance.* If there is safety staff, how does it fit into the organization? What should it do?
- *Safety committees.* Will there be committees? What will they do? Why do they exist?
- *Authority.* Who has it, and how much?
- *Standards.* What rules will the company abide by?

These questions may provide some insight to the reader who is contemplating writing a safety policy. There is one more procedure to be followed: To be effective, the policy must be signed by management. It may be conceived and written by the safety professional, but it *must* be published under the name of the executive who is responsible for all other production activities. It is often important that the *top* executive issue safety policy, as safety affects *all* departments in the organization—not just production.

Additional Executive Functions

Just signing policy is clearly not enough. It must spell out in detail exactly the role of the executive in the safety program. For instance, it should require that the responsible executives:

- Sign and issue safety policy.
- Receive information regularly on who is and who is not performing in safety by some predetermined criteria of performance.
- Initiate the positive or negative rewards to immediate subordinates (middle managers).

On a day-to-day basis this may be the only executive input needed in the program. At times, of course, policy decisions might be necessary, but these would be handled as any other management policy decision might be handled.

One way of informing the executive of what is going on and of explaining exactly what you want done is your annual report. Every safety manager ought to utilize this tool. If management does not ask for such a report, it should be submitted anyway. It would certainly be peculiar for management to appoint a person to handle the staff safety function (whether full or part time) and then never ask what the company is getting for the money spent.

Whatever the safety person wants management to know should go into this annual report. For example, the report might answer these questions:

- How did we fare last year? (Give results expressed in management's terms.)
- What did we accomplish last year? How are we stronger than we were before?
- What are our objectives for next year? How will we be stronger at this time next year?
- What do we need from the executive?

This report is crucial to a safety professional's relationship with management. Future goals included in the report tell management the direction you intend to take. If management approves, it is committed to that direction.

Management thinking tends to support the concept of delegation of responsibility to the lowest possible level. While this concept is excellent, there are some real hazards in the concept when applied to safety. When all responsibility is delegated in safety, it leaves upper levels of management with little to do about safety, other than the traditional role of signing policy. Too often this lack of involvement on a regular basis can be construed by the individuals in the organization as saying that safety is unimportant to the executives, that it has a low priority, and that it is not important enough for executives to spend their time on.

It is therefore extremely important that safety responsibilities not be delegated totally to the lowest possible level. All levels of management, including the top level, must retain some specific predetermined tasks that are visible—that can be seen by all workers in the organization as an indication of the importance of safety to that executive.

As indicated earlier, the role of upper management is threefold. It must

- Ensure the performance of subordinate managers (middle managers) in safety—an *accountability* function.
- Ensure the quality of that subordinate safety performance—a *quality control* function.
- Personally engage in some specific safety related function that can be seen as a demonstration of management priorities—a *visibility* function.

THE STAFF SAFETY ROLE

The job of staff safety specialist is very much a self-defined one. The duties will vary, depending on the size of the organization, the number of locations, the operations themselves, the people above the safety specialist and in line management, the problems presently facing the company, the other staff people and specialists available, and where the safety specialist fits into the organization.

In 1963 the executive committee of the American Society of Safety Engineers initiated a plan designed to identify the type of work that safety personnel should be doing. This was part 1 of a three-part major project; it described the scope and functions of the safety professional's position. Part 2 developed a curriculum or formal course of study, leading to a university degree, which would prepare the safety professional to perform the functions described in part 1. Part 3 established procedures leading to the acquisition of some form of certification or registration as a means of demonstrating competence in the field. Following are the results of part 1 of the project.

The major functions of the safety professional are contained within four basic areas. However, application of all or some of the functions listed below will depend upon the nature and scope of the existing accident problems, and the type of activity with which he is concerned.

The major areas are:

— Identification and appraisal of accident- and loss-producing conditions and practices, and evaluation of the severity of the accident problem.
— Development of accident prevention and loss-control methods, procedures, and programs.
— Communication of accident and loss-control information to those directly involved.
— Measurement and evaluation of the effectiveness of the accident and loss-control system and the modifications needed to achieve optimum results.

Exhibit 4.24 briefly outlines the major activities the safety manager will engage in. First of all structured systems of measurement must be established so that accountability can be fixed and so that rewards can be applied properly to the right people and at the right time in order to

---

1. Measuring safety performance, using
   Results Measures
   Activity Measures
   Audits
   Records
   Statistics
   Inspections
   Sampling
2. Safety program development, including
   Orientation
   Training
   Supervisory department
   Motivation
   Gimmicks
   The Selection process
   Medical Controls
3. Being a technical resource
   In investigation
   On standards and regulations
   On consumer products
   On new equipment purchases
4. Being a Systems Analyst

Exhibit 4.24 Safety staff's role.

reward or reinforce the desired behavior. Some of these systems might incorporate estimated costing, sampling, rating effort (SCRAPE) performance measurements as described in the next chapter. Secondly, acting as a programmer overseeing those aspects of the safety program that are not completely under personal purview, the safety professional must ensure that safety is included in orientation, that safety training is provided where needed, that safety is a part of supervisory development, that things are done that help to keep the organization's attention on safety, that safety is included in employee selection and in the medical program. Thirdly, the safety professional must function as a technical resource: know how to investigate in depth, know where to get technical data, know the standards, know how to analyze new products, equipment, and problems. And fourthly, the safety professional must function as a systems analyst, searching for whys and whethers.

Exhibit 4.25 says much the same thing. It suggests that the safety manager should concentrate initially on five key areas and find answers to questions in these five areas. These questions, or similar ones, might suggest those weaknesses in the system that would be most important to begin working on.

As staff, the safety manager has no responsibility for the safety record or the results, but is responsible for activities which help the line achieve its goals in safety. At first, the safety

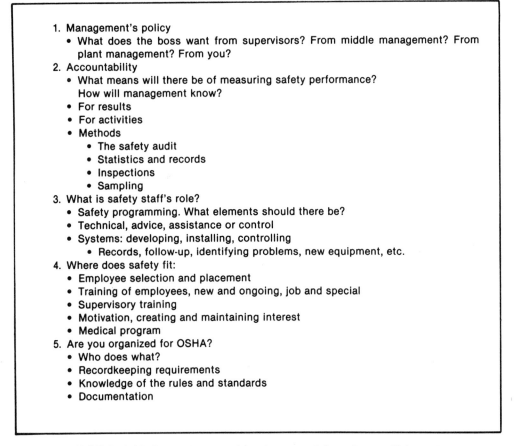

1. Management's policy
   • What does the boss want from supervisors? From middle management? From plant management? From you?
2. Accountability
   • What means will there be of measuring safety performance? How will management know?
   • For results
   • For activities
   • Methods
      • The safety audit
      • Statistics and records
      • Inspections
      • Sampling
3. What is safety staff's role?
   • Safety programming. What elements should there be?
   • Technical, advice, assistance or control
   • Systems: developing, installing, controlling
      • Records, follow-up, identifying problems, new equipment, etc.
4. Where does safety fit:
   • Employee selection and placement
   • Training of employees, new and ongoing, job and special
   • Supervisory training
   • Motivation, creating and maintaining interest
   • Medical program
5. Are you organized for OSHA?
   • Who does what?
   • Recordkeeping requirements
   • Knowledge of the rules and standards
   • Documentation

Exhibit 4.25 Areas to consider in organizing the staff function.

specialist has no authority over the line. He or she may have a great deal of influence, but this is quite different from authority. How much influence or how much power the safety manager represents will depend on the organization and on the personality of the individual in the line position. The safety specialist in any organization obtains results by using one or two methods: (1) making a recommendation to an executive in the line chain of command and that executive issues an order or (2) obtaining acceptance for suggestions voluntarily from line supervisors without taking the chain-of-command route. More often than not, the safety specialist achieves desired results by the second route and uses the first route only for rare emergencies.

Most line managers realize that the safety specialist does have stronger influence than that shown on the organization chart.This staff specialist is an expert in the field, has certain status, has management's interest and backing, and, if worst comes to worst, has some influence on management's appraisal of the line manager, and hence on that line manager's future. So, although the safety specialist has no authority, he or she is not without power.

So we see a staff job that is pretty well self-defined and whose influence depends on personalities as well as location.

It may be stated in the policy that, in certain situations, the safety specialist must be consulted, give approval, or can even temporarily step in and assume command (stop the operation). This granting of temporary authority to the staff safety specialist in industry is common and right. Even so, we must keep in mind that basically safety staff works through influencing the line, not by directing it. Any time that safety must step in and assume direction of the line, it indicates a failure in that the line has not taken care of the situation first.

Where to Install Safety

To whom should the staff safety specialist report? Should the specialist be staff to a line manager or staff to a staff executive? These questions have been debated for years, and we are sure of only one thing—that there is no one right answer. It depends on the organization and the personalities of the people involved. There are, however, some criteria that can be used in assessing the right place for safety in an organization.

Any discussion of organization charts must be qualified by the observation that charts do not reflect the give-and-take of powerful executives. Whether safety should report to line or to staff might depend on a more rational process of structuring the organization to achieve goals, or it might depend only on personalities.

The location of safety within the organization must reflect the fact that root causes of accidents exist at all levels, in all departments, and in all functions. Interrelated causes can be in maintenance, purchasing, tool-crib control, or selection of personnel. Therefore, safety seeks to exert a control, or at least an influence, on every department head, every function, and every supervisor.

We can begin to ask some questions about company A in Exhibit 4.26. Presumably, if the works manager demands accident control from superintendents, the safety director could be effective in the area controlled by the superintendents. But can the safety director exert effective influence on people who are wrecking company vehicles or on people in the R & D laboratory? If the works manager is apathetic about safety, could the safety director even reach the superintendents? How would it be possible to establish liaison and understanding with the treasurer, who buys the insurance and who best realizes the dollar cost of accidents? Note that personnel is a staff function to the works manager, not to the whole organization. Could personnel exert standards of competence in the hiring of salespeople? What, if any, difference

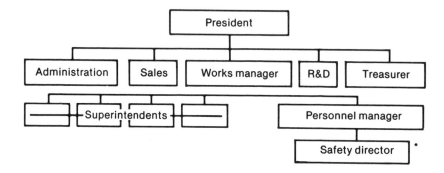

Exhibit 4.26 Organization chart, company A.

would it make if safety were assigned to report directly to the line works manager instead of the staff personnel manager?

In company B (Exhibit 4.27) safety reports to a staff function, industrial relations, which is staff to the whole organization, not to just one line department. Presumably, the safety director, through his or her executive boss, now has a channel for communicating with the R & D department (or the sales department, purchasing, maintenance). In any case, the boss is close to the ear of the chief executive. This proves nothing yet about where safety should be installed, but we have arrived at a principle: *The voice of the safety director is the voice of the boss.* Who that boss is, how far the boss's voice reaches, and whether the boss speaks for safety are among the factors that limit the safety director's effectiveness. Is the boss in line or staff? If staff, to whom is the boss staff? These are factors to contend with. We cannot say that safety must of necessity report high or low, staff or line. But we can now present criteria for determining how safety is allocated in the organization:

— *Report to a boss with influence.* In part this is a personal evaluation; in part it is an evaluation of the structure of the organization. If the boss is line, does his or her line authority encompass the hazards to be controlled? If staff, can the boss's voice reach an executive whose command will buttress necessary action?

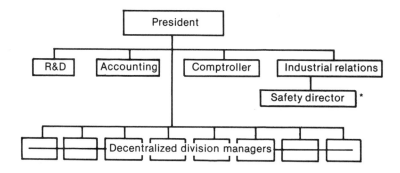

Exhibit 4.27 Organization chart, company B.

— *Report to a boss who wants safety.* Problems arise when a chief executive wants results but the voice of the safety director is muffled by an immediate boss who is concerned with other problems.

— *Have a channel to the top.* Management properly sets the priorities between production results and safety results, between sales expansion and elimination of unsafe driver-sales-people, between security of confidential research and the prying eyes of the safety professional. This is not to say that safety must be placed in the upper echelons, but it does assert that all parts of the organization must have a channel to the upper echelons. Too often the only channel is a bypass, with all its frictions.

— *Perhaps install safety under the executive in charge of the major activity.* The safety function in this case serves as staff to that executive. This obviously eliminates the "channel to the top." Nonetheless, if the acute need is in the shop, let the safety specialist work with the shop executive. If the acute need involves truck operation, let the transportation executive handle the problem with the safety director as staff assistant. Influence in other departments may be weak, but management has directed its control to the spot that hurts.

These criteria serve more to assess present structure than to determine where safety should be located. The latter depends on management goals. This is illustrated in Exhibit 4.26 and Exhibit 4.27. In company C (Exhibit 4.28) note the allocations for fire, security (guards), and health. This arrangement makes sense when we realize that this company manufactures a non-toxic but flammable household product, but has recently gone into insecticides and other products of high toxicity.

Fire control is concentrated where it is needed, rather than as a part of either safety or security. Other functions of security (theft, vandalism, sabotage, confidential information) demand little management attention, and the security function becomes simply one of "guarding." Safety is staff to a line executive. Note "health," however. Here management must know

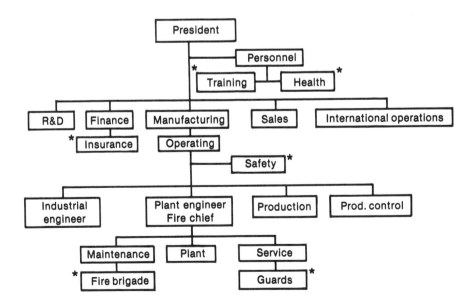

Exhibit 4.28 Organization chart, company C.

what problems will arise with any new product. Health and control of health hazards are situated to serve top-echelon decision-making. In short, the chart reflects management control of its problems and a sense of management goals.

This principle is further revealed in company D (Exhibit 4.29). Here, the left branch portrays an East Coast home office. The right branch portrays a number of locations across the United States, including a California operation. It sells "brainpower," primarily on government con-

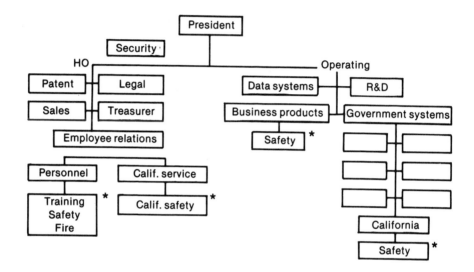

Exhibit 4.29 Organization chart, company D.

tracts, but it also has a manufacturing operation called "business products." The chart, of course, does not reveal the reason for the multiple niches of safety, but it reveals something. The California operation needs safety on the spot, but 3,000 miles away it also needs safety as part of "California service" of the home office. The niche for security is understandable when we realize that a breach of security could jeopardize the existence of the whole company.

Organization structure begins with management objectives and goals. The tug and pull of contending executives change and distort the structure, but even this reflects management goals or accommodation of goals. Though business affairs cannot be constrained or described in neat little boxes, no organization is totally unstructured.

Structure reflects grouping of activities, but the logic of grouping is the logic of goals, not the logic of words. Safety has been effectively allocated into line or into staff, grouped with fire or security, or the legal department, or maintenance, or personnel, or even the training department. The practice of other organizations is a poor guide to follow.

The Part-Time Specialist

What has been said in this chapter refers not only to the full-time safety professional but also to the personnel specialist, the line manager, the insurance manager, and others with a part-time safety responsibility. The safety professional is still staff and works through influencing, not directing. The safety professional's niche in the organization still affects his or her effectiveness.

There are, of course, far more part-time safety specialists in American industry today than full-time safety professionals. The role should be no different from that of the full-time safety professional; the function is still that of advising and assisting the line organization in its job of loss control. Hence, anything that has to do with loss control concerns the part-time specialist as much as it does the full-time professional.

The problem is that when safety is a part-time function, it is too often a "tacked-on" function; it is the secondary job, never the primary one. To cast safety as a secondary function is to admit that safety is less important than other functions. In most cases, management's policy will say otherwise. Perhaps here again the part-time safety specialist should share the fault for not holding management to its own stated policy when the time squeeze is on.

The outside consultant can give valuable assistance to the part-time safety specialist. Too often the consultant wants to help but is not used properly by the part-time safety specialist. Whether this person is from the government or the insurance field or is a private consultant, develop an action plan. If used properly in the areas of their specialty, consultants can end up as staff assistants to part-time specialists. This often becomes the equivalent of a full-time function. Consultants often specialize in those areas in which the part-time specialist needs help the most: supervisory training, industrial hygiene, cost analyses.

Often the primary reason the part-time specialist is so much less effective than the full-time safety professional is simply lack of knowledge. This is hardly excusable, for ample safety information is available.

OSHA (and other federal and legal) Responsibilities

After defining the safety responsibilities of the job, the safety director must define the OSHA responsibilities. In terms of OSHA compliance, the safety director assumes a quite different role, that of the corporate OSHA expert—the only one who "knows" the standards. Probably the reason for this is that the standards are written in such a way that the line organization perceives them as being much more technical and difficult to understand than they really are. As a result, the safety specialist is not able to ask line managers to ensure that their departments are in compliance.

Staff Safety Effectiveness

William English, former Director of Loss Control for Marriott Restaurant Operations, wrote recently in the *National Safety News* on the effectiveness of the safety manager. He quoted Peter Drucker as saying efficiency is doing things right, but effectiveness is doing the right things. Therefore, the effective safety manager is the one who is able to analyze needs, coordinate the formulation of remedies, sell the remedies to the organization, and help the organization implement the prescribed programs. These functions require judgment, planning, and enough industrial experience to understand the organization and its function.

It is no doubt true that the most effective safety managers are the ones who in fact do the right things (that is advise, develop solutions, communicate those plans to the line, and monitor results), and the effective safety managers are the ones who concentrate on doing the things that the line managers refuse to do.

The Safety Professional's Job

Dr. Bill Tarrants describes the safety professional's role like this:

> The contemporary approach of the safety professional is a unique application of measurement and evaluation, scientific principles and concepts, engineering design and analysis, accident problem appraisal, safety program management, environmental study and analysis, safety education and promotion, communications and language arts, and various techniques intended to motivate human behavior within acceptable limits of safe performance. The safety professional's consideration of people, the safe design and operation of systems involving people, and the manner by which equipment and machinery failures and errors in human performance are analyzed and various counter-measures are prescribed and applied, requires a fundamental analytical approach that is essentially different from the more traditional academic disciplines. In fact, there are some who believe that the work of the safety professional is so different from the traditional disciplines that it should be considered as a separate and unique academic field of study.
>
> The contribution of the safety professional is in the management decision-making process. Problems in this area deal with the optimum utilization of people, materials, equipment, and energy to achieve the objectives of an organization. The managers need factual information along with well defined alternatives and predicted consequences to help them recognize and solve existing or potential accident problems. The safety professional collects, analyzes, and arranges this information in such a way as to fulfill management's decision-making needs. He assists various levels of management by originating and developing policy recommendations, operating plans, programs, measures and controls which will permit the effective, accident-free use of human and material resources.
>
> The authority of the safety professional is that of knowledge and the soundness of the information he provides. His success is a function of his ability to gather well-documented facts based on valid and reliable measures of safety effectiveness. In one sense, we might define accidents as *local system failures*. The safety professional, by perfecting various techniques of measurement and applying appropriate measurement tools, is able to provide the information necessary for the prediction and control of these local system failures. The safety professional's primary skill is the ability to perceive problems, measure loss potential, and deliver sound information accompanied by well-conceived action alternatives so that managers can properly weigh the safety aspects of their management decisions.

The Safety Committee

The industrial safety movement has been blessed with (and cursed by) an organizational phenomenon that has affected no other segment of industry. This is the safety committee. The marriage of industrial safety and the committee approach is, to this day, so strong that in some states companies are literally forced into forming safety committees through codes or through insurance rating plans. Government publications urge committees.

How and why committees were initially used in safety is a mystery. They have not been used to the same extent in any other management function. Many organizations utilize committees in their executive ranks, but safety has few executive committees and many supervisor and worker committees.

Although many competent professionals use committees, some extensively, no competent professional depends on committees to do the safety job alone. At best, committees provide training and motivation for their members. At worst, they are a total waste of time.

Whether or not an organization wishes to use safety committees is strictly its own decision, but to say that committees are essential to safety results is ludicrous. They may help the safety director; they may not. This will depend on many things.

If management takes care of the basic management functions of preparing policy, fixing accountability, training supervisors and employees, and selecting and motivating employees well, then safety committees probably are not necessary.

On the other hand, if management *does not* effectively carry out these functions, then using safety committees certainly will not ensure results. In this kind of situation it is doubtful whether having committees will help at all.

Although safety committees are extremely common in industry today, it is difficult to imagine a situation to which there would be a better solution than the more normal, more effective one of good management-directed control. If safety committees are used, the safety professional would do well to examine closely the workings of those committees. For effective operation these rules are essential:

— Define the duties and responsibilities clearly.
— Choose members in view of these duties.
— Provide any necessary staff assistance.
— Design procedures for prompt action.
— Choose the chairperson carefully.

THE ROLE OF THE EMPLOYEE

The employee role has also changed over the years. In the early years under Heinrich thinking, the employee was at fault. In later years under our classical management methods, the worker had to be directed, motivated, and controlled. In the human relations era, the worker had to be made happy. In the situational management period, the worker had to be handled as an individual. Under cultural assessment, the worker had to live in an environment that was good for him.

In most of the above the worker had no role in safety; perhaps in the later years there was a little piece of involvement or participation, whatever those terms meant to management.

Today, however, the role of the employee is becoming different from any previous era. As we evolve towards true participative management, the employee role enlarges and becomes two-fold:

1. Totally responsible for his or her own actions.
2. Totally responsible for (or at least is a full partner in) running the safety system.

In the next chapter we'll look at what drives performance (the carrying out of the role) of each level in the organization.

REFERENCES

American Society of Safety Engineers. "Scope and Functions of the Professional Safety Position," Chicago, 1966.

English, W. "What Does It Take to be an Effective Safety Manager," *National Safety News,* Chicago, September, 1980.

National Safety Council. *Accident Prevention Manual for Industrial Operations,* National Safety Council, Chicago, 1974.

Petersen, D. *Safety Supervision,* Aloray, Goshen, NY, 1984.

Petersen, D. *The OSHA Compliance Manual,* McGraw-Hill, New York, 1975.

Pigors, P. and F. Pigors. "Let's Talk Policy," *Personnel,* July 1950.

Savles, L. *Managerial Behavior,* McGraw-Hill, New York, 1964.

Strode, M. "How to Motivate Employees to Work 2,000,000 Safe Man Hours," presented before the Pulp and Paper Section of the 1955 National Safety Congress.

Tarrants, W. "Preparation, Growth and Development in Safety," *Professional Safety,* Oakton, IL, May 1981.

The Traffic Institute. "Organization: Simplified Definitions of Staff," Northwestern University, Evanston, IL, 1962.

Weaver, D.A. and D.C. Petersen. "Criteria to Niche Safety," *Industrial Security,* August 1966.

Weaver, D.A. "How to Conduct TOR Analysis," Employers Insurance of Wausau, Wausau, WI, 1967.

Drivers

This chapter **is** about drivers—those elements that drive performance. Many of the elements of traditional **safety** programs will not be covered in this chapter because they simply are not drivers.

WHAT DOESN'T WORK AND WHY

Let's look at some of the traditional safety elements and see why they are not drivers.

Manuals

Classical management theory says every function should have a manual and that the manual will drive performance. Current thinking suggests that any competent manager quickly figures out what he or she must adhere to and which pages can be ignored. They simply pay attention to those dictates the boss measures; the rest they can comfortably ignore (until something happens and that probability is usually rather small).

Manuals do not drive performance. Measurement does.

Job Descriptions

Classical management suggests a job description ensures performance. It doesn't. Usually what we do each day is quite different from the job description.

Rules and Regulations

Seldom do rules and regulations govern behavior. If they make sense they might be a factor (Have you looked at your rule book lately?). One organization recently reviewed their safety rule book and sorted out their "musts" from their "like-to-haves," reducing a book of hundreds of pages to one page of "cardinal sins." Another organization with a rule book of over 600 safety rules found procedures in the book that were eliminated in 1933. The fact that nobody noticed this for 50 years indicated how often the rule book was looked at and used.

Training

Training imparts knowledge and skills. Seldom does training require the use of that knowledge or those skills. Training does *not* beget performance at any level.

Awareness Campaigns

Do awareness campaigns work? We don't know, but an educated guess would be "probably not." What are we trying to make those workers aware of? Current theory says that most workers know the difference between safe and unsafe behavior; they are aware. They choose to work unsafely for reasons other than lack of knowledge.

Gimmicks

Gimmicks, or incentives, are used almost universally. We are fairly certain through research that they do not markedly influence or change behavior.

What Does Drive Performance?

In Chapter 3, Principle 5 states that "the key to line safety performance is management procedures that fix accountability." This is simply a restatement of a key management principle.

For well over 50 years we have been preaching the principle of line responsibility in safety work, and yet there are still supervisors today who say, "Safety is the safety director's job" or "If that's a safety problem, take it up with the safety committee." Worse yet, in many companies, when an accident occurs, it goes on the safety specialist's record instead of on the record of the line supervisor in the department where it occurred. Thus, instead of simply preaching that the line has responsibility, we should have been devising procedures to *fix* such accountability. When a person is held accountable (is measured) by the boss for something, he or she will accept the responsibility for it. If not held accountable, he or she will not accept the responsibility. Effort will be put forth in the area in which the boss is measuring.

The attitude of the majority of supervisors today lies somewhere between total acceptance and flat rejection of comprehensive accident prevention programs. Most typical is the organization in which line managers do not shirk this responsibility but do not fully accept it either and treat it as they would any of their defined production responsibilities. In most cases their "safety hat" is worn far less often than their "production hat," their "quality hat," their "cost control hat," or their "methods improvement hat." In most organizations, safety is not considered as important to the line manager as many, if not most, of the other duties that he or she performs.

On what does a manager's attitude toward safety depend? It depends on abilities, his or her role perception, and effort (see Exhibit 5.1). All are important and a manager will not turn in the kind of performance we want unless we take all three into account.

Two basic factors determine how much effort a person puts into a job: (1) his or her opinion of the value of the rewards and (2) the connection the person sees between effort and those rewards. This is true of a manager's total job, as well as of any one segment of it, such as safety.

The Value of Rewards

Rewards for safety performance are no different from rewards for performance in any other area where management asks for performance from the supervisor. While our model focuses primarily on positive rewards (peer acceptance, subordinate approval, enhancing the likelihood of promotion, merit salary increases, higher bonus, intrinsic feelings of accomplishment, pat-on-

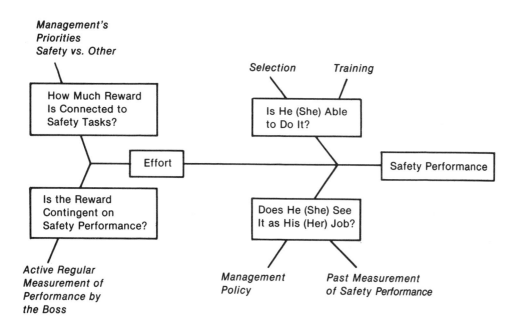

Exhibit 5.1 Supervisory safety performance model. (Adapted from the Lawler and Porter model).

the-back, compliment from the boss), it also could mean negative reward (chew-out, lower and harder pat-on-the-back, reprimand).

The main difficulty in developing a reward system is not in determining what the rewards will be for performance but rather in determining when the reward should be given, when the supervisor has "earned" it. For in safety we have precious few good measurement tools that tell us when a supervisor is performing.

The manager looks at the work situation and asks, "What will be my reward if I expend effort and achieve a particular goal?" If the supervisor considers that the value of the reward which management will give for achieving the goal is great enough, he or she will decide to expend the effort.

"Reward" here means much more than just financial reward. It includes all the things that motivate people: recognition, chance for advancement, increased pay. Most research into supervisory motivation today indicates that advancement and responsibility are the two greatest motivators.

The Effort-Rewards Probability

In assessing whether rewards really depend upon effort, managers asks the following kinds of questions:

— Will my efforts here actually bring about the results wanted, or are factors involved that are beyond my control? (The latter seems a distinct possibility in safety.)
— Will I actually get that reward if I achieve the goal?

— Will management reward me better for achieving other goals?

— Will it reward the other manager (in promotion) because of seniority, regardless of my performance?

— Is safety really that important to management, or are other areas more crucial to it right now?

— Can management really effectively measure my performance in safety, or can I let it slide a little without management's knowing?

— Can I show better results in safety or in some other area?

The line managers unconsciously asks these questions and others before determining how much effort to expend on safety. They must get the right answers before deciding to make the effort needed for results. Often line managers decide that their personal goals would be better achieved by expending efforts in other areas, and too often their analysis is correct because management *is* rewarding other areas more than safety.

Changing this situation is the single greatest task of the safety professional. Change can be achieved by instituting better measurement of line safety performance and by offering better rewards for line safety achievement.

Ability

Job performance does not depend simply on the effort that managers expend. It depends also on the abilities they bring to the task, both inherent capabilities and specialized knowledge in the particular field of endeavor. In accident prevention, this means that we must ensure through training that line managers have sufficient safety knowledge to control their people and the conditions under which these people work. In most industries lack of knowledge is not a problem, for line managers usually know far more about safety than they apply. Managers can achieve remarkable results on their accident records merely by applying their management knowledge, even if they have little safety knowledge. If a manager does not have adequate safety knowledge, the problem is easily handled through training.

Role Perception

Role perception is even more important than ability. Line managers' perceptions of their safety role determine the direction in which they will apply their efforts. Lawler and Porter describe a good role perception as one in which the manager's views concerning placement of effort correspond closely with the views of those who will be evaluating his or her performance.

In safety, role perception has to do with whether line managers know what management wants in accident control and with whether they know what their duties are. In role perception, the safety professional should search for answers to some questions about the organization and about each line manager in it. These questions concern the content and effectiveness of management's policy on safety, the adequacy of supervisory training, company safety procedures, and the systems used to fix accountability.

Exhibit 5.1 suggests that performance in safety is determined by four things being in place:

1. Is the required performance clearly defined? This was discussed for various levels and positions in the last chapter.

2. Is the accountability system in place? Here we are looking at how the person is measured and rewarded.
3. Does each person know how to do what is expected?
4. Is the perceived reward enough to capture attention and ensure performance?

Supervisory Performance

Let's start by looking at one level, the first line supervisors. What drives performance? At this point it is still simple. What drives performance is the perception of what the boss wants done; their perception of how the boss will measure them, and their perception of how they will be rewarded for that performance. To restate what the research shows, these things dictate supervisory performance:

— What is the expected action?
— What is the expected reward?
— How are the two connected?
— What is the numbers game (how measured?).
— How will it affect me today and in the future?

Accountability Systems

Any accountability system which defines, validly measures, and adequately rewards will work. Here are some examples.

SCRAPE

SCRAPE is a systematic method of measuring accident prevention effort. Most companies measure accountability through analysis of results. Monthly accident reports at most plants suggest that supervisors should be judged by the number and cost of accidents that occur under their jurisdiction. We should also judge line supervisors by what they do to control losses. SCRAPE is one simple way of doing this. It is as simple as deciding what supervisors are to do and then measuring to see that they do it.

The SCRAPE rate indicates the amount of work done by a supervisor and by the company to prevent accidents in a given period. Its purpose is to provide a tool for management which shows before the accident whether or not positive means are being used regularly to control losses.

The first step in SCRAPE is to determine specifically what the line managers are to do in safety. Normally this falls into the categories of (1) making physical inspections of the department, (2) training or coaching people, (3) investigating accidents, (4) attending meetings of the workers' boss, (5) establishing safety contacts with the people, and (6) orienting new people.

With SCRAPE, management selects which of these are things it wants supervisors to do and then determines their relative importance by assigning values to each. Let us suppose that management believes the six items above are the things it wants supervisors to do and believes that (1) and (2), inspections and training, are the most immediately important, followed by accident investigations and individual employee contacts, and that attending meetings and orienting new people are relatively less important at this time. Management might then assign these values:

Item	Points
Departmental inspections	25
Training or coaching (e.g., 5-minute safety talks).	25
Accident investigations.	20
Individual contacts	20
Meetings	5
Orientation	5
Total	100

Depending on management's desires, the point values can be increased or decreased for each item.

Every week each supervisor will fill out a small form (see Exhibit 5.2) indicating activity for the week. Management, on the basis of this form, spot checks the quality of the work done in all six areas, and rates the accident prevention effort by assigning points between 0 and the maximum.

For example, in department A the supervisor makes an inspection and makes six corrections. The safety director later inspects, finding good physical conditions. Supervisor A rates the maximum of 25 points.

In department B the supervisor makes an inspection but no corrections. The plantwide inspection, however, indicates that much improvement is needed. Supervisor B might get only 5 points for making the inspection but doing a poor job of it.

In Department C there were five accidents. Only one individual lost time, and supervisor C turned in only one investigation, getting only 5 points for effort.

Department _____ Week of _____ Points _____

(1) Inspection made on _____   # corrections _____   _____
(2) 5-minute safety talk on _____   # present _____   _____
(3) # accidents _____   # investigated _____
          Corrections _____
          _____   _____
(4) Individual contacts:
          Names _____
          _____
          _____
          _____
(5) Management meeting attended on _____   _____
(6) New men (names):   Oriented on (dates):
          _____   _____
          _____   _____
          _____   _____   _____

Exhibit 5.2 SCRAPE acitvity report form.

In department D there are 43 employees, but only 3 were individually contacted during the week. This might also be worth only 5 points.

Management decides relative values by setting maximum points, and also sets the ground rules about how maximum points can be obtained.

Each week a report is issued (see Exhibit 5.3).

SCRAPE can provide management with information on how the company is performing in accident prevention. It measures safety activity, not a lack of safety. It measures before the accident, not after. Most importantly, it makes management define what it wants from supervisors in safety and then measures to see that it is achieving what it wants. SCRAPE is a system of accountability for activities.

Safety by Objectives

Historically our safety programs have failed in some essential elements. Many programs are far from producing behavior which could be considered goal-directed. Responsibilities, even with written policy, are often unclear. Participation in goal setting and decision making is almost nonexistent. Feedback and reinforcement is slow and often not connected to the amount of effort expended in safety (especially when the number or severity of accidents is the measuring stick). Planning is minimal and while results are often measured, freedom of decision or control is seldom left to the lower levels. And finally, imagination and creativity are rare commodities in most safety programs. The principles of MBO, adapted to safety programming (SBO), can overcome some of these failings.

These are the steps of SBO:

1. *Obtain management-supervision agreement* (with staff safety consultation) on objectives. In the installation stages of an SBO program the agreement will emphasize not only results objectives but also activities objectives. Initially, then, the agreement reached will be on strategies and objectives (what means, tools, and resources to be used as well as results). Once under way, only objectives are agreed to.

Week of							
	Activity						
Department	Inspect (25)	5-min talks (25)	Acc inv (20)	Ind cont (20)	Meet atten (5)	Orient (5)	Total rate (100)
A	25	15	20	15	5	5	85
B	5	10	20	5	5	5	50
C	25	10	5	5	5	5	55
D	15	25	20	20	-	5	85
E	10	5	-	-	5	-	20
F	20	20	15	5	-	-	60
Average	17	14	13	8	3	3	58

Exhibit 5.3 SCRAPE weekly report.

2. *Give each supervisor an opportunity to perform.* Once agreements are completed, leave supervisors alone to proceed with their action plans. Require only progress reports.
3. *Let them know how they are doing.* With quantified objectives (either result or activity objectives must be quantified) give regular, current, and pertinent feedback so they can adjust their plans when they see the need.
4. *Help, guide, and train.* Both management and safety staff fulfill this role. Safety staff provides the technical and safety technical expertise while management provides the managerial help when asked, guiding when indicated, and training at the outset.
5. *Reward them according to their progress.* This requires a reward system that is geared to the progress made toward agreed upon objectives. The various managerial rewards should be used: pay, status, advancement, recognition.

The SBO approach has been installed in a variety of different industries: brewing (Adolph Coors Company), chemical (DuPont Corp.), railroading (Frisco, Union Pacific, Santa Fe, Chicago Northwestern), paper industry (Hoerner-Waldorf), and others. The results, of course, are not uniform. There are some successes and some less successful, as much depends on the installation, the commitment of management, and the meaningfulness of the objectives that were set.

See Exhibit 5.4 for the results in organizations using SBO. It has, of course, been installed differently in each organization. Some have left the objective-setting areas wide open—totally up to the supervisors and their bosses. Others limited supervisory choices to certain allowable strategies. One organization's management spelled out twelve areas that were felt to be areas of

Type Organization	Size	Type	Results
Food processor	15,000 employees	Delimited, with both mandatory & optional	36% and 40% red. yrs. 1 & 2
Food processor	10,000 employees	Delimited, with both mandatory & optional	30% and 25% red. yrs. 1 & 2
Contractor	3,000 employees		67% red. in 2 years
Brewery	15,000 employees	½ limited ½ no limits	64% red. 3 years
Oil Production	3,000 employees		50% red. 1st year
Railroad	4,000 employees	Limited	50% red. 1st year
Metal Manufacturer	3,000 employees	Select from a list of 12	50% red. 1st year

Exhibit 5.4 Safety by objective results.

slippage in the safety program. Most have required tasks and additional optional ones that a supervisor can select from a menu and then set goals in those selected.

For the most part, SBO works and continues to work once installed. One of the above companies reported a 75 percent reduction in the frequency rate in the first three months of the program. One reported a six-month claim savings of $2.3 million. One reported a 67 percent reduction in frequency continuing after three years.

Exhibit 5.5 shows an example of the definition of accountabilities for first line supervisors in two different organization. Exhibit 5.6 shows a weekly safety report.

Middle Management

The basic performance model (see Exhibit 5.1) applies here also. Middle managers also must have clearly defined roles, valid measures of performance, and rewards contingent on performance sufficient to get their attention.

The performance at this level is critical to safety success. The middle managers (the persons to whom the first line supervisors report) are by far more a key than the supervisors, for they make the system run or allow it to fail. As stated earlier, the middle manager's role is threefold:

— To ensure subordinate performance.
— To ensure the quality of that performance.
— To do some things that visually say safety is important.

Exhibit 5.7 shows two organizations' definition of tasks, measures, and rewards for this level.

Notice how the measurement is spelled out (discussed in the next chapter) as is the definition of the reward to be used (the performance appraisal system). What drives middle management safety performance is a clear perception of what tasks and activities are expected, that the performance will be validly measured, and of what rewards are contingent on that performance. Exhibit 5.8 is a sample department manager safety report.

Top Management

At the executive level we have a somewhat different ball game. While the basic performance model (Exhibit 5.1) may be similar, it is certain some inputs are more important at the top. The basic model might look more like the model in Exhibit 5.9.

The individual traits (genuine executive interest) become a large factor and the role perception box (believing he or she has a role) becomes crucial.

Clearly visibly demonstrated management committment is essential. But what drives performance? The classic answer is money—and that answer is probably less true than we think. Different executives are driven to safety performance for different reasons. Here are some:

— An honest caring for the employee welfare.
— Realizing that safety is an integral part of the quality of work life improvement.
— Realizing that safety and improving culture are closely related.
— Realizing that safety builds a healthy climate.
— Realizing that safety management, productivity, management, and quality management are one and the same.
— Realizing that error free performance is the real answer to bottom line success.

ABC COMPANY

ACCOUNTABILITIES — FIRST LINE SUPERVISOR

- ACCOUNTABILITIES:
  - Accident Investigation
  - Continuous Departmental Inspection
  - Employee Communications
  - Employee Training
- OPTIONAL ACTIVITIES
  - One-to-One Contacts
  - Group Meetings
  - Stop Program
  - Safety Circles
  - Positive Reinforcement
- PERFORMANCE MEASUREMENT:
  - 100%, Based on Activities
- PERFORMANCE WEIGHTING:
  - 20% of Total Performance Appraisal

XYZ COMPANY

ACCOUNTABILITIES — FIRST LINE SUPERVISOR

GENERAL

The key accountability of the First Level Manager is to carry out the tasks defined below.

TASKS

Required tasks are:
- Hold a period safety meeting with all employees.
- Include safety status in all work group meetings.
- Inspect department weekly and write safety work orders as required.
- Have at least five one-to-one contacts regarding safety with employees each week.
- Investigate injuries and accidents in accordance with Managing Safety guidelines within 24 hours.

In addition, in agreement with Department Head:
- Select at least two other tasks from a provided list (see Section 4, Subject 4.8) and agree on what measureable performance is acceptable.
- Report on these activities weekly.

WEEKLY SAFETY REPORT

The First Level Manager shall prepare and distribute a Weekly Safety Report in accordance with the format shown in Exhibit 5.6.

MEASURE OF PERFORMANCE
- Successful completion of tasks.

REWARD FOR PERFORMANCE
- Safety will be listed as one of the key measures on the Accountability Appraisal Form.

Exhibit 5.5 Sample SBO definitions of accountability.

```
┌─────────────────────────────────────────────────────────────────────┐
│           FIRST LEVEL MANAGER'S WEEKLY SAFETY REPORT                   │
│   FROM: _____ TO: _____ WEEK ENDING: _____    │
│   1.  WORK GROUP SAFETY MEETING                                        │
│       DATE: _____ SUBJECT: _____     │
│       _____     │
│       _____     │
│                                                                       │
│   2.  DEPARTMENT SAFETY INSPECTION                                     │
│       DATE: _____ FINDINGS: _____     │
│       _____     │
│       _____     │
│                                                                       │
│   3.  ONE-TO-ONE CONTACTS                                             │
│       EMPLOYEE: _____ DATE: _____     │
│       EMPLOYEE: _____ DATE: _____     │
│       EMPLOYEE: _____ DATE: _____     │
│       EMPLOYEE: _____ DATE: _____     │
│       EMPLOYEE: _____ DATE: _____     │
│       EMPLOYEE: _____ DATE: _____     │
│   4.  INJURY STATUS                                                    │
│       NAME: _____ DATE: _____    │
│       INJURY DESCRIPTION: _____     │
│       _____     │
│       _____     │
│                                                                       │
│   5.  OTHER SAFETY TASKS                       % COMPLIANCE            │
│          DESCRIPTION         ACTION              TO GOAL               │
│       _____     │
│       _____     │
│       _____     │
│       _____     │
│                                                                       │
│   REPORT DISTRIBUTION: STAFF II MANAGER                               │
│                        EMPLOYEE RELATIONS MANAGER                      │
└─────────────────────────────────────────────────────────────────────┘
```

Exhibit 5.6 First level manager weekly safety report.

— Saving dollars.
— Fear of lawsuit.
— Fear of criminal liability.
— Seeing a serious incident or fatality in the organization.

There is probably more interest and commitment at the executive level today than we have ever seen before. Perhaps all of the above are behind this interest and commitment. Perhaps the Bophal, Chernobyl and Challenger incidents explain some of it. The safety professional who does not take advantage of this interest is missing a major opportunity.

While interested today, the executive typically does not have the foggiest idea of what to do to make safety happen. This is the safety professional's job—to spell out the role and to spell out the system.

Exhibit 5.10 shows how some companies identify tasks, measures, and rewards at the plant manager level, and Exhibit 5.11 looks at the definition at an executive level.

Exhibit 5.12 shows the types of things we are talking about when we discuss rewards at the supervisory and management levels. Several of the items on the list are of exceptional importance and will be mentioned separately.

ABC COMPANY

DEPARTMENT MANAGER

- ACCOUNTABILITIES:
  - — Assure Supervisory Performance by Receiving and Reacting to Reports
  - — Audit Performance Through Spot Checks
  - — Maintain Departmental Budget
  - — Visibly Participate in Programs
  - — Develop Safety Management Knowledge and Skills in Subordinates
- OPTIONAL ACTIVITIES
  - — Participate in Audits
  - — Participate in Inspections
  - — Initiate One-to-One Contacts
  - — Create Ad HOC Committees
  - — Support Recognition Programs
- PERFORMANCE MEASUREMENT:
  - — 25-50%, Numbers
  - — 50-75%, Audit & Activities
- PERFORMANCE WEIGHTING:
  - — 20% of Total Performance Appraisal

XYZ COMPANY

DEPARTMENT MANAGER

GENERAL

The key accountability of the Department Manger is to ensure that the plans and programs of the XYZ Company Safety System are carried out in their area.

TASKS

- ■ Review reports from their area on task accomplishments and act accordingly.
- ■ Assess task performance defined for subordinate Managers and feedback as appropriate.
- ■ Engage in some self-defined tasks that can readily be seen by the work force as demonstrating a high priority to employee safety.
- ■ Develop safety management knowledge and skills in subordinate managers.
- ■ Make one-to-one safety contacts with hourly employees.
- ■ Participate in department safety inspections.

PERIOD SAFETY REPORT

Department Managers shall prepare and distribute a period safety report in accordance with the format shown in Exhibit 5-8.

MEASURES OF PERFORMANCE

- ■ Safety audit results for area(s) of accountability.
- ■ The 13 period rolling total injury frequency record for area(s) of responsibility.

REWARD FOR PERFORMANCE

- ■ Safety will be listed as one of the key measures on the Accountability Appraisal Form.

Exhibit 5.7 Sample middle management safety accountability requirements.

DEPARTMENT MANAGER PERIOD SAFETY REPORT

FROM: _____ TO: _____ PERIOD ENDING: _____

1. FUNCTION RESULTS/ANALYSIS (INCLUDES: PRIORITES, ACTIONS, INJURIES, EF-
   FECTIVENESS, ETC.)

2. ACCOUNTABILITIES

   DESCRIPTION   ACTION   % COMPLIANCE
                                                                 TO GOAL

   ONE-TO-ONE

   INSPECTION

   SELF-DEFINED
   TASKS

   REPORT DISTRIBUTION: PLANT MANAGER
                                      PLANT SAFETY MANAGER

Exhibit 5.8 Department manager period safety report.

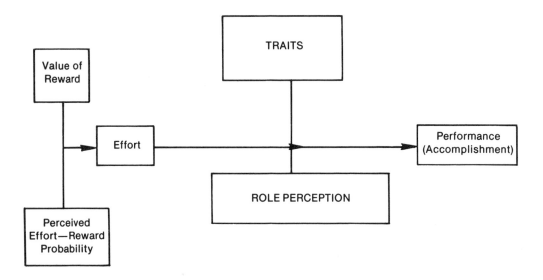

Exhibit 5.9 Top management performance model.

Performance Appraisals

Utilizing performance appraisals as a part of the reward structure for safety performance is essential in almost all organizations. If safety performance (validly measured by an activity measure) is not considered meaningful in the performance appraisal system a clear message is sent that says safety is not a high enough priority to be considered when judging performance. It must be included and preferably at a clearly spelled out percentage.

The Regular Numbers Game

Most organizations have clearly spelled out performance numbers that are looked at regularly (daily, weekly). These numbers are of such importance that management wants them that often because they are the indicators of corporate health. They are usually production numbers (tons of steel out, pounds of product) or quality numbers (scrap rate, rejects, flights on time) or other similar numbers.

If safety performance is not a part of this reporting system another clear message is transmitted which says safety performance is simply not important enough to track. Valid measures of safety performance *must* be an integral part of the numbers game–the reporting system. It states that top management wants to know if safety performance is on track.

Staff Safety Drivers

The performance model of Exhibit 5.1 might be amended again to describe what drives effective staff safety performance. It might look like Exhibit 5.13.

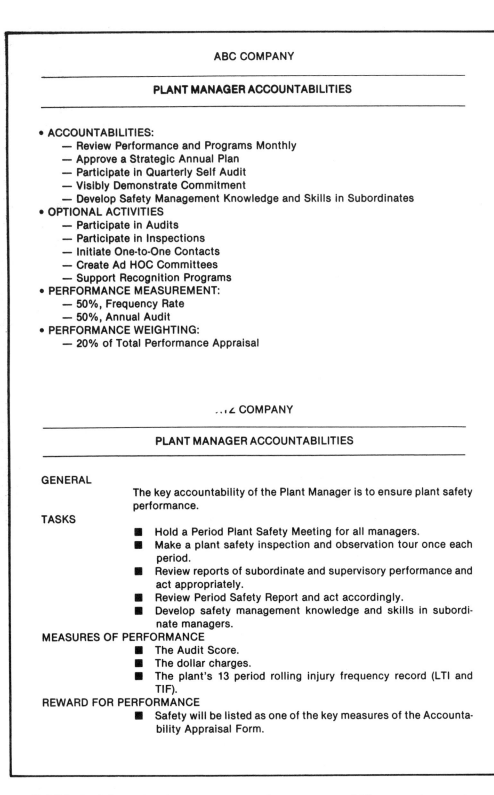

ABC COMPANY

PLANT MANAGER ACCOUNTABILITIES

- ACCOUNTABILITIES:
  - Review Performance and Programs Monthly
  - Approve a Strategic Annual Plan
  - Participate in Quarterly Self Audit
  - Visibly Demonstrate Commitment
  - Develop Safety Management Knowledge and Skills in Subordinates
- OPTIONAL ACTIVITIES
  - Participate in Audits
  - Participate in Inspections
  - Initiate One-to-One Contacts
  - Create Ad HOC Committees
  - Support Recognition Programs
- PERFORMANCE MEASUREMENT:
  - 50%, Frequency Rate
  - 50%, Annual Audit
- PERFORMANCE WEIGHTING:
  - 20% of Total Performance Appraisal

...₂ COMPANY

PLANT MANAGER ACCOUNTABILITIES

GENERAL

The key accountability of the Plant Manager is to ensure plant safety performance.

TASKS

- Hold a Period Plant Safety Meeting for all managers.
- Make a plant safety inspection and observation tour once each period.
- Review reports of subordinate and supervisory performance and act appropriately.
- Review Period Safety Report and act accordingly.
- Develop safety management knowledge and skills in subordinate managers.

MEASURES OF PERFORMANCE

- The Audit Score.
- The dollar charges.
- The plant's 13 period rolling injury frequency record (LTI and TIF).

REWARD FOR PERFORMANCE

- Safety will be listed as one of the key measures of the Accountability Appraisal Form.

Exhibit 5.10 Sample plant manager safety accountability requirements.

XYZ COMPANY
AREA VICE PRESIDENT

GENERAL

The key accountability of the Area Vice President is to ensure individual subordinate plant performance.

TASKS

■ Conduct at least a semiannual audit of each plant under their jurisdiction. (Audit details per Section 4 or as self-designed.)
■ Review the Period Safety Report and act appropriately.
■ When in plants, execute appropriate tasks that will visibly demonstrate to employees that safety is a high priority and specifically to the Area Vice President.

MEASURES OF PERFORMANCE

■ Achievement of lost time injury frequency goals
■ Achievement of total injury frequency goals
■ Achievement of audit score goals for individual subordinate plants.

REWARD FOR PERFORMANCE

■ Safety will be listed as one of the key measures on the Accountability Appraisal Form.

Exhibit 5.11 Executive safety accountability requirements.

Contrived On-The-Job Rewards				Natural Rewards
Manipulatables	Visual and Auditory	Tokens	Social	Premack
Desk accessories	Office with a window	Money Stocks	Friendly greetings Informal recognition	Job with more responsibility
Wall plaques	Piped-in music	Stock options	Formal acknowledgement of achievement	Job rotation
Company car	Redecoration of work environment	Profit sharing		Work on personal project on company time
Watches	Company literature		Invitations to coffee/lunch	
Trophies	Private office			Use of company machinery or facilities for personal projects
Commendations			Solicitations of suggestions	
Rings/tie pins			Solicitations of advice	
			Compliment on work progress	
			Recognition in house organ	
			Pat on the back	
			Smile	
			Verbal or non-verbal recognition or praise	

Exhibit 5.12 Classifications of rewards.

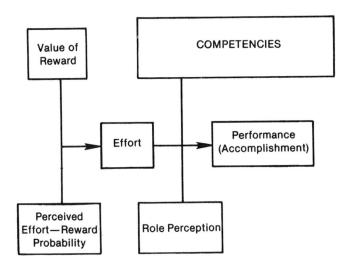

Exhibit 5.13 Staff safety performance model.

The only changes we've made is to replace the abilities and traits box wilh one labelled "competencies," and made the box much bigger to indicate its importance.

Competencies, as defined by Will Jenkins of Mobil Oil Company, are the characteristics that are shown by research to be related to superior performance. They are what superior performers are more likely to do more often and more completely for better results. They may be motives, traits, aptitudes, self-image, knowledge, or skills. Exhibit 5.14 shows the relationship between competencies and tasks.

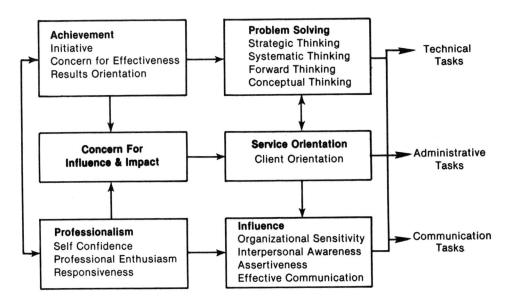

Exhibit 5.14 Relationship between competences and tasks.

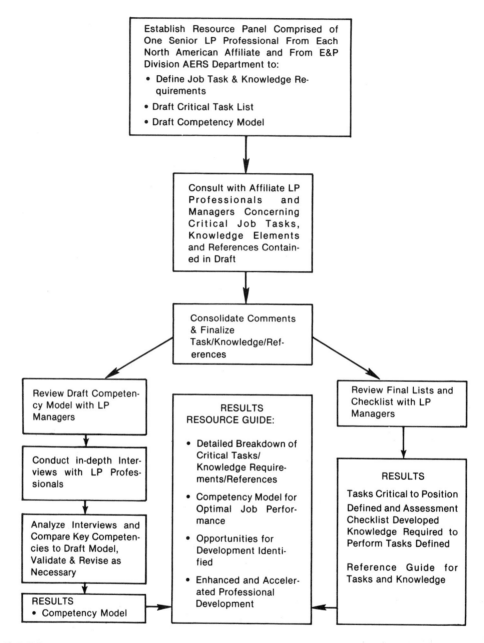

Exhibit 5.15 The process used to identify competencies, critical tasks, and development alternatives.

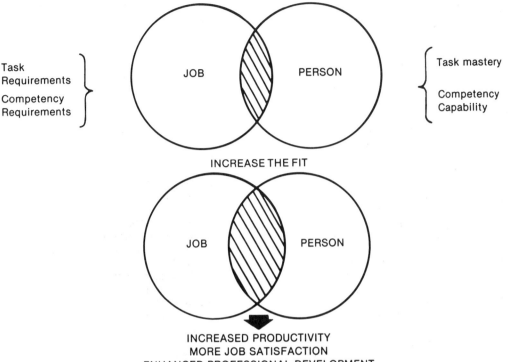

Task
Requirements

Competency
Requirements

JOB   PERSON

Task mastery

Competency
Capability

INCREASE THE FIT

JOB   PERSON

INCREASED PRODUCTIVITY
MORE JOB SATISFACTION
ENHANCED PROFESSIONAL DEVELOPMENT

Exhibit 5-16 Job-Person fit.

Job   X------X
Requirements
Profile

Person ⊙—⊙
Competency
Profile

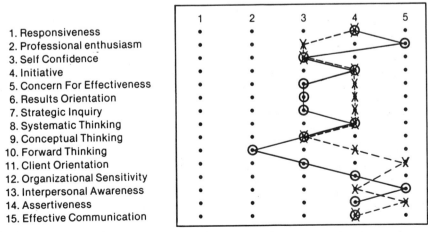

1. Responsiveness
2. Professional enthusiasm
3. Self Confidence
4. Initiative
5. Concern For Effectiveness
6. Results Orientation
7. Strategic Inquiry
8. Systematic Thinking
9. Conceptual Thinking
10. Forward Thinking
11. Client Orientation
12. Organizational Sensitivity
13. Interpersonal Awareness
14. Assertiveness
15. Effective Communication

Exhibit 5.17 Job-Person profile (loss prevention).

I. PROFESSIONALISM
  1.0 Responsiveness
    1.1 Responds to requests and projects with a sense of urgency
    1.2 Spends whatever time or effort required to meet deadlines
    1.3 Anticipates and provides the correct amount of detail in work
  2.0 Professional Enthusiasm
    2.1 Displays enthusiasm for the job and profession
    2.2 Seeks opportunities to expand and apply technical knowledge, skills and experience
  3.0 Self Confidence
    3.1 Expresses confidence in own abilities to complete challenging assignments
    3.2 Trusts own judgement
    3.3 Takes calculated risks to achieve objectives
    3.4 Makes independent decisions
    3.5 Accepts responsibility for failures and mistakes
II. ACHIEVEMENT
  4.0 Initiative
    4.1 Takes initiative to identify and solve problems
    4.2 Looks for ways to expand LP services
    4.3 Takes action to ensure best possible outcome
    4.4 Persists in overcoming obstacles
  5.0 Concern for Effectiveness
    5.1 Identifies areas where improvement can be realized
    5.2 Acts according to management and project priorities
    5.3 Seeks ways to reduce costs and increase effectiveness
    5.4 Takes pride in doing job well
  6.0 Results Orientation
    6.1 Seeks opportunities to have a positive impact on Company LP goals or to enhance the reputation of the department
    6.2 Analyzes costs and benefits of developed recommendations
    6.3 Sets challenging and realistic goals and standards
    6.4 Expresses pride and enthusiasm about meeting functional goals
III. PROBLEM SOLVING
  7.0 Strategic Inquiry
    7.1 Identifies the correct information needed to accomplish the task
    7.2 Formulates the right questions to ask
    7.3 Seeks the advice of experts or informed sources to get information
  8.0 Systematic Thinking
    8.1 Develops a well defined systematic approach to the problem
    8.2 Evaluates the merits of acceptable alternatives
    8.3 Utilizes past knowledge and experience to deal with current problems
  9.0 Conceptual Thinking
    9.1 Articulates how activities or problems relate to a larger framework
    9.2 Sees connections that are not obvious to others
    9.3 Draws conclusions from independent events or pieces of information
  10.0 Forward Thinking
    10.1 Anticipates future problems and events
    10.2 Relates future needs to analysis of past events and experience
    10.3 Develops contingency plans and solutions in case things go wrong
    10.4 Develops innovative or unconventional solutions
IV. SERVICE ORIENTATION
  11.0 Client Orientation
    11.1 Views the line and staff organizations as "clients"
    11.2 Demonstrates satisfaction in recommending or implementing solutions to the problems and concerns of the "client"
    11.3 Articulates specific benefits for client in implementing a recommendation.
    11.4 Does not let personal feelings or frustrating interfere with needs of the "client"
V. INFLUENCE
  12.0 Organizational Sensitivity
    12.1 Uses understanding of how the organization works to accomplish a specific objective

Exhibit 5.18 E&P Division loss prevention competency model.

12.2 Identifies the key groups or individuals with whom to
establish/maintain interface
relationships

12.3 Anticipates reactions to
recommendations or advice and
is prepared to address them pro-
fessionally

13.0 Interpersonal Awareness

13.1 Solicits input from those who
will be affected by the project

13.2 Anticipates the points of view
and reactions of others

13.3 Changes influence strategies
when required

13.4 Observes and interprets non-
verbal cues and behavior

14.0 Assertiveness

14.1 Confronts other abouts problems or controversial issues

14.2 Persistently seeks ways to
persuade others on important
issues

14.3 Resolves disagreements by
seeking mutually acceptable
solutions

15.0 Effective Communication

15.1 Makes written and oral
presentations in a clear, concise
and organized manner

15.2 Makes effective use of support-
ing data and other resources to
enhance proposals and
recommendations

Exhibit 5.18 Continued.

Exhibit 5.15 describes the process Jenkins uses to get a better job-person fit in the professional safety position (Exhibit 5.16). Each position has a defined profile and potential safety
professionals can be tested for the competencies that fit (see Exhibit 5.17).

What are the competencies that correlate with safety professional success? In Mobil, Jenkins
has found those listed in Exhibit 5.18 to be the essential competencies.

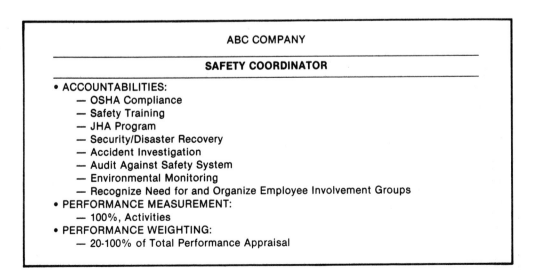

ABC COMPANY

SAFETY COORDINATOR

• ACCOUNTABILITIES:
  — OSHA Compliance
  — Safety Training
  — JHA Program
  — Security/Disaster Recovery
  — Accident Investigation
  — Audit Against Safety System
  — Environmental Monitoring
  — Recognize Need for and Organize Employee Involvement Groups
• PERFORMANCE MEASUREMENT:
  — 100%, Activities
• PERFORMANCE WEIGHTING:
  — 20-100% of Total Performance Appraisal

Exhibit 5.19 Sample safety coordinator accountability measurements.

Besides the competency box in the performance model for staff safety, there remain the other boxes pertaining to measurement and reward. Exhibits 5.19 and 5.20 show how two organizations spell these out. Exhibit 5.21 shows a sample report.

Chapter 6 looks at the element of the driving system that has been mentioned over and over again in this chapter—measurement.

XYZ COMPANY

PLANT SAFETY MANAGER

GENERAL

The key accountability of the Plant Safety Manager is to advise and assist the Plant Manager in ensuring plant safety performance.

TASKS

- Analyze the plant systems to determine weaknesses that could lead to injuries and loss.
- Develop action plans based on analyses.
- Communicate plans to those in line organizations who must implement them.
- Monitor results with Employee Relations generated reports and with other gathered data.
- Ensure that training is provided where needed.
- Assist in the Plant Manager's Period Safety Meeting.
- Coordinate the Safety Committee.
- Assist and train supervisory staff in the options available to them.
- Ensure OSHA compliance.
- Make plant ergonomic analyses (optional).
- Conduct quarterly safety inspections.
- Ensure that new employees receive training.
- Identify safety qualifications for job classifications.

PERIOD REPORT

The Plant Safety Manager shall prepare and submit a period report as shown in Exhibit 5-20.

MEASURE OF PERFORMANCE

Appraisal by Plant Manager as to task achievement.

REWARD FOR PERFORMANCE

- Safety achievements will be listed as individual items on the Accountability Appraisal Form.

Exhibit 5.20 Sample plant safety manager accountability measurements.

```
┌──────────────────────────────────────────────────────────────────────┐
│                   PLANT SAFETY MANAGER PERIOD REPORT                    │
│                                                                        │
│  FROM: _____ TO: _____ PERIOD ENDING: _____  │
│  1. TRAINING ACCOMPLISHED                                              │
│                                                                        │
│                                                                        │
│     INJURY DESCRIPTION                                                 │
│                                                                        │
│                                                                        │
│  2. % SAFETY ACTION COMPLIANCE BY DEPARTMENT                           │
└──────────────────────────────────────────────────────────────────────┘
```

	MATERIALS	MAINTENANCE	PRODUCTION	SANITATION	QUALITY ASSURANCE
ONE-TO-ONE					
DEPARTMENT AUDITS					
WORK GROUP SAFETY MEETINGS					

3. DESCRIBE OTHER PSM ACTIVITIES

    E.G. DEPARTMENT SAFETY AUDIT UPDATES, AWARENESS PROGRAMS,
          PLANT SAFETY INSPECTION, ERGONOMIC ANALYSIS,
          PLANT MANAGER SAFETY MEETING, OTHER MEETINGS ATTENDED

    REPORT DISTRIBUTION.   PLANT MANAGER
                           FILE

Exhibit 5.21 Sample plant safety manager report.

REFERENCES

Jenkins, W. *LPPD Resource Guide,* Mobil Oil Company, Dallas, 1987.

Lawler, E. & L. Porter. *Managerial Attitudes and Performance,* Irwin-Dorsey, Homewood, IL, 1968.

Measurement

In Chapter 5, the concept of measurement was introduced as part of the discussion of account-ability. To hold someone accountable, we must know whether he or she is performing well, so we must measure that person's performance. Without measurement, accountability becomes an empty and meaningless concept.

Thus, starting with what we propose as the single most important factor in getting good line safety performance—accountability—we find that we are really talking about ways to measure the line manager better. And measurement has been our downfall in safety for years.

Measurement has been discussed over and over again and we have yet to come up with a meaningful measure of safety performance. Perhaps our inability to create these needed measures is one reason for our lack of good safety performance. We have not devoted much time to good measures of lower management's safety performance and this might account for our inability to get good safety performance from line organization.

For the line manager, "to measure is to motivate." Although this statement might have sounded a little ridiculous 20 years ago, I believe that it expresses a profound truth, at least in terms of the safety performance of line organizations. Managers react to the measures used by the boss; they perceive a task to be important only when the boss thinks it is worth measuring.

Having perceived the importance of measurement in obtaining good safety performance, we then hit our biggest snag: What shall we measure? Should we measure our failures as demonstrated by accidents that have occurred in the past? If this is in fact a good measure, as has historically been believed (for that is what we usually measure), then what level of failure should be measured? We can measure the level of failure we call "fatalities." Fatalities are used to measure our national highway traffic safety endeavors. Is the measure of fatalities, then, a "good" measure? Obviously, we cannot answer this question until we examine the size of the unit being measured. Fatalities could be a good measure if we are assessing the national traffic safety picture, but it would be a little ridiculous in the case of a supervisor of ten factory workers. Such a supervisor might well do absolutely nothing to promote safety and still never experience a fatality in his or her department. Obviously, measuring fatalities would make little sense in this case.

Unfortunately, our traditional frequency rate is not much better than fatalities in the example above when we use it as a measure of supervisory safety performance. It measures a level of failure somewhat less than a fatality (an injury serious enough to result in a specified amount of time lost from work), but the fact remains that a supervisor of ten workers can do absolutely nothing for a year and attain a zero frequency rate with only a small bit of luck. By rewarding such a supervisor, we are actually reinforcing nonperformance in safety. He or she learns that it is not necessary to do anything in order to get a reward. While this may sound a little ridiculous, it accurately describes what is going on in many safety programs today.

If fatalities (or frequency rates) are a poor measure of supervisory performance, what is a good measure? Or, more important, what is wrong with fatalities as a measure? Perhaps measuring our failures is not the best approach to use in judging safety performance. After all, this is not the way we measure people in other aspects of their jobs. We do not, for instance, measure line managers by the number of parts the people in their departments failed to make yesterday. And we do not measure the worth of salespeople by the number of sales they did not make. Rather, in cases like these we decide what performances we want and then we measure to see whether we are getting them.

What would be a good measure of supervisory safety performance? More important, what set of criteria can we develop for measuring supervisory safety performance? Or the safety performance of the corporation? Or our national traffic safety performance? Or anything else related to safety?

Even a brief look at the problem of measurement shows us that we need different measures for different levels in an organization, for different functions, and perhaps even for different managers. What is a good measure for one supervisor of ten people may not be a good measure for another, much less for a plant superintendent or the general manager of seven plants and 10,000 people. What might be a good measure for the supervisor of a foundry cleaning room may be inappropriate for use in judging the effectiveness of OSHA.

CRITERIA

The development of criteria for good safety measures is certainly not an easy task. We have been grappling with it for years, and the most noted theorists and scholars in safety have been writing on the problem since the late 1950s. From their writings (notably those of Dr. Tarrants, Dr. Rockwell, and Dr. Grimaldi), general criteria for safety measures can be distilled.

We need more than a list of general criteria, however. For instance, such a list usually includes an item called "statistical reliability," which has to do with whether a measure tends to fluctuate wildly when there has not been much change in the system being measured. The criterion of statistical reliability would obviously be useless for measuring a first-line supervisor's safety performance. A supervisor who does nothing for safety (zero performance level) could have a perfect record or a miserable record. At the supervisory level, statistical reliability is next to impossible to achieve with any measure—the data base is simply too small.

The following sections describe criteria for different levels and different functions. These were developed by relating what has been written in the field since the 1950s to different organizational levels. Much of the thinking in these sections has come from discussions held with, and papers written by, graduate students in safety management at the University of Arizona. You, the reader, may agree or disagree with some of the items included in each list of criteria. The important thing is that you decide on some criteria of your own that you believe in, for without criteria we cannot do a good job of developing better measures.

Safety-Measurement Criteria for Employees

Few measures of safety performance are used at the lowest organizational level, that of the employee. Those which are used are essentially individual assessments of performance. They seem to be supervisory techniques, rather than measures. Nonetheless, we can consider a few possible criteria for a measure of employee safety performance:

- It should be so constructed that it can be used to affect the employee's rewards (appraisal, promotions, bonuses).

— It should be constructed in such a way that it recognizes or can be used to recognize safe performance (rather than unsafe performance).
— If possible, it should be self-monitoring.

Safety-Measurement Criteria for First-Line Supervisors

Measurement is more crucial at the supervisor level, and the measure (which is the motivator here) must do many more things than at the employee level. The following are a few criteria:

— It should be flexible to encompass individual managerial styles and different strategies that supervisors use to get things done.
— It should give swift and constant feedback.
— It should be able to be used to judge promotability.
— It should get the supervisor's attention.
— It should measure the presence of safety activity, not only its absence (as indicated by accidents).
— It should be sensitive enough to indicate when effort has slowed.
— It should provide an alert, showing that something is wrong.
— It should be understandable to those at both the supervisory and upper-management levels.
— It should provide recognition of supervisors' efforts.
— It should allow for creativeness.
— It should be valid; that is, it should measure what it was intended to measure. If you want performances from supervisors such as accident investigations, inspections, and training, the measure should tell whether you are getting these performances. It should not measure only failures (accidents) as an indication of whether you are getting the desired performances.
— It should be mainly performance-oriented.
— Insofar as possible, it should be self-monitoring.
— It should be meaningful.

Obviously, no one measure will meet all the above criteria. Thus there will have to be some trade-offs, and it will be necessary to devise and use measures that meet as many of the criteria as possible. Later, we shall examine some traditional and nontraditional measures to see how well they meet these criteria.

Safety-Measurement Criteria for Middle and Upper Management

Many of the same criteria apply at the upper levels of management as at the lower levels, although some new ones become important. A measure of performance at the upper- and middle-management level should meet the following criteria:

— It should be flexible enough to allow for individual managerial styles and strategies.
— It should be capable of giving swift and constant feedback.
— It should be able to be used in judging promotability.
— It should get the attention of those in middle and upper management.
— It should measure the presence, as well as the absence, of safety.
— It should be sensitive to change (able to alert management to new situations and problems).

- It should be understandable to both the middle manager and those in top management.
- It should be built in such a way that it can offer recognition for performance.
- It should allow for creativeness.
- It should be valid.

Safety-Measurement Criteria for Corporate Management

The measure used at this level indicates how well the company is doing. It is used only internally, within the organization. It can indicate the progress of the entire organization and could be construed as constituting a judgment of the president and controlling officers. Such a measure is not intended for use in comparing the performance of one corporation with that of another (different criteria will be suggested for this type of measure). The following are the most important criteria for a corporate measure of safety performance for internal uses only:

- It should be valid; that is, it should measure what it was designed to measure.
- It should be statistically reliable; it should not fluctuate without reason.
- It should be objective.
- It should be meaningful to management.
- It should be quantifiable.
- It should be sensitive to changes and problems.
- It should ensure input integrity.
- It should be primarily results-oriented.
- It should be able to be computerized.
- It should point up weaknesses in the system and thus make it possible to take preventive action.

Since we are dealing here with larger numbers of people, we can demand certain things of this measure that we could not expect of others: statistical reliability, quantifiability, input integrity, and the ability to be computerized, for example. These are ideal criteria for a measure and at this level of the organization we can ask for them.

Safety-Measurement Criteria at the Staff Level

A measure of staff safety performance should meet the following criteria:

- It should be valid.
- It should be understandable to management.

Additional Criteria

The following additional criteria for a safety measure should be added to all the above lists:

- It should have a good cost-benefit ratio.
- It should be administratively feasible.
- It should be practical.

AVAILABLE MEASURES

There are two categories of available tools: (1) activity (or performance) measures and (2) results measures. As a general rule, in selecting measuring devices, use only activity measures at the lower managerial levels, primarily activity measures (with some results measures) at the middle-upper management levels, and reserve the pure results measures for the executive level. With rare exception (like Safety Sampling), this rule of thumb gives us statistical validity and is extremely important.

Exhibit 6.1 illustrates the safety measures available. They fall into three categories: (1) activity (did the supervisor do what was supposed to be done?); (2) results before the accident (are things better around here because of what this supervisor has done?); and (3) results after the fact of the accident (how many did we have?).

Exhibit 6.2 shows the variety of choices in determining which measure to use.

As indicated in Exhibit 6.2, we can use either activity measures or results measures to determine performance, and we can use them at the supervisory, the managerial or the system-

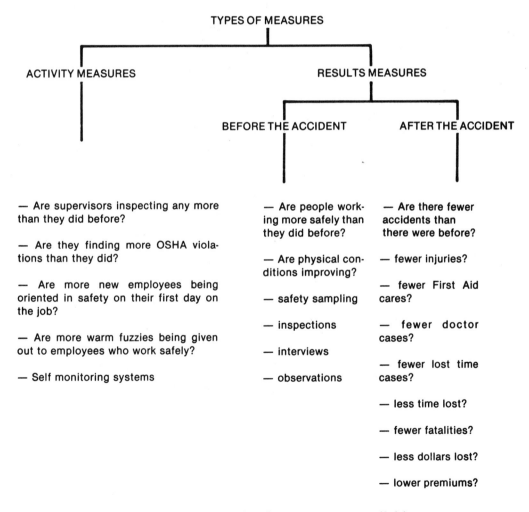

Exhibit 6.1 Types of safety measures available.

ACTIVITY		
SUPERVISOR For: Objectives Met # Inspections # Quality Investigations # Trained # Hazard Hunts # Observations # Quality Circles	**MANAGER** Objectives Met Use of Media # Job Safety Analyses # Job Safety Observations #One-on-Ones # Positive Reinforcement Group Involvement	**SYSTEM-WIDE** Audit —Questionnaires — Interviews
RESULTS		

FOR:
SUPERVISORS
Safety Sampling
Inspection Results
	MANAGERS
	Safety Sampling
	Inspection Results
	Safety Performance
	  Indicator
	Estimated Costs
	Control Charts
	Property Damage
		SYSTEM-WIDE
		Safety Sampling
		Safety Performance Indi-
		  cator
		# First Aid or Frequency
		# Near Misses or
		  Frequency
		Property Damage
		Frequency-severity Index
		Estimated Cost Control
		  Charts

Exhibit 6.2 Activities and results measures for supervisors, managers and system-wide safety programs.

wide levels, provided we use some caution in measurement selection. At the supervisory level the activity measures are most appropriate, whether it be the number of inspections, number of people trained, or number of observations made. These activity measures are equally appropriate at the managerial levels, and can even be used at the system-wide levels (through audits or questionnaires). All activity measures are extremely well suited to MBO (SBO) approaches (were the objectives reached?). Using these measures in SBO assumes that the original objectives were well written (realistic, rifled, measurable, and under the direct control of the objective setter).

Results measures can also be used at all levels, as long as extreme care is used at the lower levels. The traditional safety measures such as frequency rate or severity rate cannot be used at the lower levels except over long periods of time and then probably only as a quality check.

Measures at the Supervisory Level

Performance Measures

We start our examination of supervisory measurement by looking at what the supervisor does to get results and by determining whether the supervisor actually does them.

Performance measures have certain distinct advantages:

— They are flexible and take into account individual supervisory styles. We do not have to use the same measure for all supervisors. We can allow each supervisor to select performance and levels of performance and then measure these performances and levels. Performance measures are excellent for use in MBO approaches.

- They give swift feedback, since most require supervisors to report their level of performance to the boss. (They are also often self-monitoring.)
- They all measure the presence, rather than the absence, of safety.
- They usually are simple and thus administratively feasible.
- They are clearly the most valid of all measures.

Certain newer safety approaches require performance measures. SCRAPE and SBO mentioned in the last chapter require performance measures to be effective. Occasionally safety and management people resist using these activity measures, particularly in an MBO process. Somehow, somewhere some people ended up with the mistaken belief that MBO really meant MBR (management by results), and MBR suggests that objectives have to be results objectives. It is extremely counter productive to look at the process with that approach. Leonard Sayles of Columbia University Graduate School of Business clarifies:

> Another casualty of the current major recession is the simpleminded belief that top management can manage decentralized units solely by their results, profitability or bottom line.
> For years, this seemed like such a sensible way to manage because it appeared to avoid upper management's second-guessing of division heads and to encourage the latter to take responsibility for tough decisions and trade-offs. The rewards (or punishments) for division heads would come in direct proportion to the profitability of their units. The ideal incentive system for managers appeared to be: *Pay for results.* The main role of top management was to handle strategic and financial planning while monitoring the performance of individual units.

PROFITS AND RESPONSIBILITY

But this has proven to be a simplistic view of management and motivation. Top management has more functions (except in a pure conglomerate) than simply overseeing a race among independent division heads. It's also clear that those managers, in turn, have greater responsibilities than can be measured by annual or quarterly profits. With increasing numbers of bankruptcies and unexpected, severe losses, these responsibilities are being unveiled.

Profits are *not* a leading indicator. They don't tell you where the business is going, only where it has been. Over-emphasis on immediate profits can encourage sacrificing longer-run product development and innovation.

In fact, there is a great deal of evidence that powerful incentives, such as tying pay and promotion to earnings, drive other criteria out of the decision-making process. Even when management proclaims that it wants safety or long-run considerations factored in, these get lost when the temptation to make a strong showing for the current accounting period becomes compelling enough.

Furthermore, profits may not always be profits. Through "creative accounting"— shifting sales from one period to another or misevaluating inventory—managers can *appear* to be doing better than they really are. The business pages of the press recently have been filled with stories about well-known and presumably well-managed companies that have been fooled by key divisions. In many such cases, the firms' auditors pick up these "tricks" too late.

Top management needs to know *more* than how to read a gauge of profit figures or even return-on-investment numbers or other kinds of performance measures. It needs technical expertise to determine whether lagging performance is due to external problems, over which managers have no control, or to poor decisions by those same managers.

Even good and properly reported results can mirror dumb luck or being in the right place at the right time. In successful organizations, top managers have substantial knowledge of operations. When they don't (often because of an acquisition), those units are the ones most likely to get into trouble.

There are other reasons against telling division heads to act as though they're in business for themselves. In most companies, divisions are dependent in varying degrees on other company units in both production and marketing. They share resources, help each other's performance and cooperate for economies of scale.

But all these get ignored when a decentralized company gives its divisions too free a rein. Under these circumstances, unit managers scream if they have to take the products of a sister division (rather than a cheaper substitute sold in the marketplace), and they fight with each other in full view of potential customers. Even International Business Machines Corp. discovered it had pushed decentralization too far when it found that several of its divisions had developed product lines that were competing with each other.

The trend of the future for a slow-growth business world is a *return to greater centralization* and an abandoning of those easy management schemes that involve managing *only* by results.

Some of the performance measures at the supervisory level are listed in Exhibit 6.3 along with some of the results measures often used.

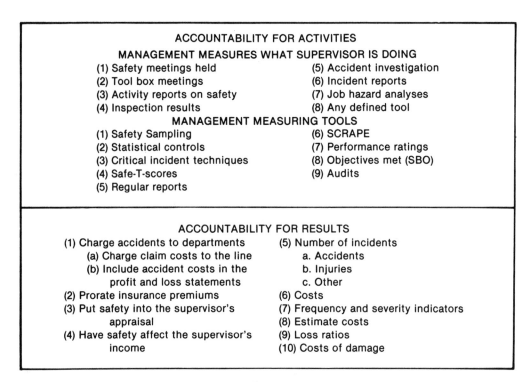

ACCOUNTABILITY FOR ACTIVITIES

MANAGEMENT MEASURES WHAT SUPERVISOR IS DOING

(1) Safety meetings held	(5) Accident investigation
(2) Tool box meetings	(6) Incident reports
(3) Activity reports on safety	(7) Job hazard analyses
(4) Inspection results	(8) Any defined tool

MANAGEMENT MEASURING TOOLS

(1) Safety Sampling	(6) SCRAPE
(2) Statistical controls	(7) Performance ratings
(3) Critical incident techniques	(8) Objectives met (SBO)
(4) Safe-T-scores	(9) Audits
(5) Regular reports	

ACCOUNTABILITY FOR RESULTS

(1) Charge accidents to departments	(5) Number of incidents
(a) Charge claim costs to the line	a. Accidents
(b) Include accident costs in the	b. Injuries
profit and loss statements	c. Other
(2) Prorate insurance premiums	(6) Costs
(3) Put safety into the supervisor's	(7) Frequency and severity indicators
appraisal	(8) Estimate costs
(4) Have safety affect the supervisor's	(9) Loss ratios
income	(10) Costs of damage

Exhibit 6.3 Supervisory performance and results measures.

Results Measures

There is a simple rule of thumb for results measures: don't use them. There is only one exception to this rule, safety sampling, discussed earlier. The only possible way that results measures should be used at this level is to use them over a long period of time as a quality check (is supervisor doing the right things?).

Measures at Upper Management Levels

Performance Measures

While we begin to emphasize results measures more at the middle-management level, some performance measures should be retained here also. Performance measures at this level are simply measures of what middle managers do—of whether they perform the tasks that it is necessary for them to perform. Thus everything we have said about supervisory performance measures applies also at the middle-management level.

However, supervisory performance and middle-management performance should be quite different, necessitating some changes in the measures used at this level. Normally, we want the middle managers to get their subordinates (supervisors) to do something in safety. Thus we can measure them to see whether they meet with their supervisors, check on these supervisors, or monitor the quality of the supervisors' work.

Audits

A measure of performance (and results) that enters the scene at the upper- and middle-management levels (management of a location, for instance) is the audit of safety performance. Many of these are discussed in Chapter 10, but we will take a quick look at a few here.

Safety Audit Factor Evaluation Report (SAFER)

One example of a safety audit is SAFER, used by Grumman Aerospace Corporation and developed by Robert J. Mills, corporate safety engineer. Following is his description of SAFER:

> Safety is essentially a line management function, and local operational managers must accept the responsibility to implement corporate policy and procedures. Local management can be assisted in its task of evaluating the work environment by the Safety Audit Factor Evaluation Report (SAFER), which acts as a guide for the self-evaluation and audit. Within the line organization, supervisors must accomplish predetermined safety objectives and are encouraged to treat safety objectives and production efficiency as inseparable.
>
> The corporate safety manager, working with local managers and safety supervisors, renders a consultative service to each location and provides technical support essential to the success of the local program.
>
> Corporate policy directives prescribe the structure, functions, and procedures for the local safety organization. These broad policy guidelines provide the balanced program which is essential in responding to the requirements of local, state, and federal regulations.

The SAFER audit report is a guide for reviewing and checking key factors in the current loss-prevention activities on a broad category evaluation of "none," "minimal," and "adequate." The evaluation is done annually by the local plant manager and safety supervisor. Items which are evaluated as less than adequate should be incorporated into the safety objectives for the coming year. This exercise enables the local plant safety supervisor to review the status of past objectives, as well as project new items to be worked on to resolve current problems.

The report creates a very definite involvement and commitment action, interrelating the local plant manager and his safety supervisor. Their combined efforts are then stated as an official status report of what they think has to be done, and it enables their superiors at divisional and corporate headquarters to review, concur, or advise on the merits of their conclusions.

Another plus is that the completed safety objectives can also be presented as evidence of "good faith" in regard to complying with the provisions of the OSHA law. The following is a cross-sectional sampling of the eight major sections of the SAFER report, which contains 77 specific items to be rated within the local plant safety program activities. A guide is also furnished to assist local plant management in developing meaningful objectives from the SAFER report. The guide contains more detailed explanations and definitions of the corresponding elements in the SAFER report. Beneath the specific items are questions lettered (a), (b), (c), etc., contained in the guide, which are to be considered before a rating is given to each specific item. (See Exhibit 6.4).

This overall plan with the described management control tools and support material, properly administered, has a proven success record with various large corporations. Obviously, it cannot be implemented and bring improved changes overnight. You cannot go from 25 percent loss control to complete control immediately. A corporate safety manager is a quality improvement specialist and must be an integral part of management. The big problem in management today is determining what level of reliability and efficiency is necessary. Any system is a number of parts and subparts working in unison to support the whole. Each function in a management system can be evaluated for excellence. The avoidance of performance error is the key to treating the cause instead of the symptom. Therefore, a corporate safety manager properly handling the described program will not only reduce losses but also improve performance.

Chevron's Review

Another example of the audit of safety performance is the safety program and evaluation used by Chevron Oil Company. It has the following purposes:

- To determine "soft spots."
- To gain insights into employee attitudes regarding the effectiveness of the safety program.
- To give local management direction in making the program more effective.
- To assist top management in determining how to improve its support efforts.

The Chevron method is slightly different from some of the other audit approaches. First, a review team is formed consisting of a staff safety engineer and one or two persons responsible for coordinating or implementing the safety program of the organization being reviewed. The next step is to administer a questionnaire to all employees and supervisors.

The major sections of the Safety Audit Factor Evaluation Report (SAFER) are:

A. ORGANIZATION (five items listed)

Ext. Item #3. Safety committees effectively organized at all levels . . . . 3 □ □ □ *None Minimal Adequate*

(a) Are all levels of management involved in regularly scheduled safety meetings?

(b) Do first-line supervisors conduct safety meetings with their employees routinely?

(c) Is relevant safety information exchanged between employees and management?

B. ADMINISTRATION (18 items listed)

Ex. Item: #1. Local management actively participates in safety effort . . . . 1 □ □ □

(a) Plant/facility manager or representative are part of management safety committee?

(b) Plant/facility manager reviews all significant accidents?

(c) Plant/facility manager actively supports efforts of supervisors on safety matters?

C. CONTROL OF PHYSICAL HAZARDS (17 items listed)

Ex. Item: #3. Approved type personal protective equipment available and used . . . . 3 □ □ □

(a) Equipment conforms to established standard: ex. Respiratory protection has Bureau of Mines approval; hard hats meet ANSI Z89.1; eye protection meets ANSI Z87.1; safety footwear meets ANSI Z41.1

(b) Areas requiring the use of protective equipment are properly posted?

(c) Personal protective equipment is routinely examined for defects and indications of deterioration?

(d) Employees are fully knowledgeable as to why equipment is required?

D. CONTROL OF OCCUPATIONAL ENVIRONMENTAL HAZARDS (13 items listed)

Ex. Item: #1. Atmospheric surveys conducted periodically (i.e., air, dust, gases) . . . . 1 □ □ □

(a) Plant monitored for concentration of materials identified in OSHA Standard 1910.93 and accurate records of results maintained)

(b) Where measurements show concentrations in excess of established limits, engineering controls initiated to reduce the level?

(c) Records of environmental tests and samples available for review? (Records maintained for 20 years)

E. INVOLVEMENT AND DEVELOPMENT (10 items listed)

Ex. Item: #4. Safety indoctrination given to all new or transferred employees . . . . 4 □ □ □

(a) Each new employee given a copy of the general plant/facility safety rules?

(b) Job safety analysis reviewed with supervisor?

(c) All relevant departmental operating procedures reviewed with supervisor?

(d) Employee given apprentice status until he demonstrates capability to accomplish job assignment unassisted?

Ex. Item: #7. First aid techniques taught to all employees . . . . 7 □ □ □

(a) Selected individuals given thorough first aid training and respond to plant/facility emergency?

(b) All employees instructed in techniques of emergency life saving?

F. MOTIVATION (10 items listed)

Ex. Item: #1. Posting of safety information adequate in volume and strategically located . . . . 1 □ □ □

(a) Information as required by OSHA posted in a place accessible to all employees?

(b) Accident prevention signs posted as required by OSHA Standard 1910.145?

(c) Excerpts from general safety committee meeting minutes posted for employees?

G. ACCIDENT EVALUATION AND REPORTING (11 items listed)

Exhibit 6.4 SAFER outline.

Ex. Item #4. Plant/facility management carries on in-depth analysis for reasons of incidents/accidents..... 4 ☐ ☐ ☐

(a) Medical data evaluated and correlated to ascertain any pattern to health problems?

(b) Statistical analysis conducted on factors related to accidents to determine and pattern?

(c) Production systems analyzed to eliminate hazards?

(d) Employee work schedule evaluated for casual factors?

Ex. Item #7. Records maintained of all accidents and illnesses .... 7 ☐ ☐ ☐

(a) Maintain records per OSHA (29 CFR 1904) for 5 years.

(b) Notation made in employee medical records (certain records must be kept for 20 years).

(c) File of accident investigation reports maintained for statistical evaluation?

NOTE: Each major section has space for local management to include any additional items they feel are necessary or appropriate.

Projected Safety Objectives: After the Safety Audit Factor Evaluation Report (SAFER) is completed, review all items rated "None" and "Minimal" and decide which specific items need the most attention. A list of these items constitute the safety objectives that need priority attention and should be stated in terms of specific programs or procedures to be implemented. For example:

Safer
Ref. No.

	Priority	Project
C—Item #4. Develop and Institute Hazard Operation Permit System for all hazardous operations. System to be coupled with existing work order system.	1	Completion Date September 1

Exhibit 6.4 Continued.

Next, the review team conducts the program review by means of interviewing key management people, supervisors, and employees on a random basis; attending a safety committee meeting; reviewing all loss-prevention files, accident logs, and reports; observing people at work; tabulating and analyzing the completed survey questionnaire; making inspections of facilities; and making spot reviews of safety suggestions, operating procedures, work orders, and training programs. After this, the safety program review and evaluation is prepared with the assistance of key management people. On the basis of this evaluation, an overall profile of the safety program is constructed and ratings are determined.

The review is followed by a verbal feedback session with the responsible manager and staff. Then a written report is prepared, covering the points made in the closing conference. The report includes a transmittal letter, the safety program review and evaluation forms, and any suggestions for improving the overall effectiveness of the program. A copy of the written report is sent to the appropriate executives and managers.

Safety Performance Indicator

Another approach to measurement that incorporates both results and performances is the Safety Performance Indicator (SPI). It was developed and used by C. D. Attaway, chief safety engineer for Thiokol Chemical Corporation. Following is his description of the SPI:

The Safety Performance Indicator does not replace the standard disabling injury frequency or severity rate, but is rather a means of combining it with other pertinent and important elements in a systematic fashion to reveal why accidents occur and what management has done to control these causes. This method is geared primarily to factory, plant, or installations of a large operating or manufacturing organization and is designed to highlight those factors that control a corporation's accident experience.

An added feature of the formula is the flexibility offered by adding or removing elements. For "within the division" purposes such things as promptness or completeness of accident reporting can be made an element. Or violations of safety rules and regulations and other circumstances that impede progress can be included as factors in the formula.

The intent of the method is simply to indicate whether the performance of the safety program at an activity or operation is getting better or worse and *why*.

Explanation and Computation of the Formula

The Performance Indicator is expressed by the formula $PI = A + DL + PD + RC + MD + S$. All data used in this formula can be obtained from accident reports. These reports are evaluated as submitted, with nothing added or deleted. Accidents and their consequences which are beyond a manager's immediate resources are not included.

A. The first element in the formula is A, which denotes the total number of reportable accidents occurring on work or operations under the jurisdiction of the manager that are within the resources of the corporation to correct. (One day or more actually lost or charged by ANSI Z16.1, or accident resulting in $100 or more property damage.)

B. DL denotes the total number of days lost, including fixed charges for permanent impairments and fatalities, as the result of injuries sustained by *on-duty* personnel.

C. PD denotes the amount of property damage in dollars to property and/or equipment.

D. RC denotes reason for cause, which is each instance where a supervisory and/or physical deficiency produced the accident cause. Physical deficiencies are defined as those accident factors which stem from poor construction practice, maintenance, layout, or equipment. Supervisory deficiencies are defined as those accident factors which stem from failure of supervisors at the job level to direct, inspect, instruct, and place the working force in a manner conducive to safety operation.

These definitions make it obvious that one, two, or none may be assigned a single accident. Selection depends on which factors, if any, are predominant. For example, if a man falls from a scaffold not provided with a guardrail, a physical deficiency is assigned. If a man falls from a scafford while using the cross members as means of access when adequate access has been provided by means of suitable ladders or stairways, a supervisory deficiency is assigned. If a piece of equipment is wrecked because of brake failure while being operated at excessive speed, both a physical and a supervisory deficiency are assigned.

E. MD denotes management deficiency, each instance where the report indicates, on the part of management, an apathetic review, a poor grasp of the facts involved, and inadequate directions to correct deficiencies. A management deficiency is also assigned to a report involving a fatality, permanent impairment, or property damage in excess of $10,000 where such incidents occur as a result of a repetitive cause in the plant in a calendar year. For example, if a man loses a finger operating a power saw without a guard and afterward another man loses a finger by the same cause, a management deficiency is assigned.

If a fire occurs because of an unattended salamander stove, with resulting damage in excess of $10,000, and then another fire occurs, with cause the same and costing over $10,000 loss, a management deficiency is assigned. This is a weighty factor in the formula because it is a determined evaluation of report data in terms of application of policy and instructions issued by the general manager to top supervision. Top management support is the hub around which the division safety program revolves, and therefore it becomes a principal consideration in evaluating safety performance. When reports are evaluated, a management deficiency is not assigned when a manager corrects a deficiency.

F. S denotes surcharges, assigned in each instance in which an accident resulted in a fatal, permanent impairment and/or damage in excess of $10,000. Surcharges are also a weighty factor because such serious incidents must be effectively controlled before the safety program can be considered to be making progress. It is conceded that in some instances severity of an injury or amount of property damage is a matter of chance. However, the experience of the corporation establishes that most of our fatalities, permanent impairments, and large dollar losses are foregone conclusions if certain incidents occur.

For example, when a man falls from a considerable height, the severity of his injuries is extreme. The same holds true if he is in contact with equipment which engages a power line. When valuable property is stored without segregation of combustibles or provision of fire protection or where high explosives are processed and explosions occur, the losses are usually large. Since the severity of incidents of this nature is largely predictable and since practical controls are well publicized, the occurrence of such accidents must have special consideration when measuring performances.

G.   The computations made from the formula $A + DL + PD + RC + MD + S$ are added together for current performance and compared to a normal. The normal is computed by applying the same formula to accident reports received for three preceding years. After the accumulation of five years' data, the normal will be a five-year rolling average, i.e., the most recent five-year experience. This will tell whether the performance of the installation is better or worse, how it is contributing to the overall performance, and the reasons why the performance is better or worse.

Results Measures

Results measures can be used either before an accident occurs or afterward. After-the-accident measures might be considered measures of our failures.

Failure Measures

Failure measures tend to be generated from our injury record system.

In order that the injury records can be used for measuring the results of the line manager's safety performance, they should be set up so that:

— They are broken down by unit.
— They give some insight into the nature and causes of accidents.
— They are expressed eventually in terms of dollars by unit.
— They conform to any legal and insurance requirements.

Beyond these broad outlines, each company can devise any system that seems right for it. The "dollar" criteria is included because of the belief that a dollar measuring stick is much more meaningful to line personnel than any safety specialist's measuring stick (such as our frequency and severity rates).

Many companies today have elaborate and successful record systems that do not measure up to these criteria. Obviously, if they serve their own intended purpose, this is the most important factor.

With failure measures we can count incidents, accidents, or injuries, and we can count these at various levels of severity. For example, we can count only fatalities, only lost-time injuries at the level of days lost when compensation starts, only those resulting in one or more days lost, only those requiring the attention of a doctor, or perhaps only cases requiring first aid. We can even start counting all close calls.

What we choose to count will change our results markedly. For lower-level managment, it is best to use a measure that gives us lots of numbers. For a supervisor of ten people, a measure of close calls as well as injuries makes more statistical sense and is more meaningful than one that counts only fatalities or only lost-time injuries.

Before-the-Fact Measures

Before-the-fact measures measure the results of supervisor action before an accident occurs. For instance, a periodic inspection is made of a supervisor's work area to measure how well he or she is maintaining physical conditions. This is a measure of whether things are wrong and, if so, how many things are wrong. We can also measure how well a supervisor gets through to the people in the department by measuring the people's work behavior. Safety sampling, which is discussed later, is used for such measurements.

Some results measures can be useful:

— They can be constructed to give swift feedback.
— They can be used for promotion purposes.
— They can attract attention.
— They can measure the presence, not the absence, of safety.
— They can be sensitive to change.
— They can be understandable to those who must use them.
— They can provide recognition of good performance.

However, we must use care in selecting results measurements. Since we are judging performance by some means other than the performance itself, we must be sure that what we are looking at is fairly closely related to performance. With failure measures, this close relationship is quickly lost at lower levels of the organization. Before-the-fact measures tend to retain the close relationship much better, therefore they are the next best measure (after activity measures) for lower levels. Two before measures are the best: safety sampling and inspection. Both were described earlier. Each was mentioned as a supervisory tool, but they are also excellent measurement tools of that supervisor.

The single most important reason for making inspections or for sampling is seldom mentioned. It is:

> To measure the supervisor's performance in safety.

Perhaps if the line manager felt that this was the primary purpose of management's inspection, he or she might do a better job of making sure that nothing amiss could be found in the department.

Results Measures at Middle and Upper Management

At the middle management level we concentrate on results measures. Here again we can work with before-the-fact measures or with failure measures. The validity and reliability of our failure measures may depend upon the size of the unit we are working with. Before-the-fact results measures used to judge middle managers' performance would include safety sampling and inspections by staff safety. Failure measures can be used a bit more extensively at this level. Here our traditional indicators become more useful.

Frequency and Severity Rates

Traditionally, we have used two figures to measure company-wide safety performance: the frequency rate, which is the number of disabling injuries (lost-time cases) per-worker-hours, and the severity rate, which is the number of days lost, or charged, per-worker-hours.

As indicators of internal performance, these rates have some serious weaknesses. The validity and reliability of these measures are a function of the size of the company, or how large the data base is, but the real weakness seems to be the fact that the rates are not meaningful to people in the organization. Those in top management often do not understand the rates and may wonder why their safety people cannot talk like managers and use more meaningful measures. As indicators for use in comparing the company's progress with that of other organizations, the major difficulty has to do with input integrity. Each organization seems to compile its safety records according to its own rules. It is common knowledge that when we attempt to compare our progress with that of other organizations using these measures, we have no idea how the other organizations have kept their records. Still, frequency and severity rates are the final measures used in most record systems today. Since these measures are very weak, however, we ought to examine other available measures, such as the following:

— *Frequency-severity indicator (FSI)*. This is a combined frequency and severity rate. FSI equals the square root of the frequency rate times the severity rate divided by 1,000:

$$FSI = \sqrt{\frac{F \times S}{1,000}}$$

For example:

Frequency rate	Severity rate	FSI
2	125	0.5
4	250	1.0
8	500	2.0

— *Total cost of first-aid cases.* In this appraisal the pro rata cost per case handled in-plant would be considered. For example, say that a first-aid unit in a plant costs $10,000 to operate. Forty percent of the unit's time is used to treat first-aid cases. An average of 1,000 cases are treated; therefore, the cost is $4 per case. This example combines industrial and nonindustrial first-aid cases. If average time per case is substantially different for each of these categories, a separate average cost per case may be better.
— *Cost incurred.* This includes the actual compensation and medical costs paid for cases which occurred in a specified period plus an estimate of what is still to be paid for those cases.
— *Estimated cost incurred.* This is an estimate of cost incurred, based on averages.
— *The cost factor.* This equals the total compensation and medical cost incurred (see item 3 above) per 1,000 worker-hours of exposure:

$$\text{Cost factor} = \frac{\text{cost incurred x 1,000}}{\text{total worker-hours}}$$

— *Insurance loss ratio.* This is equal to the incurred injury cost divided by the insurance premium:

$$\text{Loss ratio} = \frac{\text{incurred costs}}{\text{insurance premium}}$$

— *Cost of property damage and public liability costs.* This is a measure of damage to property of others caused by company operations.
— *Nonindustrial disabling injury rate.* This is a measure of off-the-job safety:

$$\text{Rate} = \frac{\text{no. of injuries x 1,000,000}}{312 \text{ x no. of employees}}$$

The 312 is computed as follows:

$$
\begin{array}{rl}
7 \text{ x } 24 \text{ hours} = & 168 \text{ hours per week} \\
& \underline{40} \text{ hours of work} \\
& 128 \\
& \underline{56} \text{ hours of sleep} \\
& \overline{72} \text{ hours exposed}
\end{array}
$$

4 1/3 x 72 hours per week = 312 hours per week

The above list is only a beginning. Each organization should devise an injury record system that will measure what management wants measured.

Dollars

Many of the measures discussed above are dollar-oriented. For internal uses, dollar-related measures are perhaps the best. Dollars are understandable and meaningful to everyone in the organization, particularly those at the corporate level. When we talk dollars, we are talking management's language. What dollar indicators can be used with management? Here are some possibilities:

- Dollar losses (claim costs) from the insurance company.
- Total dollar losses (insurance direct costs) plus first-aid costs not paid by insurance.
- So-called "hidden costs."
- Estimated costs.
- Insurance loss ratio.
- Insurance premium.
- Insurance experience modification.
- Insurance retrospective premium.
- Cost factor (see Chapter 4).

Dollar losses (items 1 and 2) are good indicators. (It seems more realistic to include first-aid cost than to ignore it.) Many companies use these figures or some measurements based on them, such as cost per worker-hour.

Although hidden costs are very real, they are difficult to demonstrate. To say arbitrarily to management that they amount to four times the insurable costs is asking for trouble. If management asks for proof, you can only say, "Heinrich said so." Management wants facts—not fantasy. Without proof, hidden costs become fantasy.

All of the above techniques utilize the dollar as the measuring stick, instead of measuring number of accidents, number of days lost, or the commonly used frequency rate or severity rate. Many people today believe that the dollar is a far better measuring stick than any other in safety, and many companies are beginning to utilize it effectively. In the past, it was difficult if not impossible to utilize the dollar for the following reasons:

- In the case of serious accidents, actual claim costs are not available for a long period of time—in some cases, years after the accident happened.
- There is no way to convert a frequency rate or a severity rate into dollars.
- If a company operates in several states, the actual claim costs will be an unfair measuring stick because the benefits vary so much from state to state.

Today, it is possible to fix dollar accountability by using a system in which the costs are estimated by a predetermined formula based on previous costs for similar accidents. Thus, when an accident does occur, an estimate of the final cost can be readily computed using this formula. This estimate can be based on the average costs in the state where the company operates. In approximating actual costs, it is possible to use average costs per medical-only cases, costs for the time away from the job, or daily hospital charges.

Hidden costs are real. They consist of such item as:

- Time lost from work by an injured employee
- Lost time by fellow workers
- Loss of efficiency due to breakup of crew
- Lost time by supervisors

- Cost of breaking in a new employee
- Possible damage to tools and equipment
- Time lost while damaged equipment is out of service
- Rejected work and spoilage
- Losses through failure to fill orders on time
- Overhead cost while work is disrupted
- Loss in earning power
- Economic loss to employee's family
- At least 100 other items of cost which may arise from any accident

The actual direct cost of a work injury—compensation and medical benefits—often is not established until long after the injury occurs, especially for the more severe cases. Thus, the loss statements provided by insurance carriers do not serve the purpose of effective, *current* cost evaluation. The only information that such statements can provide on relatively recent injuries consists of the amounts of medical and compensation benefits paid to date, plus the outstanding reserves. These reserves are established primarily to assure that sufficient funds are set aside for the eventual cost of the claims. It is not until the healing period has ended and the degree of disability has been positviely determined, however, that any accurate estimation of the cost can be made. And the more complicated the injury, the longer it takes to establish the final cost.

In recent years there has been an increasing demand for some method by which employers can determine promptly the approximate cost of compensable occupational injuries. Several such estimation systems are now in use in various companies in the United States. Estimated costs have proved themselves to be an effective method of obtaining cost information for line measurement.

This can be complicated or simple. One organization simply charges back to the location on the basis of $2,000 for an OSHA recordable injury, $7,000 for a lost time injury plus $500 for each day lost. As long as everybody understand the ground rules, a system like this works just as well as a complicated system.

Insurance Costs—Management pays an insurance premium for workers' compensation coverage. This is a real cost. Why not then measure the results in terms of this real out-of-pocket cost to management? The insurance premium is based on industry averages, adjusted by each company's record in the past and in some cases again adjusted by the current accident record. These adjustment factors, known as the "experience modifiers" and the "retrospective adjustment," are excellent indicators of past and present performance. Safety people ought to know, understand, and use these adjustment modifiers, since they represent the true costs of safety to management.

Direct Costs vs. Indirect Costs—There are both direct and indirect accident costs: direct costs are those medical and compensation costs paid to the claimant by the insurance company; indirect costs are those so-called "hidden" costs not covered by insurance and, in fact, not easily observed or recorded. Indirect costs include time lost by others who observed or gave help at the time of the accident and time lost by the supervisor in investigating.

Many studies have been made of indirect costs, with varying results. Each accident that occurs will, of course, have a different indirect cost and a different indirect-cost-to-direct-cost ratio. To attempt to strike averages for a company, for an industry, or for all industries is a meaningless exercise. Obviously, the ratio will vary tremendously, depending on circumstances, injury severity and other factors.

More important, any indirect costs that a safety specialist claims on the basis of some assumed ratio, even on the basis of computed past figures, really have little validity, and management knows this. To state arbitrarily that direct costs are $10,000, indirect costs are $40,000, and total accident costs are therefore $50,000 usually is not good for much more than safety preaching.

Direct costs are more meaningful. They are close to being real costs—in fact, they *are* real costs if not clouded by insurance company "reserves." Reserves constitute the insurance company's estimate of dollar losses per case, and they must be established for all serious cases. A safety professional whose estimate shows these reserves to be inaccurate can, of course, establish an estimate with his or her own figures on those cases still open, providing, of course, that management understands this approach.

Even direct costs are somewhat unreal costs to the average company, since they are paid by the insurance carrier, not by the company itself. However, the amount the company does pay out is dependent to a large extent on what these direct costs are.

This discussion is included so that the reader will better undertand how the workers' compensation insurance premium is arrived at—and hence might be able to use this information in communicating with management. In most states this premium figure is not as subject to manipulation as premiums on other lines of insurance, thus it is a more objective measure of safety performance. The workers' compensation insurance premium is determined by three rating systems:

1. Manual (or state) rating
2. Experience rating
3. Retrospective rating

MANUAL RATING—In making rates, the first step is to determine a basis of exposure. In workers' compensation, the basis is payroll. This reflects the number of employees, the hours they work or are exposed, and their pay scale. It is a measure of exposure that can be more readily ascertained and more easily verified than any other.

Workers' compensation rates are expressed in terms of dollars per $100 of payroll. Different types of work entail various degrees of hazard; therefore, in rate making, the types of work are classified to give consideration to these degrees of hazard. Through long experience and study of injury records from all types of business, a list of classifications has been developed in which the different types of work are arranged according to the degree of hazard. There are 600 to 700 classifications and each company fits into one or more of these. Of course, each company must first make sure that it is in the right classification, as each classification has its own rate. Each classification or group of similar industries actually sets its own insurance rate.

EXPERIENCE RATING—After manual rates are applied, experience rating is used to vary the company's own rates, depending on its experience in recent years. Under the manual rate, all industries of the same type (such as machine shops) pay exactly the same rate, regardless of their own accident record. Obviously, in any group of machine shops, some will have a good safety record, and some will have a poor one. Experience rating attempts to change this so that the company with a good safety record pays less than the company with high losses by making a statistical comparison between what losses *occurred* in a particular company during the past three years and what losses *were expected to occur* during that period in a machine shop of that size.

RETROSPECTIVE RATING—After experience rating, the company may select a retrospective rating plan of insurance instead of a regular plan. In a retrospective plan the amount of premium paid will depend on the amount of losses that occurred this year (not in the past three years). Retrospective adjustments are made on top of, after, or in addition to experience rating.

The experience modification cannot be escaped by the company—it must pay that rate.

Retrospective rating is a gamble—the company decides on partial self-insurance or, rather, on cost-plus insurance. The premium paid under this plan depends on the losses that the company will sustain this year. In retrospective rating, the company will pay for the administration of the insurance company, its losses, the use of the insurance company's claim department, and taxes—all subject to a minimum and a maximum.

The higher the losses during the year, the higher the insurance premium. If management succeeds in keeping losses low, the insurance premium will be low. At the same time, it must pay a penalty if its losses go over the break-even point.

As safety specialists, we should be aware of these insurance costs and of the different plans that set the costs. We should know the manual rate, the experience modifier, and whether the company is on retrospective. Management looks at these insurance costs, so safety people ought to be very familiar with the figures and utilize them in their communications with management.

Some companies choose to prorate their insurance premiums. When an accident occurs, the insurance company pays the direct costs of that accident. However, the insurance premium that the company pays is influenced by the costs of the accidents. If a department or a location is charged a specified amount of the insurance premium based on its percentage of the total accident costs, this will be a more accurate dollar measuring stick.

Let us assume, for example, that the XYZ Manufacturing Company has three plants; plant A has 2,000 employees, plant B has 500 employees, and plant C has 5,000 employees. Prorating the insurance premium suggests that, regardless of size, the plants will be charged the premium in proportion to their losses. The XYZ Manufacturing Company insurance premium is $100,000. Thus:

	Accident loss record	Premium charged
Plant A	$15,000 or 30%	$30,000
Plant B	$20,000 or 40%	$40,000
Plant C	$15,000 or 30%	$30,000

It will be difficult for plant B to show a profit this year because of its large accident losses.

Internal Measures

Thus, many measures can be used to judge corporate progress and performance in safety. Results (failure) measures such as dollar-oriented measures, insurance measures, and estimated cost measures, and before-the-fact results measures such as inspections and safety sampling are options. OSHA code violations, discovered either by the safety specialist or by compliance officers, can also be used as measures. In short, to judge a firm's internal safety performance, there is a wide variety of measures from which to choose. Often top executives, rather than safety staff choose the measures that will be used. There is nothing wrong with this, although one would hope that safety staff might encourage corporate executives to find more meaningful measures than the traditional ones.

Comparison Measures

It is less easy to find measures for use in comparing one company's progress with that of others. Most safety professionals agree that traditional methods of doing this are ineffective and inaccurate.

Perhaps, however, the real question is whether such measures are needed. Does a comparison to other organizations mean anything anyway? A company's safety record reflects many things: hazards, controls, employee morale, the climate and style of the company. It is difficult to see how or why a company should compare itself with others when all the items that go into the making of each company's record are different. It would seem that the only important thing is whether the company is getting better or worse in each period measured.

Past Performance as a Standard

The past performance of any group is the best standard to use as a guide for its present performance. The people in the group understand it and will accept it, whereas they are often reluctant to accept an arbitrary standard set by someone outside the group. Often they believe that competing against a different group is unfair, since no two groups face the same challenges, hazards, or situations. Management is interested principally in its own company and whether it is improving.

When past performance is used as a standard, it is often difficult to know whether the changes that have occurred are truly significant, reflecting a different supervisory performance, or whether they are merely chance happenings—random fluctuations. There is a technique now in use which helps to detemine this.

Safe-T-Score

The Safe-T-Score is based on a statistical quality control test which is used to examine the means of two groups of comparable data for significant differences. In statistics it is usually referred to as the "t test." Here is its formula:

$$\text{Safe-T-Score} = \frac{\text{frequency rate now} - \text{frequency rate past}}{\sqrt{\dfrac{\text{frequency rate past}}{\text{million worker-hours now}}}}$$

The Safe-T-Score is a dimensionless number. A positive Safe-T-Score indicates a worsening record; a negative Safe-T-Score indicates an improved record over the past. Here is how the Safe-T-Score can be interpreted:

- If the Safe-T-Score is between +2.00 and –2.00, the change is not significantly different. The variation is due to random fluctuation only.
- If the Safe-T-Score is over +2.00, the record is significantly worse than it was in the previous period. Something has gone wrong.
- If the Safe-T-Score is under –2.00, the record is significantly better than it was in the last period. Something has changed for the better.

For example, to compare the all-accident frequency rates of two locations of a company, each against its past record, start with this information:

Location X	*Location Y*
Last year— 10 accidents 10,000 worker-hours	Last year— 1,000 accidents 1,000,000 worker-hours
Frequence rate = 1,000	Frequency rate = 1,000
This year— 15 accidents 10,000 worker-hours	This year— 1,100 accidents 1,000,000 worker-hours
Frequency rate = 1,500	Frequency rate = 1,100

The frequency rate for location X has increased by 50 percent; that for location Y has increased by only 10 percent. Are these figures significant? Has something gone wrong at one or both locations? By using the Safe-T-Score, these questions can be answered.

Perception Surveys

One of the more recent and one of the best measurement tools is the perception survey. A number of questionnaire approaches have been used for years to dipstick employee morale, climate, etc. Recently several organizations have carried this to the point where they have developed questionnaire approaches and methodology so precise that the results correlate much more closely with the accident indicator than any audit or other measuring stick. These are discussed in more detail in Chapter 10.

REFERENCES

Cook, Ken. "Safety Sampling," Ken Cook Lectron Company, Milwaukee, WI, 1963.

Grimaldi, John V. "The Measurement of Safety Engineering Performance." *The Journal of Safety Research,* September 1970.

Martin, J. A. "Large Plant Safety Program Management," *Journal of the ASSE,* May 1963.

Pollina, Vincent. "Safety Sampling," *Journal of the ASSE,* August 1962.

Rockwell, Thomas, and Vivek Bhise. "Two Approaches to a Non-Accident Measure for Continuous Assessment of Safety Performance," *The Journal of Safety Research,* September 1970.

Sayles, L. "Limitation of managing by results," *Boardroom Reports,* 1983.

Tarrants, William. "A Definition of the Safety Measurement Problem," *The Journal of Safety Research,* September 1970.

Tarrants, William. "Applying Measurement Concepts to the Appraisal of Safety Performance," *Journal of the ASSE,* May 1965.

Part III

Proactive System Elements

Changing Behavior

Chapter 7 focuses on employee behavior. While this book's focus is the systems approach to safety management, not the behavioral approach, we would be totally remiss not to include at least this one small section on the behavioral approach to safety results. This is an integral part (perhaps the most important single part) of the management process.

We believe it is important that the management system deal with behavior—it is the primary cause of accidents. It is important also that it deal with behavior in ways that make sense—too often it has not.

A companion book, *Safety Management—A Human Approach, Second Edition,* covers the behavioral approach to safety management. So this chapter is primarily an overview of the elements in the safety system that deal with employee behavior.

In this brief overview we examine the following influences on employee behavior:

— Selection.
— Training.
— Supervision.
— Options to influence behavior.
— Where discipline fits in.
— Where gimmicks fit it.

SELECTION

Perhaps the logical starting point is to select people who will work safely—to screen out those "unsafe" workers and hire the "safe" ones. Obviously this is easier said than done. In the early years of safety, selection was one of the biggest tools. We believed we could select those applicants that would be safe workers and screen out the unsafe. It is not that simple. Here's why:

— Proneness—It used to be believed that certain applicants were "accident prone", that is, with permanent susceptibility to accidents. But three problems emerged:

  1. Theorists began to question the concept.
  2. No one ever figured out how to test for "proneness."
  3. Such testing probably would be illegal.

Current theorists seem mixed as to the concept. In this country we tend to reject the proneness thinking, while in others it is still believed. Probably the other countries are more right than we are for research has shown there is a link between personality types and risk taking

141

(thus leading to more accidents) and between what people have had to cope with and having accidents (a temporary susceptibility).

Research has shown that:

— Accident proneness is real but only in a tiny percentage of people, and is probably a system of maladjustment. We probably cannot afford to find those people.
— Accident susceptibles are an ever-shifting group. Finding them will not help much because next year the susceptibles will be a different group.

Whether or not applicants can be tested to find the susceptibles is also questionable. The typical selection process consists of the items shown in Exhibit 7.1.

THE PROCESS OF SELECTION

Biographical Data	Tests
Application	Physical examination
Interviews	Qualifications
References	Defects
Other sources: credit bureaus,	Skill tests
schools, driving record, agencies	Job knowledge tests
	Psychological tests

Exhibit 7.1 The selection process.

As the exhibit shows, there are two basic sources of information about an applicant: biographical data and test results.

The value and validity of most of these selection devices have been fairly well subjected to research in the field of occupational psychology and that research leads to these conclusions:

— Job knowledge and skill tests can be administered effectively, provided the job criteria can be established in these areas.
— Similarly, physical exams can be helpful, provided physical job criteria are available.
— Interviews are generally invalid, not because the applicant is lying, but rather because the interviewer's biases and stereotypes invalidate them.
— Checks to previous employers (if we have established a good working relationship) will provide an applicant's past history (assuming this is indicative of his or her future, which is not and probably cannot be proved).

In addition, there are the further selection constraints of law. Most tests are illegal in the U.S. under the Equal Employment Opportunity laws, unless they have been statistically validated to job success at your company.

TRAINING

When training is done systematically by the organization, as opposed to the portions of the training process occurring daily by supervisors with their people, we can say it consists of the organization going through the following process:

1. Finding out where people are in terms of their current knowledges and skills.
2. Finding out where people should be in terms of the behaviors required to perform the job safely.
3. Figuring out a systematic way to provide the difference.

The above puts a concentration in the training task of defining training needs. In safety training, historically concentration has been in other areas—on training method, on content, on other things. Learning theorists tell us to spend the bulk of our time in defining needs, for if we do a good job of this, all else falls easily into place. The assessment of training needs involves a three-part analysis.

1. Organization analysis should uncover the resources and objectives of the company which relate to training.
2. Operations analysis should define corporate jobs and tasks.
3. Manpower analysis explores the human dimensions of attitudes, skills, and knowledge as they relate to the company and the employee's job.

Organization analysis involves a study of the entire organization: its objectives, its resources, the allocation of these resources in meeting its objectives, and the total environment of the organization. These things largely determine the training philosophy for the entire organizatinn.

Job analysis for training purposes involves a careful study of jobs within an organization in a further effort to define the specific content of training. It requires an orderly, systematic collection of data about the job, similar to job safety analysis procedures. Other ways are also available for job analysis.

— Observations. (Is there obvious evidence of unsafe acts or poor methods? Are there incidences on the part of individuals or groups that reveal poor personnel relationships, emotionally charged attitudes, frustrations, lack of understanding, or personal limitations? Do these situations imply training needs?)
— Management requests for training of employees.
— Interviews with supervisors and top management personnel to accumulate information about safety problems, as well as interviews with employees concerning safety.
— Group conferences with interdepartmental groups and safety advisory committees to discuss organizational objectives, major operational problems, plans for meeting objectives, and areas in which training could be of value.
— Comparative studies of safe versus unsafe employees to underline the basis for differentiating successful from unsuccessful performance.
— Questionnaire surveys.
— Tests or examinations of safety knowledge of current employees; analyses of safety sampling.
— Supervisors' reports on the safety performance of employees.
— Accident records.
— Actually performing the job.

Manpower analysis focuses on individuals and their performance on the job as it relates to safety. The performance analysis chart shown in Exhibit 7.2 helps define individual training needs. This performance analysis chart was devised by Dr. Phil Brereton of Illinois State University. The chart helps to identify whether or not the person does in fact have a training problem at all. Too often we jump to training as the solution to almost any problem. While

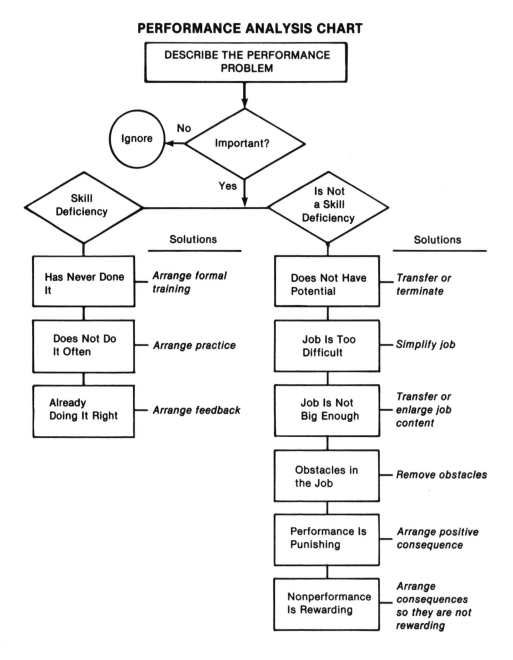

PERFORMANCE ANALYSIS CHART

Exhibit 7.2 The Brereton model.

training is highly important, it is the solution only when the person's problem is the result of a lack of knowledge or skill. If the problem is the result of a skill deficiency, training might be the answer (see the left branch of the chart).

Many times, however, the problem is not the result of a skill deficiency, in which case the right branch of the chart applies, leading to other solutions.

Finally, what is usually missing from the training design is evaluation. The evaluation of training is not simple. We try to determine what changes in skill, knowledge, and attitude have taken place as a result of training. We also try to determine how these skills, knowledges and attitudes contribute to organizational objectives. We suggest evaluation measures in four broad categories:

1. Objective-subjective.
2. Direct-indirect.
3. Intermediate-ultimate.
4. Specific-summary.

— *Objective-Subjective.* The major distinction between an objective and a subjective measure is its source. A measure is objective if it is derived from overt behavior. If the measure represents an opinion, a belief, or a judgment, it is subjective.
— *Direct-Indirect.* A measure is classified as direct if it measures the behavior or the results of behavior of the individual. An indirect measure assesses the action of an individual whose behavior can be measured only by its influence on the actions of others. For instance, supervisory training effectiveness is usually measured indirectly: Did his or her department have a better safety record?
— *Intermediate-Ultimate.* An intermediate measure might be the supervisor's test grade—an ultimate measure of the results back on the job.
— *Specific-Summary.* Somewhat related to the problem of intermediate and ultimate measures of training outcome is the problem of measures which are used as an index of successful performance of a specific phase of a job or as an index of the degree of performance of the total job against its potential contribution to organizational goals. Should the effectiveness of the supervisory inspection training program be measured by the number of code violations or by improvement in the department's accident record?

These four categories of measures are not mutually exclusive. A rating scale could be a subjective, intermediate, direct, and specific measure. It also could be a subjective, direct, ultimate, and summary measure. Regardless of their type, measures must have certain characteristics if they are to be used in studying training outcomes. These characteristics are relevant to job criteria and corporate goals, reliability (consistency), and freedom from bias.

Too often in safety training there simply is no measure of any kind to evaluate whether or not the training has accomplished its objectives—or if it has accomplished anything. There should be. Participants in the training process, regardless of organizational level, should have a measurable improvement in some skill or behavior, and management should insist that the training be evaluated in terms of improvement of the skill or in terms of measurement of behavior change.

SUPERVISION

The third element in behavior control, and probably by far the most important, is supervision—what happens every day between the worker and his boss. This is what affects the behavior most. What the supervisor says and does; how measures and rewards are presented give workers signals as to what is important and what is not. Supervisors have a vast array of ways to send these messages as indicated below:

BEHAVIOR CHANGE OPTIONS

In the early years of safety it was believed that employee behavior could change through education and enforcement. Period. Early texts taught that.

Historical roots in safety suggest that industrial safety programs should consist of the "3 Es"—engineering, education, and enforcement. This historical belief has led safety programs to concentrate on physical conditions, training, and rule enforcement. Of the three, two have to do with behavior control, and thus safety's traditional emphasis on behavior control uses only those two.

Sixty to seventy years of research in organizational behavior, organizational and industrial psychology, and other fields suggest that the above traditional safety approach has been off base. Organizational psychologists today are quite clear that education and enforcement of safety rules are relatively weak as behavior changing techniques, and perhaps the weakest of those available to management. They suggest that other approaches are much more effective.

Those approaches can be put into these three categories:

1. *Approaches that attempt to change worker behavior directly.* These approaches typically will be called Behavior Modification or Positive Reinforcement approaches.
2. *Approaches that attempt to change a worker's attitude* in the belief that behavior will follow.
3. *Approaches that attempt to build a psychological environment* that makes it more likely that the worker will choose to work safely (motivation).

MOTIVATION

What motivates people (or, rather, what causes people to motivate themselves) at various levels? Exhibit 7.3 attempts to describe the process of motivation. At various levels of the organization, people are "turned on" and "turned off" by different things. These "key motivators," listed in the left-hand column, are based on certain theories from the behavioral sciences (as indicated in the right-hand column).

Employee Level

At the employee level, the primary motivating influence is peer pressure, or pressure from the informal group.

Other key influences on employee attitudes and behavior (actions) are whether the job is important and meaningful to the employee, the degree of employee involvement and participation in the important aspects of the job, and how much recognition is given for good performance. These insights come from the thinking of Frederick Herzberg, Chris Argyris, Rensis Likert, and other influential management theorists.

Conflict Theory

Argyris's conflict theory provides safety managers with insights into why people commit errors.

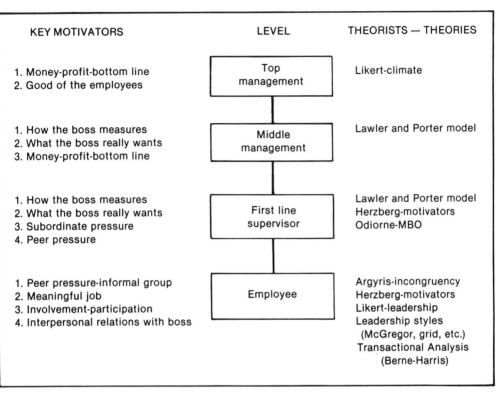

Exhibit 7.3 Motivators at various levels.

First, Argyris examines these principles of management:

— *Work Specialization.* American management has made a science of specialization. Ideas inherent in this assumption are:

  — That the human personality will behave more efficiently if more specialized.
  — That one best way to define the job can be found.
  — That any individual differences in the human personality may be ignored.

— *Chain of Command.* Following the logic of specialization, the planners created a new function (leadership), the primary responsibility of which is to control, direct, and coordinate the interrelationships of the parts and to make certain that each part performs its objective adequately. They then must be motivated to accept direction, control, and coordination of their behavior. The leader, therefore, is assigned formal power to hire, discharge, reward, and penalize the individuals in order to mold their behavior in the pattern of the organization's objectives. This makes the individuals dependent on, passive toward, and subordinate to the leader. As a result, the individuals have little control over their working environment.

— *Span of Control.* The principle of span of control states that administrative efficiency is increased by limiting the span of control of a leader to no more than five or six subordinates whose work interlocks. By keeping the number of subordinates to a minimum, great emphasis is placed on close supervision.

Argyris concludes that there are some basic incongruencies between the growth trends of a healthy personality and the requirements of formal organization. The individual in our culture tends to develop along specific trends as follows:

- From a state of passivity as an infant to a state of increasing activity as an adult.
- From a state of dependence on others as an infant to a state of relative independence as an adult. Relative independence is the ability to "stand on one's own two feet."
- From being capable of behaving in only a few ways as an infant to being capable of behaving in many different ways as an adult.
- From having erratic, casual, shallow, quickly dropped interests as an infant to possessing a deepening of interests as an adult.
- From having a short-time perspective as an infant to having a much longer time perspective as an adult.
- From being in a subordinate position in the family and society as an infant to aspiring to occupy at least an equal and/or superordinate position relative to peers.
- From having a lack of awareness of the self as an infant to having an awareness of the control over the self as an adult.

These characteristics are descriptive of a basic developmental process along which the growth of individuals in our culture may be measured.

In the work environment, because of management principles, people are:

- Provided minimal control over their workaday world.
- Expected to be passive, dependent, subordinate.
- Expected to have a short-time perspective.
- Induced to perfect and value the frequent use of a few superficial abilities.
- Expected to produce under conditions leading to psychological failure.

All of these characteristics are incongruent to those which healthy human beings desire. They are much more congruent with the needs of infants in our culture.

Argyris then suggests that the conflict leads to these escapes for the workers:

- Leaving the organization.
- Climbing the organizational ladder.
- Manifesting defense reactions such as daydreaming, aggression, ambivalence, regression, projection, and so forth.
- Becoming apathetic and disinterested toward the organization, its makeup, and its goals. This leads to such phenomena as reducing the number and potency of the needs they expect to fulfill while at work, and goldbricking, setting rates, restricting quotas, making errors, cheating, slowing down, accidents, and so on.
- Creating informal groups to sanction the defense reactions and the apathy, disinterest, and lack of self-involvement.
- Formalizing the informal group.
- Evolving group norms that perpetuate the above behavior.
- Evolving a psychological "set" against the company.

Then, because of the employees' actions, many managements tend to respond to this behavior by:

- Increasing the degree of their pressure-oriented leadership.

— Increasing the degree of their use of management controls.
— Increasing the number of "pseudo" participation and communication programs.

These three reactions by management actually compound the dependence, subordination, and so on, that the employees experience, which, in turn, cause the employees to increase their behavior, the very behavior management desired to curtail in the first place.

Is there a way out of this circular process? Argyris suggests the basic problem is the reduction in the degree of dependency, subordination, and submissiveness experienced by the employee in the work situation. It can be shown that job enlargement and employee-centered (or democratic or participative) leadership are elements which, if used correctly, can go a long way toward helping the situation. Close supervision leads subordinates to become dependent on, passive toward, and subordinate to the leader. Close supervision also leads to the problems and behavior described. Obviously, close supervision and tightened control is not the answer to the problem.

Certainly Argyris's incongruency theory has a great deal of meaning in safety. As employees adapt to frustration and conflict in their work situation and manifest defense reactions such as daydreaming and aggression, these surely influence the accident record. As employees exhibit the various forms of escape, our safety record is surely influenced to a very marked degree. When we see these things, our reaction is very much as Argyris indicated:

1. We increase the degree of pressure to enforce safety rules (three-day layoff).
2. We increase the degree of control. We watch even closer—step up our activities.
3. We increase the number of "pseudo" participation and communication programs (posters, contests, gimmicks).

This is real. We have all seen exactly these reactions toward a working population that "does not care about their own safety."

What should we do in safety programming? Argyris suggests we should consider programs that reduce the degree of dependency, subordination, and submissiveness. We should build safety programs that incorporate the concepts of job enrichment, participation, and employee-centered leadership.

Argyris proposed "leveling." The use of group decision-making and supervision means that the boss does not necessarily make all decisions alone. The emphasis here is on the involvement of people in the decision-making process to the extent that their perceptions of problems are sought, their ideas on alternative solutions are cultivated, and their thoughts on implementing decisions which have already been made are solicited.

Motivation-Hygiene Theory

The second behavioral theory is Herzberg's motivation-hygiene theory. While Chris Argyris and other behavioral scientists came to a conclusion that job enrichment is a necessity, it was actually Frederick Herzberg who first coined the term and expressed the principle. Herzberg's interest in job enrichment grew out of his discovery of what might be called his *dual factor theory of job satisfaction and motivation.*

Herzberg uses the term *hygiene* to describe such things as physical working conditions, supervisory policies, the climate of labor-management relations, wages and various "fringe" benefits. In other words, hygiene includes all the various factors through which management has traditionally sought to affect motivation. Herzberg chose the term hygiene to describe these factors because they are essentially preventive actions taken to remove sources of dissatisfaction from the environment, just as sanitation removes the potential threats to health from the physical environment. His research has shown that when any of these factors are deficient, employees are quite likely to be displeased and to express their displeasure in ways that hamper the organization, for example, through grievances, decreased productivity, or even strikes. But when the deficiencies are corrected, even though productivity may return to normal, it is not likely to rise above that level.

Herzberg uses the term *motivation* to describe feelings of accomplishment, professional growth, and professional recognition that are experienced in a job that offers sufficient challenge and scope to the worker. These factors seem capable of producing a lasting increase in satisfaction, and with it, an increase in productivity above "normal" levels. This has been found to be true in a wide variety of jobs and organizational settings.

Dissatisfiers seem to be such items as pay, benefits, company policies and administration, behavior of supervision, working conditions, and other factors that are generally peripheral to the task. Though traditionally thought of by management as motivators of people, these factors are actually more potent as dissatisfiers. High motivation does not result from their improvement. Dissatisfaction does result from their deterioration. Motivators are such items as achievement, recognition, responsibility, growth, advancement, and other matters associated with the self-actualization of the individual on the job. Job satisfaction and high production are associated with these motivators, while disappointments and ineffectiveness were usually associated with the dissatisfiers. A challenging job which allows a feeling of achievement, responsibility, growth, advancement, enjoyment of work itself, and earned recognition motivates employees to work effectively. Factors which are peripheral to the job (work rules, lighting, coffee breaks, titles, seniority rights, wages, fringe benefits, and the like) dissatisfy workers. These could be considered tease items which will temporarily please but which will be long-range motivational failures. In other words, the factors in the work situation which motivate employees are different from the factors that dissatisfy employees. Motivation stems from the challenge of the job through such factors as achievement, responsibility, growth, advancement, work itself, and earned recognition. Dissatisfactions more often spring from factors peripheral to the job.

Likert's Theory

The third theory is that of Rensis Likert, whose studies concerned the effect of the supervisor-employee relationship on productivity. Among Likert's findings are:

— The tighter the supervisor's control over the employee, the lower the productivity.
— The more the supervisor watches and supervises the worker, the lower the productivity.
— The more punitive the supervisor is when the employee makes a mistake, the lower the productivity.

In short, Likert's research indicates that if we wish to promote safety we should not try to "control" employees. Controlling employees causes them to work less safely.

Today management accepts these behavioral theories, and certainly they can help safety directors increase the acceptability and productivity of a safety program.

Coping with the Group

Each employee is an individual, and also an integral part or a member of a group. Each manager must manage a crew as different individuals, but also as a group. As in chemistry where elements combine together to make other substances with entirely different properties, individuals combine together to produce a group which has entirely different properties. A group has a distinct personality of its own. We have to recognize the group properties as well as the individual's properties. Each group makes its own decision and sets its own work goals. These may be identical to management's goals, or they may be different. The group also sets its own safety standards, and it lives by *its* standards, regardless of what management's standards are, and regardless of what the OSHA standards say.

These group pressures put more pressure on workers than do the standard procedures which are written in the manuals. In safety, work-group pressures and group norms are perhaps the single most important determinant of worker behavior. To restate our original thesis: the group sets its own safety rules, and they live by their rules, not management rules.

The safety program then must not only speak to the individual, it must also attempt to understand the group's safety norms and to influence those group norms so that they are safety-oriented norms. The safety program must help to build strong work groups with goals that coincide with safety goals.

The Employee Motivation Model

Exhibit 7.4 depicts the employee motivation model.

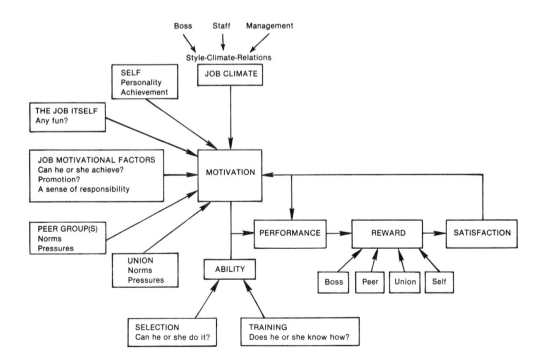

Exhibit 7.4 The motivation reward satisfaction model.

There are a number of influences over the employee. The supervisor must determine which can be altered and which cannot. This model of worker performance attempts to depict the various motivational influences upon workers, which summed together determine what their performance will be—whether or not they will perform in a safe or an unsafe manner.

ATTITUDES

There are a number of current theories about attitude and change or development today that might be helpful to the safety professional. Traditionally safety professionals have used preaching and teaching as the primary methods of developing or changing a worker's safety attitude. The current theories suggest that the whole process of attitude change is considerably more complex than we formerly realized.

These theories are discussed in depth in *Safety Management—A Human Approach, Second Edition*. The theories are well tested and validated. The difficulty is in how to use them. There are two problems in using attitude theory in practical safety:

1. Attitudes cannot be measured. All we can do is observe a behavior and infer it means an attitude.
2. Whether or not attitude can be changed is dependent upon many things:
   - The other person's perception of us to start with.
   - How he or she reacts to the message we are sending.
   - His or her current situation.

The bottom line is that perhaps we are better off pursuing either of the other means.

BEHAVIOR MODIFICATION

Behavior modification is not new. The basic process involves systematically reinforcing positive behavior while at the same time ignoring or exercising negative reinforcements to eliminate unwanted behavior. There are two primary approaches used in behavior modification programs. One is an attempt to eliminate unwanted behavior that detracts from organizational goal attainment and the other is the learning of new responses. In safety, the objective is to eliminate unsafe acts.

The second major goal of behavior modification is to create acceptable new responses to an environmental stimulus.

The basic concept of behavior modification is the systematic use of positive reinforcement. As is indicated in the above, the results of using positive reinforcement are improved performance in the area to which the positive reinforcement is connected. The concept is based on the simple formula

$$B = f(C)$$

which means that a person's behavior is a function of the consequences of past behavior. If a person does something and immediately following the act something pleasureable happens, he or she will be more likely to repeat that act. If a person does something and immediately following the act something painful occurs, he or she will be less likely to repeat that act again (or will ensure not being caught in that act next time).

Positive reinforcement is gaining wider and wider acceptance in industry. It has been used to increase productivity and quality, improve labor relations, meet EEO objectives, and reduce absenteeism.

A number of recent experimental programs show the potential of positive reinforcement in safety. Exhibit 7.5 outlines a number of these. As the Exhibit shows, the results are impressive, with reductions in unsafe behavior ranging from 8 to 350 percent with no negative results.

Perhaps behavioral influences can be illustrated by Exhibit 7.6 from *Analyzing Safety Performance,* by the author. An employee comes to the organization with certain influences from the past which are dictating his or her behavior. These values are tested in the present situation, resulting in a "current motivation." This, coupled with abilities, becomes current behavior. We then attempt to influence that behavior with a safety system. That system consists of antecedents and consequences.

Antecedents (things come before the behavior) are rules, regulations, group norms, climate, and so on. Consequences are what that person experiences after his or her behavior. While both are important, consequences are infinitely more important than antecedents. The rule (an antecedent) is only important insofar as it predicts what will happen following the act (consequences). Every employee knows which rules to follow and which to ignore after a short time on the job, learning from the reactions of the boss and others each day.

RESEARCHER	YEAR	TYPE REWARD	MEASURES USED	RESULTS OBTAINED	PERCENT IMPROVEMENT
1. Komaki, Barwick & Scott	1978	Praise & Feedback 3-4x/wk	% Safe Behaviors	70% up to 95.8% 77.6% up to 99.3%	37% 29%
2. Komaki, Heinzman & Lawson	1980	Feedback Daily	% Safe Behaviors	34.4% up to 68.4% 70.8% to 92.3%	98% 30%
3. Krause	1984	Feedback	Unsafe Behaviors Lost Time Accidents Severity Rates		80% 39.3% 39.2%
4. Sulzer-Azaroff & DeSantamaria	1980		Accident Freq. Accident Freq. Hazards Hazards	15 to 0 45 to 33	100% 27% 29% 88%
5. Felliner & Sulzer-Azaroff	1984	Praise & Feedback	% Safe Behaviors	78% up to 86% 79% up to 85%	10% 8%
6. Petersen	1983	Praise & Feedback	Safety Sampling	Two Railroad Divisions for Six Mos. with Control Groups	40% 49%
7. Rhoton	1980	Praise & Feedback	Violation	1-4 per mo. to 0	100%
8. Hopkins, Conrad & Smith		Feedback & Money	% Safe Behaviors Housekeeping Rate Housekeeping Rate	60% up to 100% 20 to 90 45 to 100	67% 350% 122%
9. Chokar & Wallin	1984	Feedback	% Safe Behavior	65% up to 81% 81% up to 95%	20% 17%
10. Uslan & Adelman	1977	Praise	Injury Frequency		50%

Figure 7.5 Safe Behavior Reinforcement.

These behaviors take place in a physical environment which improves (or doesn't) depending upon the corporate emphasis on it.

Some of these are influences over which we have some control, while others are not. For those that are not, it still is important to recognize that they are in fact influences.

Influences Beyond our Control

For all practical purposes all of the past influences are outside of our control, as are probably the following current influences: personality, attitudes, values, individual differences, union. That is not to say that there are not possible actions about these influences, but it does mean that they probably cannot be effectively changed through any managerial system. However, some managerial systems might be built to soften their effect.

For instance, building a management system that better helps a supervisor understand the worker's individual personality, the attitudes and the individual value systems certainly helps to soften the effect of these things on safety performance.

One example suggests that a system be provided to supervisors (after some training in the concepts) that requires them to look at each individual supervised and make a quick evaluation in terms of personality type and thus accident risk, and choose a leadership style that would be most appropriate for the that individual.

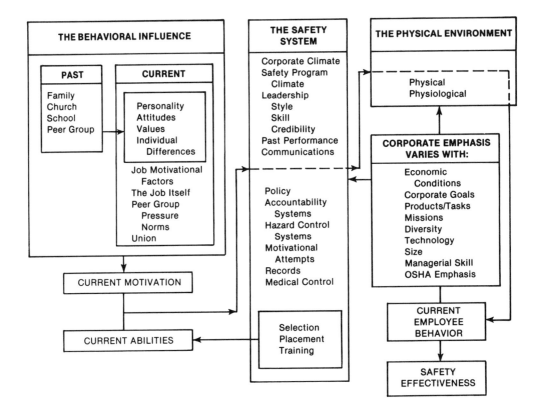

Exhibit 7.6 Behavioral factors affecting safety effectiveness.

Influences Within Our Control

Those influences within our control are considerably stronger influences than the ones beyond our control. These are the job motivational factors, the job itself, peer groups, organizational climate, and boss styles.

These influences were explained in some detail when the theories of Argyris, Herzberg, and Likert were explained along with some discussion of the group influence. Improvement here comes from improving the motivational field of each employee through some systematic method, probably administered by each employee's immediate manager. A start at this can be accomplished by providing each supervisor with some simple approaches to enable them to look at each of their employees.

DISCIPLINE

Up to now we have not even mentioned enforcement and discipline; one of the cornerstones of traditional safety. Where does it fit in? Or does it? Perhaps the answer is "only as a last resort."

Research shows that positive reinforcement is much more powerful than negative in both behavior building and behavior maintaining. The experiments mentioned earlier in positive reinforcement illustrate the power of positiveness.

This does not mean discipline should be eliminated from our safety repertoire. It remains a part, and an important part, but it is used seldom and only as a last resort. The day of the 600-page rule book and issuing of tickets for violations is probably over. It is in fact counter productive because it creates a "them vs. us," a "police vs. policed" environment exactly the opposite of the team building approaches of today.

GIMMICKS

An integral part of traditional safety is gimmickry: contests, incentives, posters, hoopla. Does all this fit? Early on this writer became disenchanted with gimmickry for a very simple reason: they only worked sporadically—usually didn't, but occasionally did. Thus the comments here on gimmicks are biased. Gimmicks simply do not fit any place in safety theory; they make little sense in management theory and even less sense in behavioral theory:

- In management theory, performances are clearly defined—to measure validity of performance and to make rewards consistent with their importance. Gimmicks define no performance; they usually measure with invalid measures (number of accidents) and reward with trivialities (plaques, tee shirts, jackets, steak dinners).
- In behavioral theory, a reward should be contingent upon performance to affect behavior and the reinforcement must be contiguous (right now) to the behavior. Gimmicks are neither contingent or contiguous to any behavior.

In brief, gimmicks are irrelevant to behavior. This is not to say they should be eliminated. Please recognize that to eliminate gimmicks in most circumstances is to incur the wrath of almost everyone, from the old safety director to the lowest rated worker, for gimmicks are a satisfier. They do not affect behavior, but to eliminate them is to remove something expected. They maintain morale but they do not shape behavior in most normal workers.

BRINGING IT TOGETHER

The behavioral aspect of safety is by far the most complex and confusing of all of the aspects. One way of looking at it is shown in Exhibit 7.7. It suggests that success (productivity or safety) is a function of culture and structure—people interacting together in a situation (within a culture and with a structure) producing certain results. Those results are influences by all four:

1. Culture.
2. Structure.
3. Population.
4. Situation.

All of this interaction results in conflict, usually much conflict (hopefully with cooperativeness).

What does all this mean? It means that in structuring an organization's safety system you must consider:

— What is the culture of the organization?
— What is the structure of the organization?
— What is the population of the organization?
— What is the current situation?

First-Line Supervisory Level

At this level of the organization there are major differences in what seems to motivate employees. First-line supervisors have moved up a step in the organization and they are now in a totally different situation psychologically. Whereas they used to be motivated primarily by peer pressure, this is now much less important. Members of the old peer group are now supervisors themselves. The group is less strong and its pressure is less important. Now there is more interest in the boss and in his or her desires, for pleasing the boss means retaining the new position of power over others and perhaps even being promoted again to a position of more power and larger financial gains. Thus the first-line supervisor's primary motivational force is the boss—how he or she is being measured by the boss, how to demonstrate good job performance to the boss, and how to find out what the boss really wants. The measures the boss uses are not always a good indicator of what the boss really wants, however, as many of them have been chosen by management. Also, priorities shift and change, without there being a concurrent change in the measures the boss uses. Thus first-line supervisors strive to find out what is important to the boss and they react to those items first.

It is true that first-line supervisors also react to pressures from other directions. Since they must live with their people—their subordinates—they do react to them and to the pressure they exert. The first-line supervisor is motivated first by what the boss wants (as indicated by the measure used and by the boss's expressed wishes) and second by the members of the group. The first-line supervisor must know these people well enough to know what they want and need; success on the job depends upon it.

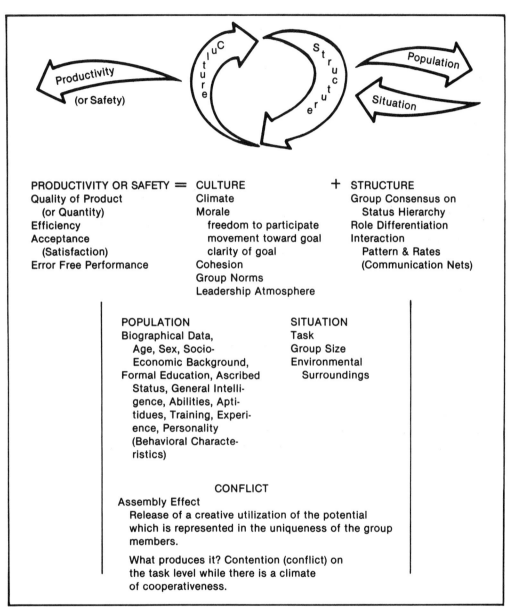

PRODUCTIVITY OR SAFETY = CULTURE + STRUCTURE
Quality of Product   Climate   Group Consensus on
   (or Quantity)   Morale   Status Hierarchy
Efficiency   freedom to participate   Role Differentiation
Acceptance   movement toward goal   Interaction
   (Satisfaction)   clarity of goal   Pattern & Rates
Error Free Performance   Cohesion   (Communication Nets)
                                      Group Norms
                                      Leadership Atmosphere

POPULATION   SITUATION
Biographical Data,   Task
   Age, Sex, Socio-   Group Size
   Economic Background,   Environmental
Formal Education, Ascribed   Surroundings
   Status, General Intelli-
   gence, Abilities, Apti-
   tidues, Training, Experi-
   ence, Personality
   (Behavioral Characte-
   ristics)

                    CONFLICT
Assembly Effect
   Release of a creative utilization of the potential
   which is represented in the uniqueness of the group
   members.

   What produces it? Contention (conflict) on
   the task level while there is a climate
   of cooperativeness.

Exhibit 7.7 Elements involved in safety (or productivity).

These motivational pulls are described well by a number of behavioral scientists and management theorists. The model of supervisory performance devised and tested in full by Edward Lawler and Lyman Porter is one of the most descriptive; it is discussed in Chapter 5.

Middle Management Level

What motivates middle managers is very similar to what motivates first-line supervisors, with one major exception. While they are very much interested in the measures of their performance used by the boss and in the desires and wishes of the boss, middle managers are especially re-sponsive to measures relating to the dollar—to profitability, which is so important to managerial thinking. The Lawler and Porter model discussed in Chapter 5 is also relevant at this level, the only difference being that results measures are probably better used as indicators of performance and effort at this level than at the lower level.

Top Management Level

What motivates top management? Top management people are highly interested in money, in profitability, in the bottom line; they must be, for that is their primary responsibility in the organization. However, those in top management often are also deeply interested in the good of the people who work for them. Although not many theories relate directly to the top levels of the organization, Rensis Likert's work on corporate climate is relevant here. After making an in-depth study of the impact of leadership styles on productivity, Likert began studying the impact of total corporate style on productivity.

Executives are generally deeply interested in building a good climate within their companies. While profitability must remain a top priority to them, most executives are almost as interested in building a good corporate climate, for they know future profitability depends on this.

REFERENCES

Argyris, Chris. *Personality and Organization,* Harper & Row, New York, 1957.

Herzberg, F. *Work and the Nature of Man,* The World Publishing Company, Cleveland, 1966.

Likert, R. *The Human Organization,* McGraw-Hill Book Company, New York, 1967.

Petersen, D. *Safe Behavior Reinforcement,* Aloray, Goshen, NY, 1988.

Porter, L., and E. Lawler. *Managerial Attitudes and Performance,* The Dorsey Press and Richard D. Irwin, Inc., Homewood, IL, 1968.

Chapter Eight

Changing Physical Conditions

The first step in the professional safety task is analysis. As indicated previously, analysis can be broken down into three categories—the behavioral, the safety system (or managerial), and the physical. This chapter looks at the physical environment and at how we can analyze our current situation.

Physical environment has traditionally been looked at more than the other two categories we'll examine. Since the beginning of industrial safety in the early 1900s heavy attention has been paid to the control of physical conditions. There is much information that we will not attempt to repeat here; rather we will look at some systematic approaches to physical condition control.

INSPECTIONS

Inspection was and is one of the primary tools of the safety specialist. Before 1931 it was virtually the only tool, and from 1931 to 1945 it was still the one most used. Until 1960 it was the primary tool of many outside service agencies and even today it remains the primary (and sometimes the only) tool of some safety professionals.

Today, however, one key question should be asked by every safety specialist engaged in inspection: "Why am I inspecting?" The answers to that question dictate how, when, and where to inspect. For instance, inspecting in order to unearth physical hazards only means looking only at things. However, inspecting to pinpoint both physical hazards and unsafe acts also includes people. Unfortunately, most inspections today are of the former kind, rather than the latter.

If the primary intent is to detect hazards not seen before, inspection differs from those where the primary interest is in checking on the inspections the department supervisor has made. If the intent is to detect hazards only, they can be corrected immediately by going directly to the maintenance department and reporting any deficiencies. If the intent is to audit the supervisor's inspection, the findings will be used to instruct and coach the supervisor so that future inspections will improve.

Many articles have been written on safety inspections and many have asked the question: Why inspect? Some typical answers have been

- To check the results against the plan.
- To reawaken interest in safety.
- To reevaluate safety standards.
- To teach safety by example.
- To display the supervisor's sincerity about safety.
- To detect and reactivate unfinished business.

- To collect data for meetings.
- To note and act upon unsafe behavior trends.
- To reach firsthand agreement with the responsible parties.
- To improve safety standards.
- To check new facilities.
- To solicit the supervisor's help.
- To spot unsafe conditions.
- To measure the supervisor's performance in safety.

It is generally agreed that the conditions and the people are the responsibility of the line supervisor. Thus responsibility for the primary safety inspection must also be assigned to the supervisor. "Primary safety inspection" means the inspection intended to locate hazards. Any inspections performed by staff specialists, then, should be only for the purpose of auditing the supervisor's effectiveness. Hence the results of inspection become a direct measurement of safety performance—effectiveness. It is important to look behind the acts and behind the conditions when inspecting, and ask, Why are these here? The answer to this question may lead back to the department supervisor—it may even lead to some other system weakness within the company— but the question should be explored fully and answered.

For instance, if an unsafe ladder is discovered, the inspector should immediately ask such questions as, Why is this ladder here? Why was it not uncovered by our ladder inspection procedure? Why did the line supervisor allow it to remain here? Answers to questions such as these begin to get at the true causes of accidents.

CHECKLISTS

The line supervisor is often provided with a checklist to use for the primary inspection. This approach does not encourage an effort to trace back the symptoms to their true causes. This type of checklist should be reworked into a form that requires determination of some of the causes of the symptoms that have been unearthed.

Checklists are common aids to inspectors. They range from a simple list of what to look for (see Exhibit 8.1) to more complex forms to record the results of the inspection (Exhibit 8.2, 8.3). Checklists can over general areas (Exhibit 8.2) or technical areas (Exhibit 8.3, 8.4, 8.5). Other methods include job safety analysis and hazard hunts, both discussed in Chapter 4.

SYSTEMS SAFETY

Systems safety is still a somewhat unused discipline by the industrial safety professional. There are, no doubt, concepts in systems safety that could be applied usefully to industrial safety.

Industrial safety and systems safety start from a common base—the desire to save lives and property. Yet systems safety is oriented toward analysis and improvement of hardware (that is, systems), and industrial safety directs itself to people more and more each year. Both start from a common base, but they start with different fields of endeavor. Industrial safety strives primarily to control accidents to employees on the job. So far, systems safety has worked mainly in the area of product safety in the aerospace and automotive fields.

Industrial safety engineers operate in a fixed manufacturing situation. They work in the midst of hazards that have often been there for a long time, many of which are accepted by production as a necessary component of their way of operating. They must work within that

A SAFETY INSPECTION INVENTORY

Planned inspection means knowing two things. First, foremen have to know what to look at. Second, they have to know what to look for.

The "Look At" Guide

The things that must be "looked at" fall into twenty-four broad categories. For any given area, not all of these categories necessarily apply, nor are all of equal importance.

1. Atmospheric Surroundings: Hazardous conditions of dust, gases, fumes, sprays, etc.

2. Chemical Substances: All liquids and solids that are toxic in nature.

3. Containers: All objects for storage of materials, i.e., barrels, boxes, bottles, cans, etc.

4. Conveyor Systems: Any system that transports materials to or from operations or processes.

5. Electrical Conductors and Apparatus: Wires, cables, etc.; switches, controls, transformers, lamps, batteries, fuses, etc.

6. Engines and Prime Movers: Sources of mechanical power, excluding electric motors, i.e., steam engines, gas engines, etc.

7. Elevators, Escalators, and Manlifts: Require no further explanation.

8. Firefighting Equipment: All firefighting equipment, plus sprinklers, fire plugs, etc.

9. Guards and Safety devices: All removable and fixed guards, and safety devices, or attachments, excluding personal protective apparel.

10. Hand Tools, All Kinds: Equipment that is held or carried when in use.

11. Hoisting Equipment: Cranes, derricks, power shovels, dredges, air hoists, hydraulic jacks. etc.

12. Flammables and Explosives: Require no further explanation.

13. Machinery and Parts Thereof: Power equipment that processes or modifies materials, i.e, agitators, grinders, forging presses, pulverizing machines, drilling machines, etc.

14. Materials, Raw or Processed: Materials and supplies essential to production of finished product, excluding materials elsewhere classified.

15. Mechanical Power Transmission Systems: Shafts, bearings, gears, pulleys, drums, cables, belts, sprockets, ropes, chains, etc., when used to transmit power.

16. Overhead Structures and Equipment: Any structural part or equipment that may fall from above.

17. Personal Protective Apparel: Goggles, gloves, aprons, leggings, etc.

18. Pressure Vessels, Boilers, and Pipes: Objects subject to internal pressure from compression of liquids or gases.

19. Pumps, Compressors, Blowers, and Fans: Objects that move or compress liquids, air, or gases.

20. Shaftways, Pits, Sumps, and Floor Openings: All types of openings into which persons may stumble or fall.

21. Walking or Standing Surfaces: Floors, aisles, stairs, platforms, ramps, roads, scaffolds, ladders, etc.

22. Warning and Signal Devices: Direct communication systems such as radio, telephones, buzzers, bells, lights, etc.

23. Vehicles and Carrying Equipment: Trucks, cars, trains, motorized carts, and nonmotorized equipment for transporting materials.

Exhibit 8.1 A safety inspection inventory.

24. Miscellaneous: Other potentially-hazardous objects or conditions that do not fall into the above categories.

The preceding twenty-four categories amount to a general answer to the question of what to "look at" in a foreman's safety inspection area. Now, what are the conditions that must be "looked for?" The conditions listed below will not cover all of the possibilities, but they are a guide to what to look for.

The "Look For" Guide

The separate items of the guide have been listed under five major categories.

1. Look for guarding of agents, such as:

   a. Missing or inadequate guards against being "struck by."
   b. Missing or inadequate guards against "striking against."
   c. Missing or inadequate guards against being "caught on, in, or between."
   d. Missing or inadequate guards against "falling from or onto."
   e. Lack of or faulty support, bracing, shoring, etc.
   f. Missing or faulty warning or signal device.
   g. Missing or faulty automatic control device.
   h. Missing or faulty safety device.

2. Look for structural defects and material characteristics, such as:

   a. Sharp-edged, jagged, splintery, etc., conditions.
   b. Worn, frayed, cracked, broken, etc., conditions.
   c. Slippery conditions (for gripping or walking).
   d. Dull, irregular, mutilated, etc., conditions.
   e. Uneven, rough, packed, or with holes.
   f. Decomposed or contaminated conditions.
   g. Flammable or explosive characteristics.
   h. Poisonous characteristics (by swallowing, breathing, or contacting).
   i. Corroded or eroded conditions.

3. Look for functional defects, such as:

   a. Susceptibility to breakage, collapse, etc.
   b. Susceptibility to tipping, falling, etc.
   c. Susceptibility to rolling, sliding, slipping, etc.
   d. Leakage of gases, fumes, or fluids.
   e. Excessive heat, noise, vibration, fumes, sparking, etc.
   f. Failure of agency to operate.
   g. Erratic, unpredictable performance of agent.
   h. Lack of adequate electrical grounding.
   i. Operation that is too fast or too slow.
   j. Low voltage leaks.
   k. Signs of excessively high or low pressure.
   l. Throwing off of parts, particles, materials, etc.
   m. Standard indications of need for special attention.

4. Look for ventilation, illumination, noise, such as:

   a. Noxious fumes or gases.
   b. Flammable or explosive fumes or gases.
   c. Insufficient illumination.
   d. Excessive glare from light source.
   e. Hazardous dusts or atmospheric particles.
   f. Hazardous temperature condition.
   g. Excessive noise.

Exhibit 8.1. Continued.

Exhibit 8.1 Continued.

framework, perhaps teaching the employees how to work around those hazards instead of removing them.

The first concern of the systems safety engineer is that the design be foolproof and that no one can possibly get hurt. For instance, in the case of a spacecraft, the systems safety engineer would be vitally concerned that the retro-rockets fire—on schedule—or all would be lost for the astronauts. Systems safety techniques, then, are the result of a need to eliminate any hardware malfunctions or mistakes in design that could have serious consequences. In many of these situations the only safeguard possible is to analyze or anticipate problems and then design in order to avoid them.

There are many methods of analysis in use in systems safety:

— *Gross-hazard analysis.* This is done early in the design stage. It is the initial safety analysis and it considers the overall system.
— *Classification of hazards.* Types of hazards are identified and classified with regard to potential severity.
— *Failure modes and effects.* The kinds of failures that could happen are examined and their effects are predicted.
— *Hazard criticality ranking.* The probability of occurrence of different hazards is determined and the hazards are ranked in order from most to least critical.
— *Fault tree analysis.* Fault tree analysis traces the progression of hazards.

Other types of analysis made in systems safety consider high-energy potentials, catastrophe accidents only, and maintenance considerations of system skills required for operation of a system.

Fault tree analysis was successfully applied initially in aerospace and has been used in other fields. It seems likely that the technique will be used in industrial safety in the future. The fault tree is actually a logic diagram that traces all the events that might have led to the undesired result being studied and can be constructed for any event. First, an undesired event is selected.

INSPECTION REPORT COVER SHEET

GENERAL INSPECTION

Area Inspected	Inspector	Date Inspected
No. of items carried from Previous Report	No. of items added To this Report	Total no. of items On this Report
No. of Class A items On this Report	No. of Class B items On this Report	No. of Class C items On this Report
Was a critical parts inventory reviewed before inspection? ☐ YES ☐ NO	Did you determine which are the critical jobs in the area before inspection? ☐ YES ☐ NO	Did you use a common hazard check list for this area as a guide? ☐ YES ☐ NO

General Reaction of Inspector to Conditions and Practices in this area

Distribution of initial inspection report and copies

FOLLOW UP FACTS

Progress Reports	Report Date	Hazards Remedied A	B	C	Hazards Open A	B	C	Intermediate Action Taken A	B	C
No. 1										
No. 2										
No. 3										
No. 4										

General comments on conditions and practices at time of final report

Follow up coordinated and reported by

Distribution of final follow up report and copies

DEFINITIONS

"A" HAZARD — any condition or practice with potential for loss of life or body part and/or extensive loss of structure/equipment/material.

"B" HAZARD — any condition or practice with potential of serious injury or property damage, but less severe than Class "A".

"C" HAZARD — any condition or practice with probable potential or non-disabling injury or non-disruptive property damage.

INTERMEDIATE SAFETY — immediate temporary measure/s taken to reduce the potential of accident occurrence until more suitable or permanent remedial action can be taken.

CRITICAL JOB — any job or task that has been associated with a high accident frequency or a high potential for severe loss.

COMMON HAZARD CHECK LIST — a list of the common and repetitive hazards and their locations that have been observed and recorded on previous inspections. If one does not exist, it can be created by a short review of several previous inspection reports.

CRITICAL PARTS INSPECTION INVENTORY — an actual inspection inventory sheet that includes: items inventoried, related critical parts, conditions to observe, recommended frequency of inspection and inspection responsibility. If available, it can serve as an excellent inspection guide.

Exhibit 8.2 Inspection report.

HOUSEKEEPING RATING FORM

Department . . . . Date Inspected . . . . Inspector . . . .

INSTRUCTIONS FOR FILLING OUT

Consider the maximum score for each item indicative of a perfect condition and pro-rate the item according to its worth. To obtain final rating: (1) Total the points in each column. (2) Average all totals for final rating:

RATING

> A place is in order when there are no unneccessary things about and when all necessary things are in their proper places.

DIVISIONS

	Maximum Scores							
MACHINERY AND EQUIPMENT								
a. Must be clean and free of unnecessary material or hangings.	4							
b. Must be free of unnecessary dripping of oil or grease.	4							
c. Must have proper guards provided and in good condition.	7							
STOCK AND MATERIAL								
a. Must be properly plied and arranged.	8							
b. Must be loaded safely and orderly in pans, cars and trucks.	7							
TOOLS								
a. Must be properly stored.	6							
b. Must be free of oil and grease when stored.	3							
c. Must be in safe working condition.	6							
AISLES								
a. Must be provided to work positions, fire extinguishers, fire blankets and stretcher cases.	6							
b. Must be safe and free of obstructions.	6							
c. Must be clearly marked.	3							
FLOORS								
A. Must have surfaces safe and suitable to work.	6							
b. Must be clean, dry and free of refuse, unnecessary material, oil and grease.	6							
c. Must have an adequate number of receptacles provided on them for refuse.	3							
BUILDINGS								
a. Must have walls and windows that are reasonably clean for operations in that area and free of unnecessary hangings.	3							
b. Must have lighting systems that are maintained in a clean and efficient manner.	3							
c. Must have stairs that are clean, free of materials, well lighted, provided with adequate hand rails and treads in good condition.	5							
d. Must have platforms that are clean, free of unnecessary materials and well lighted.	4							
GROUNDS								
a. Must be in good order, free of refuse and unnecessary materials.	10							
TOTAL SCORES								

Furnished through the courtesy of the Insurance Company of North America and the Pacific Employers Group

Exhibit 8.3 Rating form.

MONTHLY CRANE INSPECTION

Crane No. _____ Inspection By: _____

Date _____

ITEM	OK	REMARKS
1. Operating Mechanism _____		
2. Air or Hydraulic Systems ____		
3. Hook Deformation _____		
4. Chains _____		
5. Ropes _____		
6. Bolts, Rivets _____		
7. Shears, Drums _____		
8. Pins, Bearings, Shifts, etc. ___		
9. Brake System _____		
10. Power/Electrical Apparatus ___		

CORRECTIONS MADE:

OTHER REMARKS or RECOMMENDATIONS:

Exhibit 8.4 Technical checklist.

Then it is necessary to reason backward to visualize and identify all the ways in which it might have occurred. Each contributing factor or cause is then studied and analyzed to determine how *it* could possibly have happened. Such tracing of causes and factors can point to many different system failures that might not otherwise be noticed.

Say that the undesired event selected is a severe injury to the operator of a power press in an industrial plant. The fault tree (see Exhibit 8.6) for this event is made by listing those events which might have occurred, or which must have occurred, for the undesired event to happen. These events are connected by either AND or OR gates. An AND gate means that both events noted must be present for the event to occur. An OR gate means that either event alone can be responsible for the occurrence of the major event. Thus in Exhibit 8.6 events B, C, D, and E must all be present for the major, undesired event A to occur.

However, either event F or event G alone could cause subevent B, and either H or I alone could cause subevent C. Events N, O, or P alone could cause subevent G.

This fault tree shows that for a severe injury to occur to the operator of the power press, four things must happen:

1. The press must be operating.
2. The ram must be descending.

CHECK LIST FOR MECHANICAL POWER PRESSES

Press _____ Conducted by _____ Date _____

Instructions: **Cross out all questions not applicable. Answer all other questions "yes" or "no." If answer is "no" on back of sheet list question number and write actual discrepancy; for example, "keyswitch missing" or "counter balance pressure switch disconnected," and so on.**

Section I — Press Controls   YES   NO

   1. Is there a red stop control that overrides all other controls and immediately disengages the clutch and engages the brake?

   2. Is there a keyswitch for selecting "Off," "Inch," "Single Stroke," and "Continuous" (if applicable) functions of the press?

   3. Is the "Inch" control arranged so that it is impossible to reach the point of operation while actuating the press in this mode?

   4. Are all hand controls protected against accident actuation and at least 20" apart?

   5. If the press can be operated by either one or two persons, is the second set of hand controls either removable or controlled by a keyswitch?

   6. If capable of continuous operation, does the continuous option depend either on a keyswitch or on a detachable special operator's station?

   7. If a foot control is provided, are hand controls either removable or is selection between hand and foot control through a keyswitch?

   8. Is the foot control equipped with a substantial guard to prevent accidental actuation?

   9. Is there a lockable main power switch or circuit breaker?

  10. Is the motor start control protected against accidental actuation?

  11. Are all control circuits (except motor starters) of 120 volts or less?

  12. Do all control circuits have "ground connected" indicators and fuses in the ungrounded portions of the circuit?

Section II — Press Operation

  13. When operating press in "Single Stroke" mode with hand controls:
     (a) Is the concurrent use of both hands required to trip the press?

     (b) Is it ncessary to keep all buttons depressed until the die has closed in order to continue operation?

     (c) Is it necessary to replace buttons before initiating a second stroke or before continuing an interrupted stroke?

  14. If capable of continuous operation, is the press equipped so that a definite action by the operator is required before continuous operation will take place?

  15. Is it necessary to reactuate hand or foot controls to restart operation after the "Stop Control" has been operated?

Section III — Press Operation Interlocks

  16. If the press is not intended for continuous, automatic feed, operation, is the clutch-brake control equipped with a tandem air valve?

  17. When "single stroke" or "continuous" is selected, is the brake engaged whenever the forward direction contactor is disengaged?

  18. Are friction brakes set with retained, enclosed, compression springs?

Exhibit 8.5 Technical checklist.

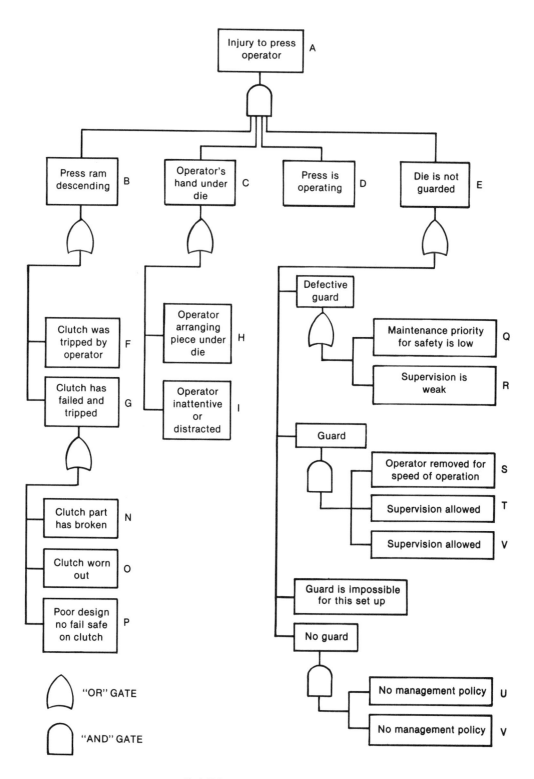

Exhibit 8.6 Sample fault tree.

3. The operator must have a hand under the die,
4. There must be an inoperative guard or no guard on the die.

Prevention lies in the elimination of one of these four events.

Each of the four is then analyzed. For the press ram to be descending, the clutch must have been tripped by the operator, or the clutch must have failed, and the press repeated without being tripped. Any of these could happen if a clutch part had broken, the clutch were worn out, or the clutch had been poorly designed, with no fail-safe mechanism built into it.

The operator's hand might have been under the die if he or she had been working under it to arrange the piece or had been distracted and was paying poor attention to the work. The die could have been unguarded if the guard had been removed, if the guard had been in some way defective, if there were no guard, or if on this particular die and setup a guard was impractical or impossible.

Each of these subevents could then be further analyzed, which might lead to such conclusions as (Q) maintenance priorities do not include press die-guard construction, (R) press department supervision is not finding defective guards (is not looking for them), (S) press operators are removing guards, (T) supervision is not enforcing the use of guards, (U) supervision is not stopping work on a press when guards are absent, and (V) management is not setting and enforcing a policy of press guarding. (Note that in this particular fault tree analysis, human failures have been included.)

The above analysis is very simple; it was chosen merely to illustrate the general idea of fault tree analysis and to indicate how it might possibly be used in industrial safety as well as in systems safety. You can no doubt visualize a complex analysis—for example, one that might be used in the aerospace or missile fields.

Fault-Tree Analysis and High-Potential Accidents

Perhaps systems safety can be best used in industry in conjunction with the high-potential accident analysis approach. Using present-day systems of locating and defining high-potential accidents or high-potential accident exposures, we can periodically and regularly identify potential high losses.

After identifying the high-potential accident, the next step would be to make an estimate of the potential dollar loss that could reasonably be expected should the accident occur. For a high-potential accident that has already happened, this estimation would be relatively simple: The type of injury would be pictured, and the loss along the lines would be estimated as described in the costing process discussed in Chapter 4. Additional cost estimates can be made of any indirect costs and property damage losses and a total estimated cost can be derived.

At some predetermined dollar loss level, a fault-tree analysis would be required. Since fault-tree analyses can be complex in some cases and hence time-consuming, this level would necessarily be high, perhaps $10,000 or $25,000. The analysis might be made by safety staff or the system could be taught to line supervision, who would then make the analysis.

Both high-potential accident analysis and systems safety are proved techniques, but both are practically unused in industrial safety.

Investigating

Accident investigation is a device for preventing additional accidents. According to the National Safety Council's *Accident Prevention Manual for Industrial Operations,* the principal purposes of an accident investigation are:

- To learn accident causes so that similar accidents may be prevented by mechanical improvement, better supervision, or employee training.
- To determine the "change" or deviation that produced an "error" that, in turn, resulted in an accident (systems safety analysis).
- To publicize the particular hazard among employees and their supervisors and to direct attention to accident prevention in general.
- To determine facts bearing on legal liability.

An accident that causes death or serious injury, obviously, should be thoroughly investigated. The "near-accident" that might have caused death or serious injury is equally important from the safety standpoint and should be investigated, for example, the breaking of a crane hook or a scaffold rope or an explosion associated with a pressure vessel.

Each investigation should be made as soon after the accident as possible. A delay for only a few hours may permit important evidence to be destroyed or removed, intentionally or unintentionally. Also, the results of the investigation should be made known quickly, as their publicity value is greatly increased by promptness.

Any epidemic of minor injuries demands study. A particle of emery in the eye or a scratch from handling sheet metal may be a very simple case. The immediate causes may be obvious and the loss of time may not exceed a few minutes. However, if cases of this or any other type occur frequently in the plant or in a department, an investigation might be made to determine the underlying causes.

SAGE ANALYSIS

One recent adaptation of systems safety is SAGE analysis as diagrammed in Exhibit 8.7. One of the primary inputs is the conducting of interviews and the holding of meetings to get a better idea from an many sources as possible as to what problems exist. Following the information gathering process from the people involved, the data is then fed into the process more typically associated with systems safety such as the logic diagrams.

PROCESS SAFETY AUDITS

A tool called the process safety audit is used at the Monsanto Company to identify and correct physical condition problems. It is described by H. Partlow and D. Schillinger in a recent article in *Professional Safety*:

> If you have an older plant, the chances are a multitude of safety sins have been committed over the years. Old operating departments, crisis changes where little thought was given to safety, and temporary installations that were never made permanent are just a few of the potential booby traps that may exist in a chemical plant. These flaws can cause staggering losses.
>
> The question is, how do you get these conditions corrected? We use a tool called a Process Safety Audit. The audit form that is used by most Monsanto plants was developed by our Corporate Safety and Property Protection Group. Each plant modifies the procedure to increase the effectiveness of the audit for a particular location or operation.
>
> Possibly the best way to tell what a safety audit consists of is to first discuss what it is *not*. It is not a hazard inspection. A hazard inspection is a necessary part

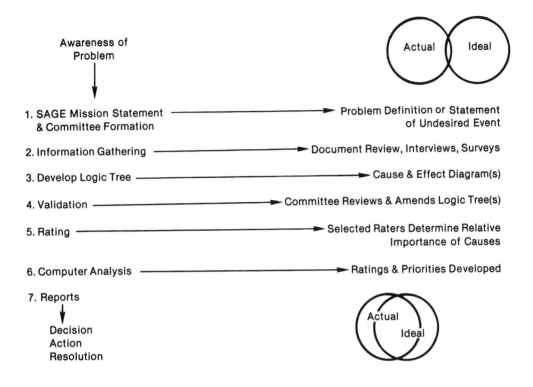

Exhibit 8.7 The seven phases of SAGE analysis.

of an effective loss prevention program, but a safety audit will ferret out many things that will never be noted in a hazard inspection. Other safety audits do no more than list control variables and operating safety problems that occur when these variables get out of control. This type of audit may be good for training purposes; but, beyond that, [it] has little use as a loss-prevention tool.

Emphasis in our audits is placed on equipment and control. Other items are also included, but the primary emphasis is placed in these areas. A good safety audit examines the safety interrelationships of equipment and processes and the safety relationship of one piece of equipment to another. For example, a vessel operating at 4,000 PSIG may discharge into a vessel with a test pressure of 200 PSIG. This vessel may operate at only 50 PSIG. Obviously, the interrelationship of the first vessel with the second imposes additional safety concerns.

In addition to interrelationships of equipment and process, items such as venting, design pressure versus relief pressure, materials of construction, access, and location are also covered.

In the control area, typical items discussed are the adequacy and dependability of existing controls, redundant control of key variables, types of control hardware used, interlocks and alarms, and raw material identification procedures.

Some of these items receive very special attention. Venting is such an item. On most complex equipment, no one can be intelligent enough to predict a course of action that will make a vent unnecessary. There is a good chance that sooner or later the venting system will stand between your plant and disaster. When it is called upon to function, it must work. In the audit, we examine certain specific characteristics of venting systems. These are listed below:

1. Is the capacity of the vent system sufficient to relieve properly under extreme conditions such as an overcharge? (Don't always design for normal conditions.)
2. Is the relieving pressure within proper limits for the equipment and process involved?
3. Does the vented material discharge in a satisfactory manner and to the proper location?
4. Is the vent strong enough to sustain the forces that it will be subjected to under emergency conditions without self-destructing?
5. Are relief devices tested and inspected on a periodic basis?

The following are typical items considered when controls and instrumentation are discussed:

1. Is the overall control scheme and instrumentation correct? (Are we controlling the right variables with the proper hardware?)
2. In the event of failure of a control loop, rather than place complete dependency on so-called "fail safe" instrumentation, do we have proper back-up systems to avoid a disaster?
3. Does the department have written verification that interlock and alarm testing are done on a regularly scheduled basis?
4. With highly automated pieces of equipment, are interlocks and alarm signals so installed that the failure of a single item cannot render the entire control system inoperative?
5. Does the department have raw material identification procedures?

The above items are typical things that are examined in a process safety audit. Typical steps performed in a plant to complete a process safety audit and a brief discussion of each step follow:

1. Selection of the process.
2. Selection of the committee.
3. Working sessions.
4. Field inspection.
5. Review with the appropriate management and assign priorities.
6. Final write-up.
7. Follow-up.

Two inputs are used to set priorities—the degree of hazard in the process and the profitability of the product. All processes are audited, and the plan is to cover each process in the plant on a biennial basis. The selection process simply helps to insure that major concerns receive priority.

A minimum of five persons [is] generally necessary to have an effective audit. The addition of other persons usually contributes to the audit. The minimum [includes] the supervisor of the unit, the operating foreman, a technical representative, a mechanical department representative, and a member of the safety department. It is helpful if the safety department representative has considerable production experience and a technical degree.

The working sessions are attended by all the members of the audit team. In these sessions the process and equipment are discussed following the normal flow pattern of the process. Equipment sheets are used for the various pieces of equipment (see Exhibit 8.8).

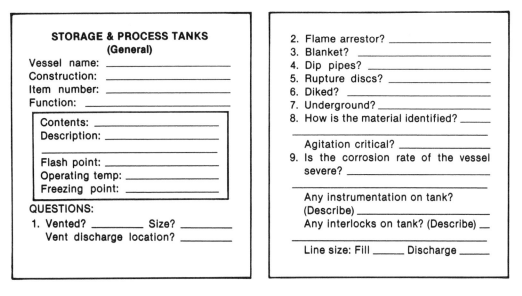

Exhibit 8.8 Sample equipment sheet.

Working sessions should last about two hours per day. This gives the members of the committee time to perform their other duties. In these sessions, deviations from standards and deviations from good safety practice are recorded. These deviations later become the recommendations of the audit. Much of the dialogue of the audit concerning safety as it applies to this process is recorded. Reading a completed audit is a useful training tool for new supervisors.

During the working sessions some information may not be known. A note is made in the audit where unknowns exist, and at the conclusion of the working sessions these notes are listed, and a field inspection is made to answer some of the unknown questions in the audit. The field inspection is also used to record any obvious safety problems which should be corrected; however, this is not its primary function.

Actually there are two reviews of the safety audit recommendations. A rough draft of all recommendations is made after the field review. This rough draft is discussed with the supervisor to reach a general agreement between the safety department and production supervision that these items should be completed.

After a general agreement has been reached with first-level supervision, a meeting is scheduled with the superintendent. At this meeting the recommendations are discussed and priorities are assigned. We use high, medium, and low for a priority system. High means completion should be within one year; medium, two years; and low, three years, or you should live so long!

Generally agreement is reached. On occasion, a superintendent will inform us that he probably will not do a certain recommendation. If this is a deviation from an accepted safety standard, the only course of action is to carry this matter to the next highest level with the superintendent's knowledge of exactly what you are going to do. It's always best to resolve a problem at the lowest level of management.

The finished report consists of a summary, recommendations, and a discussion of the process. In our company, we include maximum foreseeable and probable dollar loss information (MCA Guidelines for Risk Evaluation and Loss Prevention in Chemical Plants) in the summary. We assign priorities to each recommendation and detail equipment sheets are an integral part of the process section of the audit report.

Removing the Hazards

Once hazards have been found, what happens next? To simply say "fix them" doesn't deal with reality. In most environments there are ten times more hazards than can be fixed. It is easy to find them; it is expensive to fix them. In most companies a hazard free environment would mean two things: no production and bankruptcy. Therefore, some practical solutions must be introduced to hazard control (a concept not always clear to enforcement personnel).

Some basic decisions must be made on what to fix and what not to fix. How much money should be spent and what will the payoff be for the money spent? These are management decisions much like any other management financial decision.

There have been interesting economic treatises in this area. Notable is one by John Gleason and Darold Barnum, who looked at the management financial decision of whether or not to comply with OSHA standards. Here is an excerpt:

> analysis has indicated that expectancies of being cited for initial safety and health violations, and the fine levels if cited, are so low under OSHA that they are of little value in preventing violations of the Act. Those employers who obey the law would do so regardless of the penalties. Employers at whom the sanctions are aimed— those who will correct violations only if it is economically profitable for them to do so—are not being affected. Thus the current sanctions antagonize employers who attempt to obey the law, while having little impact on those employers who will obey the law only if it is economically profitable.

Additional excerpts from this analysis are included in Chapter 14 where OSHA compliance is discussed.

If we were to follow the above logic, it would make little sense to abate hazards (comply with OSHA standards). While economically this might be true, it is legally dangerous to follow this direction in the real world. At the other end of the decision scale is to totally abate all hazards (totally comply) regardless of cost effectiveness, which might delight enforcement officers but upset the financial wizards (who happen to sit in crucial seats in most companies). Therefore we must opt for some in-between course of action; we must prioritize our fixes.

Each organization will handle the problem differently. However, there are a few general principles that apply:

1. Set priorities. Most organizations cannot do everything at once, so decisions must be made about what comes first.
2. Schedule and assign tasks. Without planning and assigning responsibilities not much will happen.
3. Follow up to ensure that things are being done.
4. Document everything done, if not for internal reasons (to protect yourself), certainly for OSHA.

Should the presses or the exit lights be rewired first? Should collapsible handrails for the loading dock or handrails around the roof be built first? These are the kinds of decisions that someone will have to make. For each item on the list (see Exhibit 8.9) there should be a start and a finish time.

# DEPT.	VIOLATION	LAW VIOLATIONS PRIORITY	FIX SCHEDULE			
		=	6/16	6/23	6/30	7/7
306 Paint rm	Repaint lines on floor for 5,000 gal.					
307 Paint rm	Provide for bonding while pumping					
308 Paint rm	Too many barrels stored					
309 Paint rm	Open barrels in paint room					
310 QC rm	Ungrounded electric plate					
311 Pl #3	Bag sprinkler heads in booth					
312 Pl #3	Nonstandard handrails over PL					
314 Pl #3	Unguarded Ch & Sp 3rd level					
315 Pl #3	Handrail missing top PL					
316 Pl #3	Guard conveyor (unused) nip point					
317 Pl #3	Guard back of B & P on conveyor					
318 Pl #3	Ground lights on deionizing tanks					
320 Pl #3	Water on floor is excessive					
321 Pl #3	Hard hats in area					
322 Pl #3	Unguarded coupling in oven area					
323 Pl #3	Reversed polarity ext cord to PH tester					
324 Pl #3	Uninspected ladders					
325 Pl #3	Construct bar for swinging hooks					
326 Pl #3	Side shields for all hangars					
327 Pl #3	Hooks on floor					
328 Metals	Bottom B & P of fixture unguarded					
329 Metals	Oil on floor					
330 Metals	Guard back of B & P on B-5					
331 Metals	Guard back of B & P on P-11					
332 Metals	Guard back of B & P on P-12					
333 Metals	Guard back of B & P on edge trimmers					
334 Metals	Guard back of B & P on S-4					
335 Metals	Guard Ch & Sp of P-12 edge trimmer					
336 Metals	P-14 totally unguarded					
337 Metals	P-14 has improper controls					
338 Mntn.	Guard belt sander					
339 Mntn.	Guard back of B & P recip. saw					
340 Mntn.	Need dead man switches on drills					
341 Mntn.	No underslung guard radial saw					
342 Mntn.	Guard B & P, Ch & Sp on lawnmower					
343 Mntn.	Guard B & P on motorized cart					
344 Mntn.	Ungrounded fan					
345 Mntn.	Unguarded fan					
346 Mntn.	Provide approved hand rails on rolling scaffold					
347 Mntn.	Provide platform that covers top on scaffold					
348 Mntn.	Provide safe ladder to get up on scaffold					
349 Mntn.	O_2 cap off					
350 Mntn.	Guard the back of B & P portable hacksaw					
351 T & D	Guard the B & P on file					
352 T & D	Ungrounded light on vertical mill					
353 Receiving	Steel pallets stored too high					
354 Receiving	Wood pallets stored too high					
355 Receiving	RR chain down					
356 Receiving	F-7 leaking oil					
357 Receiving	Int. rails platform by truck dock					

Exhibit 8.9 One page of a maintenance list.

First assess the capabilities of the maintenance people. Can they do this work as well as their regular tasks within the required time frame? Many organizations have found they need to set up a separate safety maintenance force. It can report either to staff safety or to maintenance as long as its efforts are directed *solely* to safety items.

Not until the organization is set can priorities be established and then a schedule set. One method of setting priorities is by using a matrix.

Exhibit 8.11 is the matrix used by one organization, and it was used (1) by determining, as a compliance officer might, the degree of hazard (imminent, serious, or minor) and (2) by estimating the costliness of the change in terms of time, manpower, and money. The number in each box of the matrix—which indicates priority—was arbitrarily assigned at the outset. It could be shifted depending on management's decision, money available, manpower, and so forth. As supervisors or inspectors reported to safety personnel any physical violations they could not correct themselves, the violations were given priorities ranging from 1 to 9 and put on maintenance (task force) lists (see Exhibit 8.9). That assured early completion dates for imminent and serious hazards (particularly those of moderate or small cost). Once priorities were established, the maintenance task force scheduled each job and set a start and a finish date. The schedule became the work assignment schedule for the task force, the report to management, and the needed documentation. Updated weekly, it showed the progress in compliance.

C. Ray Asfahl offers a ten point scale (see Exhibit 8.10) to achieve the same thing.

A SAFETY PROGRAM PRIORITY SYSTEM

A slightly different system of prioritizing was used by Chester Hudlow of the North Carolina Department of Transportation. He suggests that the most common method of getting at the deficiencies in an organization is simply to start inspecting and then begin the process of hassling to get management to correct all the deficiencies you have found. Hudlow suggests that there are several things wrong with this traditional approach:

— It doesn't give priority to the deficiencies with the greatest potential for accident.
— It could bankrupt the company if all deficiencies were corrected at once.
— It does not involve the line supervisor.

A properly prepared priority list will give form and organization to safety efforts. It will provide the greatest reduction in accidents in the shortest period of time. By involving management and line supervisors, far better results should be achieved than would be possible otherwise. Last but not least, the priority list will enable management to prorate cost so it will be manageable within the framework of budgetary constraints.

Exhibit 8.11 illustrates the steps involved in preparing a safety program priority list. The list is prepared by the line supervisors and consolidated by higher-level staff, the extent and level of consolidation depending on the size and organizational structure of the company. It is suggested that a list be compiled first without regard to priority. In order to ensure that all important items appear on the list, it may be necessary for the safety professional to provide supervisors with a checklist which would be updated periodically as deficiencies are corrected.

The second step is to divide the list into three parts. List 1 will contain those items with long-term budgetary constraints that will take from one to five years to correct. Although supervisors may have to get approval on future budgets to cover some of the items on this list, they should include everything that needs to be done on this or one of the other priority lists. Supervisors may need to modify their time frame and repeat some items on priority lists at a

1	2	3	4	5
"Technical" violations; OSHA standards may be violated, but no real occupational health or safety hazard exists.	No real fatality hazard. Health hazards minor or unverified.	Fatality hazard not of real concern. Health hazards have exceeded designated action levels. OR Sound exposure action levels exceeded. (Ex: continuous exposure in the range 85-90 dBA) OR	Fatality hazard either remote or nonexistent. Health hazards characterized by occupational illnesses which are usually temporary, not permanent. Controls or personal protective equipment may or may not be required by OSHA. OR Temporary hearing damage usually will result without controls or protec-	Fatality hazard either remote or not applicable. Long-range health may be at risk; controls or personal protective equipment *advisable* or *required* by OSHA. OR Hearing damage may be permanent without controls or protection. (Ex: continuous 8-hr exposure in the range 95-100 dBA)
	Even minor injuries unlikely.	Minor injury risk exists but major injury hazard is very unlikely.	tion and a few workers may incur partial permanent damage. (Ex: continuous 8-hr exposure in range 90-95 dBA) OR Minor injuries likely, such as cuts or abrasions, but major injury risk is unlikely.	Major injuries such as amputation not very likely.

Exhibit 8.10 Category descriptions for a ten-point scale for workplace hazards.

6	7	8	9	10
Fatality hazard unlikely. Long-range health definitely at risk; controls or personal protective equipment required by OSHA. OR Hearing damage likely to be permanent without controls or protection. (Ex: continuous 8-hr exposure in the range 100-105 dBA) OR Major injury such as amputation not very likely but definitely could occur.	Fatality not very likely, but still a consideration. OR Serious long-range health hazards are proven; controls or personal protective equipment essential to prevent *serious* occupational illnesses. OR Hearing damage obviously would be *severe* and permanent without protection. (Ex: continuous 8-hr exposure in excess of 105 dBA) OR Major injury such as amputation could easily occur.	Fatality possible; this operation has never produced a fatality, but a fatality could easily occur at any time. OR Severe long-range health hazards are *obvious;* controls or personal protective equipment essential to prevent *fatal* occupational illness. OR Major injury is likely. Amputations or other major injuries have *already* occurred in this operation in the past.	Fatality likely; similar conditions have produced fatalities in the past; conditions too risky for normal operations; rescue operations are undertaken for injured workers with rescuers using personal protective equipment.	Fatality imminent; risks are grave; some employees earlier in the day have died or are dying, conditions are too risky even for daring rescue operations except perhaps with exotic rescue protection.

Exhibit 8.10 Continued.

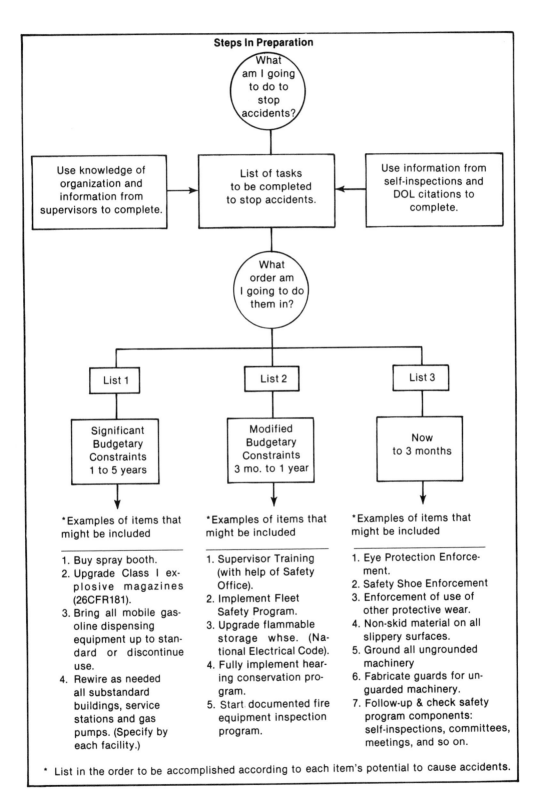

Exhibit 8.11 Priority list.

future date. Nevertheless, every deficiency should appear on a list, even though some items may prove to be very rough estimates insofar as the estimated time for correction. No item should appear on List 1 until after every reasonable and conscientious effort has been made to assign the item to List 2 or List 3.

List 2 will contain items that have a less serious budgetary constraint or that for one reason or another will require from three months to one year to correct. No items should appear on List 2 until after every reasonable and conscientious effort has been made to assign the item to List 3.

List 3 should contain most deficiencies. Normally, all items needing correction will appear on this list unless matters beyond the supervisor's control prevent it.

The TTC Hazard Rating System

D. A. Weaver, while at the Transportation Test Center (TTC) in Pueblo, Colorado, delivered another alternative method for prioritizing:

> The heart of the system can be displayed on a single sheet of paper, front and back. One side contains the hazard scenario, including the hazard coding and the hazard rating. The other side explains the coding and rating process (see Exhibit 8.12). Thus, with the scenario in hand, a line manager has all the information needed to understand the hazard being reported to him.
>
> The elements of the system can be enumerated as follows:
>
> 1. The Hazard Report or Scenario.
> 2. The Coding Process (Exhibit 8.12) which produces an assessment of severity, of level of probability, and of cost to correct the hazard. These three factors are expressed in a three letter code, such as: AAA, ACA, DCB, etc., through 64 permutations of A, B, C, & D. Severity is assessed in three factors: personnel casualty, monetary loss, and operational delay. Definition and rating of these terms can vary to suit differing operations. For instance, we chose to make "A" stand solely for potential fatality, equated with no other loss. Thus the presence of "A," as the first letter in the three letter code, instantly signalizes a potential fatality.
>
> The level of probability is assessed by one question: "How frequently is the hazard likely to be encountered?" The answer is coded A, or B, or C, or D. Thus, statistical computations of probability becomes merely a question on which everyone can readily agree. It is not, of course, probability in the statistical sense, the probability of an occurrence. It is, rather, a simple statement of exposure; the more frequently a hazard is encountered the more probable the accident. The idea needs no explanation and probability levels are readily assessed without dispute.
>
> Cost of corrective action is deliberately estimated in large steps, although, in adopting the Hazard Rating System to another operation, the steps may be large or small as best fit the needs of a particular organization. At the TTC, we established cost code "A" as less than $1,000, although most hazards can be corrected as a part of routine work at nominal cost.
>
> These three factors—severity, probability, and cost—entered the hazard coding and were expressed in a three letter code ranging from AAA to DDD. For instance, a hazard coded AAA instantly revealed three things. It was potentially fatal; it was encountered daily; and it was a low cost fix. At the opposite extreme, DDD means that injury is not likely, that the hazard is

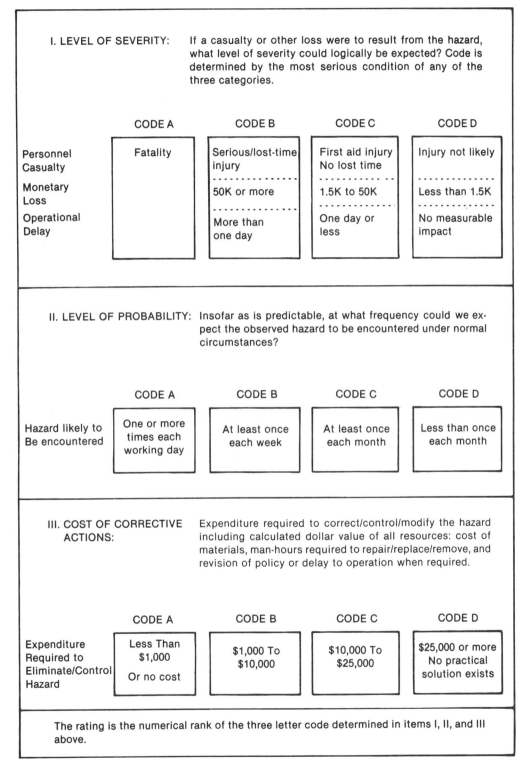

I. LEVEL OF SEVERITY: If a casualty or other loss were to result from the hazard, what level of severity could logically be expected? Code is determined by the most serious condition of any of the three categories.

	CODE A	CODE B	CODE C	CODE D
Personnel Casualty	Fatality	Serious/lost-time injury	First aid injury No lost time	Injury not likely
Monetary Loss		50K or more	1.5K to 50K	Less than 1.5K
Operational Delay		More than one day	One day or less	No measurable impact

II. LEVEL OF PROBABILITY: Insofar as is predictable, at what frequency could we expect the observed hazard to be encountered under normal circumstances?

	CODE A	CODE B	CODE C	CODE D
Hazard likely to Be encountered	One or more times each working day	At least once each week	At least once each month	Less than once each month

III. COST OF CORRECTIVE ACTIONS: Expenditure required to correct/control/modify the hazard including calculated dollar value of all resources: cost of materials, man-hours required to repair/replace/remove, and revision of policy or delay to operation when required.

	CODE A	CODE B	CODE C	CODE D
Expenditure Required to Eliminate/Control Hazard	Less Than $1,000 Or no cost	$1,000 To $10,000	$10,000 To $25,000	$25,000 or more No practical solution exists

The rating is the numerical rank of the three letter code determined in items I, II, and III above.

Exhibit 8.12 TTC hazard rating system.

encountered less than monthly, and that it is a costly fix (over $25,000). This code can now be stated numerically, as described next.

3. The Hazard Rating (Exhibit 8.13), also referred to as the priority rating table. This can readily be seen as an adaptation of the hazard pyramid attributed to Vernon Gross. The priority rating of each three letter hazard code is read directly from the table. Each three letter code (AAA #1, ADA #14, DBA #32, CCC #49, etc.) finds its unique and logical place on the 64 point scale.

The logic, by way of explanation, lies in simple addition. Assume each "A" has a value of 4. Then 4 + 4 + 4 gives AAA a value of 12, the only combination adding up to 12. Give B the value of 3, C is 2, and D is 1. Code DDD is 1 + 1 + 1 equals 3, the only combination adding up to 3. We thus establish the top and bottom of the 64 point scale.

RATING TABLE

RATING	CODE		RATING	CODE
1	AAA		33	ACD
2	AAB		34	ADC
3	ABA		35	BBD
4	BAA		36	BCC
5	AAC		37	BDB
6	ABB		38	CAD
7	ACA		39	CBC
8	BAB		40	CCB
9	BBA		41	CDA
10	CAA		42	DAC
11	AAD		43	DBB
12	ABC		44	DCA
13	ACB		45	ADD
14	ADA		46	BCD
15	BAC		47	BDC
16	BBB		48	CBD
17	BCA		49	CCC
18	CAB		50	CDB
19	CBA		51	DAD
20	DAA		52	DBC
21	ABD		53	DCB
22	ACC		54	DDA
23	ADB		55	BDD
24	BAD		56	CCD
25	BBC		57	CDC
26	BCB		58	DBD
27	BDA		59	DCC
28	CAC		60	DDB
29	CBB		61	CDD
30	CCA		62	DCD
31	DAB		63	DDC
32	DBA		64	DDD

Exhibit 8.13 Hazard rating table.

Other combinations spread logically between #1 and #64. For instance, twelve combinations add up to a value of 8, yet ABD (4 + 3 + 1) ranks 21 on the rating table while DBA (1 + 3 + 4) ranks only 33. But observe that rank 21 (ABD) is a potential fatality, encountered weekly, but costly to fix. Similar logic gives each hazard code its unique ranking on the 64 point table.

A unique feature of the system concerns hazards assigned to the safety manager. This feature is a brilliant insight of the Hazard Rating System. Safety is indeed a line function, but catastrophes result when events fall in the cracks between jurisdictions, divided responsibilities, dual authorities, and undefined and/or ambiguous responsibilities. In truth, the correction of some hazards cannot be assigned to any certain line manager. In such cases, who better to make responsible than the safety manager? It is his duty, not merely to report a hazard, but to find a handle whereby it may be corrected. Hence, certain hazards are assigned to the safety manager on the monthly action report.

Each manager, including the safety manager, receives a monthly Safety Action Report, each directed solely to a specific manager. No manager (except the Chief Executive) knows what has been assigned to other managers. Each knows only what has been assigned to him.

Safety management by confrontation almost ceases. This is the most notable result of the Hazard Rating System at the Transportation Test Center. It replaced the usual inspection report with action and deadlines imposed by the safety staff. Such reports often met with flip response. Some systems set line people at loggerheads; others ease the path of cooperation.

Mathematical Prioritizing

W. Fine has developed a method of mathematically deciding priorities. His formula weighs the controlling factors and calculates the risk of a hazardous situation, giving a "risk score" to indicate the urgency of remedial action:

Normal industrial safety routines such as inspections and investigations usually produce or reveal numerous hazardous situations which, due to limitations of time, maintenance facilities and money, cannot be corrected at once. The Safety director must then decide *which* problems he should attack first. A great aid in making this decision would be a method to establish priorities of all hazardous situations, based on the relative risk caused by each hazard. By means of such a priority system, safety personnel can allocate their own time and effort and request expenditure of funds to correct situations in proportion to the actual degrees of risk involved in the various situations. Such a priority system is created by the use of a simple formula to "calculate the risk" in each hazardous situation and thereby arrive at a *Risk Score* which indicates the urgency for remedial attention.

Another closely related problem deals with economics. When Safety comes up with a proposed remedy for a hazard, it may be necessary to convince Management that the cost of the corrective action is justified. Since most budgets *are* limited, Safety must compete with other organizations for funds for safety projects. Unfortunately in many cases, the decision to undertake a costly project depends to a great extent on the salesmanship of Safety personnel. As a result, due to a poor selling job, an important safety project may [not] be approved; or due to excellent selling by Safety, a highly expensive project may get approval when the risk may not actually be great. This difficulty is solved by use of an addition to the Risk Score formula which *weighs* the estimated cost and effectiveness of contemplated corrective action against the risk and gives a determination as to whether the cost is justified.

The Risk Score Formula is as follows:

Risk Score = Consequences x Exposure x Probability

In using the formula, the numerical ratings or weights assigned to each factor are based upon the judgment and experience of the investigator making the calculation. A detailed review of the elements of this formula follows.

The first element, *consequences,* is defined as the most probable results of an accident due to the hazard that is under consideration, including both injuries and property damage. Numerical ratings are assigned for the most likely consequences of the accident, from 100 points for a catastrophe down through various degrees of severity to 1 point for a minor cut or bruise.

CONSEQUENCES

Degree of Severity of Consequences   *Rating*

a. Catastrophe: numerous fatalities; extensive damage
   (over $1,000,000); major disruption 100
b. Several fatalities; damage $500,000 to $1,000,000 50
c. Fatality; damage $100,000 to $500,000 25
d. Extremely serious injury; (amputation, permanent
   disability); damage $1,000 to $100,000 15
e. Disabling injuries; damage up to $1,000 5
f. Minor cuts, bruises, bumps; minor damage 1

The next factor, *exposure,* is defined as the frequency of occurrence of the *hazard-event,* the hazard-event being the *first undesired event* that *could* start the accident sequence. The frequency at which the hazard-event occurs is rated from *continuously* with 10 points, through various lesser degrees down to 0.5 for *extremely remote.*

EXPOSURE

The hazard-event occurs: *Rating*

a. Continuously (or many times daily) 10
b. Frequently (approximately once daily) 6
c. Occasionally (from once per week to once per month) 3
d. Usually (from once per month to once per year) 2
e. Rarely (it has been known to occur) 1
f. Very rarely (not known to have occurred,
   but considered remotely possible) 0.5

The third factor, *probability,* is defined as the likelihood that, once the hazard-event occurs, the *complete accident-sequence of events will follow* with the timing and coincidence to result in the accident and consequences. The ratings go from 10 points if the complete accident-sequence is *most likely and expected,* down to 0.1 for the "one in a million" or practically impossible chance.

PROBABILITY

The accident-sequence, including the consequences: *Rating*

a. Is the *most likely* and expected result if the hazard-event
   takes place 10
b. Is quite possible, would not be unusual, has an even
   50/50 chance 6
c. Would be an *unusual* sequence or coincidence 3
d. Would be a *remotely* possible coincidence.
   It has been known to have happened 1
e. Extremely remote but conceivably possible. Has
   never happened after many years of exposure 0.5
f. Practically impossible sequence or coincidence:
   a "one in a million" possibility. Has never happened
   in spite of exposure over many years 0.1

Example

PROBLEM: A building at an explosive-processing laboratory contains a number of ovens which are used for environmental testing (heating) of explosive material with up to five pounds of high explosive material in each oven. This type of oven has been known to heat excessively due to faulty heat controls and thereby cause the explosives in the oven to detonate. People walk past the outside of this building. The potential hazard considered here is the endangering of persons who occasionally walk past the building.

The first step in calculating the risk is to study the situation and list the most probable sequence of events for an accident. These are as follows:

a. Several ovens are in use, each containing explosives.
b. Persons are present outside the building.
c. The thermostat of one oven fails and the oven temperature rises above the proper operating range. (This is the hazard-event.)
d. The secondary emergency shutoff control also fails to function.
e. The oven overheats.
f. The explosive detonates.
g. A passerby near the building is fatally injured by flying debris.

Factors are considered and evaluated for use in the formula:

Risk Score = Consequences x Exposure x Probability

a. *Consequences.* Considering that a fatality was most likely, the ration for *con-sequence* = 25.
b. *Exposure.* The hazard-event is the failure of the thermostat. Experience shows that this has happened before, but very "rarely." Therefore *Exposure* = 1.
c. *Probability.* Based on judgment and experience, it must be decided on what the probability is that the complete accident-sequence will follow the hazard-event, considering each step in the sequence. Consideration include the facts that all ovens have been equipped with secondary emergency shutoff controls, and that thorough maintenance procedures ensure the proper functioning of both the thermostatic controls and emergency shutoff controls. Failure of either set of controls is *quite unlikely.* For *both sets* to fail at the same time

and on the same oven, would be a very *remotely possible coincidence*. Therefore the *probability* rating is remotely possible, and *probability* = 1.

d. Substituting in the formula:

$$\text{Risk Score} = 25 \times 1 \times 1 = 25$$

The significance of this Risk Score will be seen when Risk Scores for other
hazards are calculated, using the same criteria and judgment, and then there will be
a basis for comparison of risks.

In the same manner as demonstrated in the above example, the Risk Scores for
many other hazardous situations have been calculated, *using the same criteria
and judgment*. These cases are now listed in order of their Risk Scores, or we can
say—in order of the *relative seriousness of their risks* on one sheet that is called
the *Risk Score Summary and Action Sheet* (see Exhibit 8.14).

The listing of hazardous situations in the order of the seriousness of their
risks, the higher risks first, becomes an actual priority list.

On the right side of the chart, horizontal brackets have been drawn. These are
the critical dividing lines, which signify the different zones based on the degrees
of risk, and indicate the required corrective action commensurate with the
degrees of risk.

For the hazards with the *higher Risk Scores* (in the high risk zones) the
Action Column calls for immediate corrective action. In these cases, or for any
other hazardous situation whose Risk Score is calculated to be in the high risk
zone, any operation should be stopped until something is done to lower the
risk, and get the score into a less urgent category.

The *medium risk* hazards are in the second bracketed zone, and as the Action
Column states, these cases are "urgent" and require corrective action as soon as
possible. But for these degrees of risk or urgencies we do not say—"Stop the job!"

The hazardous situations in the lowest zone of the chart are lesser ordinary
hazards which, as stated in the Action Column, should be acted upon without
undue delay, but not as emergency situations.

The *Risk Score Summary and Action Sheet* can be a very useful device:

— It establishes priorities for attention by both Safety and Management,
  since all the hazards are now listed in order of their importance. The position of any item can be lowered by corrective measures which will
  decrease any one of the factors: Consequences, Exposure, or Probability.
  For example, the consequences can be reduced by providing protective
  clothing or equipment. Better machine guarding or improved procedures
  could decrease both Exposure and Probability.

— For a newly discovered hazard, the list provides guidance to indicate its
  urgency. Once its Risk Score is calculated its urgency will be indicated by
  the *action* area in which its Risk Score places it.

— It can be used to evaluate a safety program, or to compare safety programs of
  various plants, a more realistic method than accident statistics. At any
  given time the complete chart for a plant represents the actual status of
  safety, i.e., let us say the chart shows seven "immediate actions" for emergency items; six items in the "urgent" category; and twelve "minor"
  hazards. Accomplishment of the safety program over a period of time will
  then be demonstrated by reducing risk scores and moving items downward
  on the chart, from the high risk categories into lower risk areas. For
  example, it would show progress if the number of "emergency action"
  items were reduced from seven to two, and "urgent" items from six to
  four; or if the overall average Risk Score is reduced from 140 to 115; etc.

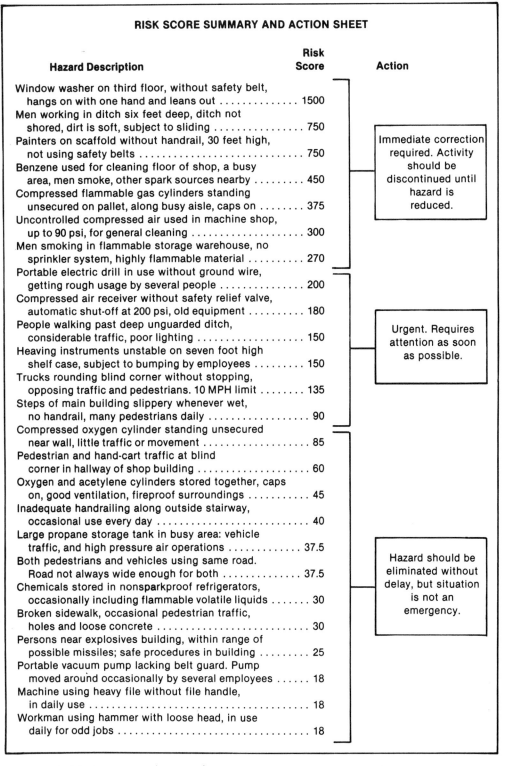

RISK SCORE SUMMARY AND ACTION SHEET

Hazard Description	Risk Score	Action
Window washer on third floor, without safety belt, hangs on with one hand and leans out	1500	
Men working in ditch six feet deep, ditch not shored, dirt is soft, subject to sliding	750	
Painters on scaffold without handrail, 30 feet high, not using safety belts	750	Immediate correction required. Activity should be discontinued until hazard is reduced.
Benzene used for cleaning floor of shop, a busy area, men smoke, other spark sources nearby	450	
Compressed flammable gas cylinders standing unsecured on pallet, along busy aisle, caps on	375	
Uncontrolled compressed air used in machine shop, up to 90 psi, for general cleaning	300	
Men smoking in flammable storage warehouse, no sprinkler system, highly flammable material	270	
Portable electric drill in use without ground wire, getting rough usage by several people	200	
Compressed air receiver without safety relief valve, automatic shut-off at 200 psi, old equipment	180	
People walking past deep unguarded ditch, considerable traffic, poor lighting	150	Urgent. Requires attention as soon as possible.
Heaving instruments unstable on seven foot high shelf case, subject to bumping by employees	150	
Trucks rounding blind corner without stopping, opposing traffic and pedestrians. 10 MPH limit	135	
Steps of main building slippery whenever wet, no handrail, many pedestrians daily	90	
Compressed oxygen cylinder standing unsecured near wall, little traffic or movement	85	
Pedestrian and hand-cart traffic at blind corner in hallway of shop building	60	
Oxygen and acetylene cylinders stored together, caps on, good ventilation, fireproof surroundings	45	
Inadequate handrailing along outside stairway, occasional use every day	40	
Large propane storage tank in busy area: vehicle traffic, and high pressure air operations	37.5	Hazard should be eliminated without delay, but situation is not an emergency.
Both pedestrians and vehicles using same road. Road not always wide enough for both	37.5	
Chemicals stored in nonsparkproof refrigerators, occasionally including flammable volatile liquids	30	
Broken sidewalk, occasional pedestrian traffic, holes and loose concrete	30	
Persons near explosives building, within range of possible missiles; safe procedures in building	25	
Portable vacuum pump lacking belt guard. Pump moved around occasionally by several employees	18	
Machine using heavy file without file handle, in daily use	18	
Workman using hammer with loose head, in use daily for odd jobs	18	

Exhibit 8.14 Hazardous situations listed in order of severity of risks.

If it is desired to compare the actual safety status of each of a number of industrial plants, it could be done simply by comparing the *average* of the Risk Scores of all the principal hazardous situations at each plant. For example, the plant with an average Risk Score of 90 would be a safer place to work than one with an average Risk Score of 120.

To determine whether proposed corrective action to alleviate a hazardous situation is justified, the estimated cost of the corrective measures is balanced or weighed against the degree of risk. This is done by integrating two additional factors into the Risk Score Formula.

The Justification Formula is as follows:

$$\text{Justification} = \frac{\text{Consequences x Exposure x Probability}}{\text{Cost Factor x Degree of Correction}}$$

Note that the numerator of this fraction is actually the Risk Score. A denominator has been added, made up of two additional elements: Cost Factor, and Degree of Correction.

The Cost Factor is a measure of the estimated dollar cost of the proposed corrective action. Ratings are as follows:

Cost	*Rating*
a. Over $50,000	10
b. $25,000 to $50,000	6
c. $10,000 to $25,000	4
d. $1,000 to $10,000	3
e. $100 to $1,000	2
f. $25 to $100	1
g. Under $25	0.5

The Degree of Correction is an estimate of the degree to which the proposed corrective action will eliminate or alleviate the hazard. Its ratings are as follows:

Description	*Rating*
a. Hazard positively eliminated, 100%	1
b. Hazard reduced at least 75%, but not completely	2
c. Hazard reduced by 50% to 75%	3
d. Hazard reduced by 25% to 50%	4
e. Slight effect on hazard (less than 25%)	6

To use the formula and make a determination as to whether a proposed expenditure is justified, values are substituted and a numerical value for *Justification* is computed. The Critical Justification Rating has been arbitrarily set at 10. For any rating over 10, the expenditure will be considered justified. For a score less than 10, the cost of the contemplated corrective action is not justified.

To demonstrate the use of the Justification Formula the same examples used in the demonstration of the Risk Score will be used.

Consider the example of the hazard to persons near a building in which explosives are processed. The corrective action that was proposed was the construction of a barricade along the outside of the building to protect passersby in event of an explosion within. The estimated cost was $5,000. Using the "J" formula:

1. The Consequences, Exposure, and Probability as already discussed were evaluated at 25, 1, and 1 respectively.
2. *Cost factor.* The estimated cost is $5,000. Therefore, based on the Rating Chart, the Cost Factor = 3.
3. *Degree of correction.* The effectiveness of the barricade to protect passersby is considered to be over 75 percent. Therefore the Degree of Correction = 2.
4. *Computation:*

$$J = \frac{25 \times 1 \times 1}{3 \times 2} = \frac{25}{6} = 4.20$$

5. *Conclusion.* The expenditure of $5,000 to construct a barricade to protect passersby is well below 10, and therefore is not justified.
6. *Further consideration.* Since the Risk Score is 25, this situation still requires attention. Review of this problem revealed that other steps could be taken to lower the risk. The Probability of the complete accident sequence occurring was considered to be remote, but it could be made much more remote (and the Risk Score halved) by administrative controls such as portable barriers and warning signs, to reduce or even eliminate the presence of passerbys in the danger zone.

The above examples demonstrate how the formula can save money. In addition it can be a highly valuable management tool. For example, immediately after a very serious accident occurs, let us say an explosion with a fatality, there is usually a tendency for managers as well as safety directors to over-react, to go to extremes in favor of safety. Judgment may become somewhat biased in favor of excessive safety measures. Such action actually hurts Safety's image in the long run, because when situations cool down and people again become rational and reasonable, the poor judgment of such projects is apparent. Therefore under excitable circumstances when costly projects are being considered and may be too hastily approved, the Justification Formula can show whether or not the measures are justified, logically and simply. This formula is a simple and positive *management tool* to help management make a proper decision.

The hazard evaluation systems presented herein can be used effectively by anyone who has sound judgment and experience in safety, with nominal training and guidance.

The Risk Score Formula is used to calculate the *relative severity of hazards.* This process establishes priorities for attention to hazards by both Safety and Management. The Risk Score Summary and Action Chart gives a quick reading or evaluation of the safety status of the organization at any time; it can show safety programs over a period of time; and it gives guidance to Safety in determining where to concentrate its efforts.

The Justification Formula provides both Safety and Management with guidance in deciding whether the cost of a proposed safety project is justified. This formula gives a solid foundation upon which Safety may base its recommendations for corrective action. Its use will assure management that safety projects which actually are not justified will not be recommended. It will therefore cause management to place greater faith in Safety and to give greater support to Safety. It will help to establish that Safety is a *good profitable business.*

For the user's convenience, all the factors used in this hazard evaluation system are given as a single chart (see Exhibit 8.15).

```
JUSTIFICATION FORMULA RATING SUMMARY SHEET
```

Factor	Classification	Rating
1. Consequences	a. Catastrophe; numerous fatalities; damage over over $1,000,000; major disruption of activites 100	
	b. Multiple fatalities; damage $500,000 to $1,000,000. . . . 50	
Most probable	c. Fatality, damage $100,000 to $500,000 25	
result of the	d. Extremely serious injury (amputation, permanent	
potential	disability); damage $1000 to $100,000 15	
accident.	e. Disabling injury; damage up to $1000 5	
	f. Minor cuts, bruises, bumps; minor damage 1	
2. Exposure	*Hazard-event occurs:*	
	a. Continuously (or many times daily) 10	
	b. Frequently (approximately once daily) 6	
The frequency of	c. Occasionally (from once per week to once	
occurrence of the	per month) 3	
hazard event.	d. Usually (from once per month to once per year) 2	
	e. Rarely (it has been known to occur) 1	
	f. Remotely possible (not known to have occurred) 0.5	
3. Probability	*Complete accident sequence:*	
	a. Is the *most likely* and expected result if the	
	hazard-event takes place 10	
Likelihood that	b. Is *quite possible,* not unusual, has an even	
accident sequence	50/50 chance 6	
will follow to	c. Would be an *unusual* sequence or coincidence 3	
completion.	d. Would be a *remotely possible* coincidence 1	
	e. *Has never happened* after many years of exposure,	
	but is conceivably possible 0.5	
	f. *Practically impossible* sequence (has never	
	happened) 0.1	
4. Cost Factor	a. Over $50,000 10	
	b. $25,000 to $50,000 6	
Estimated dollar	c. $10,000 to $25,000 4	
cost of proposed	d. $1,000 to $10,000 3	
corrective action.	e. $100 to $1,000 2	
	f. $25 to $100 1	
	g. Under $25 . . . . 0.5	
5. Degree of Correction	a. Hazard positively eliminated, 100% 1	
	b. Hazard reduced at least 75% 2	
Degree to which	c. Hazard reduced 50% to 75% 3	
hazard will be	d. Hazard reduced by 25% to 50% 4	
reduced.	e. Slight effect on hazard (less than 25%) 6	

Exhibit 8.15 Ratings of factors in hazard evaluation system.

RISK ANALYSIS

Hugh M. Douglas of Imperial Oil Ltd., of Canada, and author of several texts on loss control, has provided an interesting method of decision-making in risk analysis. The method is based upon three maxims:

   1. All risks (uncertainties of loss) can never be completely eliminated.

2. Wise use of available knowledge can reduce potential loss.
3. Efforts to reduce loss should be geared to achieve maximum cost effectiveness.

These maxims have led to a system of quantitatively identifying risks and evaluating proposed risk reduction procedures. The system uses numerical values for comparison purposes. It is based on the contents of two papers: "Mathematical Evaluations for Controlling Hazards" by William Fine and "Practical Risk Analysis for Safety Management" by G. F. Kinney.

According to Fine and Kinney, risk or uncertainty of loss imposed by a particular hazard will increase with the likelihood that the hazardous event will actually occur, with exposure to that event, and with possible consequences of that event. To calculate the risk, numerical values are assigned to each of these factors. From these, an overall risk score is computed as the product of these three separate factors. The numerical values are arbitrarily chosen, but provide a realistic but relative score for the overall risk.

The likelihood of occurrence of a hazardous event is related to the mathematical probability that it might actually occur. Likelihoods range from the completely unexpected up to an event that might be expected (see Exhibit 8.16).

LIKELIHOOD	VALUE
Might well be expected	10
Quite possible	6
Unusual but possible	3
Only remotely possible	1
Conceivable but very unlikely	0.5
Practically impossible	0.1
Virtually impossible	0.1

Exhibit 8.16 Likelihood of hazardous event rating scale.

The greater the exposure to a potentially dangerous situation, the greater the associated risk. The value of unity is assigned to the situation of a rather rare exposure, and the value of 10 is assigned to continuous exposure (see Exhibit 8.17).

Likelihood	Value
Continuous	10
Frequent (daily)	6
Occasional (weekly)	3
Unusual (monthly)	2
Rare (a few per year)	1
Very rare (yearly)	0.5

Exhibit 8.17 Value of potentially dangerous situation.

POSSIBLE CONSEQUENCE	VALUE
Catastrophe (many fatalities, or 10^7 damage)	100
Disaster (few fatalities, or 10^4 damage)	40
Very serious (fatality, or 10^4 damage)	15
Serious (serious injury, or 10^4 damage)	7
Important (disability, or 10^3 damage)	3
Noticeable (minor first-aid accident, or $100 damage)	1

Exhibit 8.18 Value of level of inquiry.

Injury or asset loss from a hazardous event can range all the way from minor injury or loss that is barely noticeable up to the catastrophic (see Exhibit 8.18).

The risk score for some potentially hazardous situation is given numerically as the product of three factors: one numerical value for likelihood, for exposure, and for possible consequences. (See Exhibit 8.19 for suggested controls attached to the risk score.)

Risk scores can be calculated graphically as shown in Exhibit 8.20. The likelihoods are listed on the first or left line of the nomograph. Exposure factors are listed on the second line and factors for possible consequences appear on the fourth line. To calculate a risk score using this nomograph, locations corresponding to each factor involved are established. Then a line is drawn from the point for the likelihood factor through that for the exposure factor and extended to the tie line at the center. A second line is drawn from this point on the tie line through that for the consequence factor and extended to the scale for the risk score.

The larger the risk score for a situation, the more effective a proposed corrective action, and the less that action costs, the greater is the justification. A quantitative index for the justification can be derived from numerical values assigned to each of its three component factors.

The effectiveness value assigned to a proposed risk reduction action is taken as unity for complete elimination of risk and as zero for an action that has no effect.

Cost and justification bear an inverse relation. A cost factor is expressed as a divisor whose numerical value increases with cost so that increased cost gives lesser justification. The justification factor provides an index for cost-effectiveness and like the risk score can be calculated graphically as shown in Exhibit 8.21. Entry to this nomograph is by three factors: one numerical value each for risk score as calculated previously, for degree of risk reduction that the proposed measure provides, and for its cost divisor. Lines through these points give both a numerical value and a descriptive term for the justification factor.

Risk Score	Risk Situation
400	Very high risk; consider discontinuing operation
200 to 400	High risk; immediate correction required
70 to 200	Substantial risk; correction needed
20 to 70	Possible risk; attention indicated
20	Risk; perhaps acceptable

Exhibit 8.19 Risk score for potentially hazardous situations.

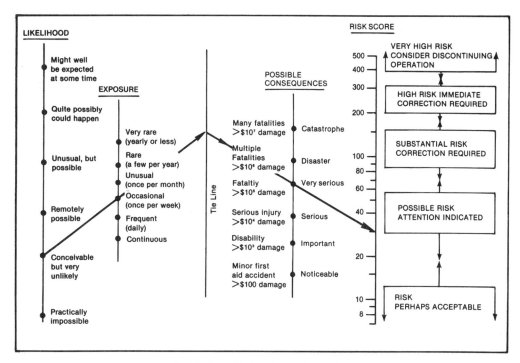

Exhibit 8.20 Risk analysis.

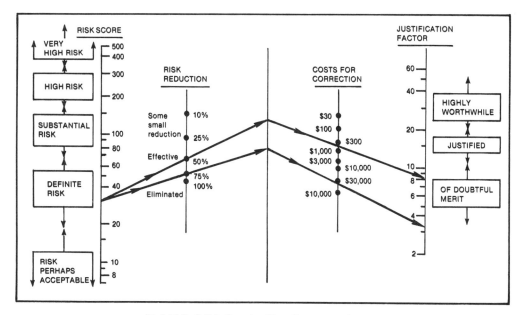

Exhibit 8.21 Cost-effectiveness factors.

FOLLOW-UP/ON-GOING PHYSICAL CONDITION CONTROL

On-going physical condition control requires that the supervisor is the key person and that middle and top management do whatever is necessary to support the key person. Therefore, this section looks generally at what makes a supervisor perform and specifically at what makes that supervisor perform to keep the department in compliance with the law. We have always felt that the supervisor was the key person in safety work, and modern thinking is in agreement, although it places additional emphasis on the "key chain-holders."

Just what is it that a supervisor's performance in safety depends on? It depends on abilities, perception of role, and effort expended. All are important, and a supervisor will not turn in superior performance unless all three are taken into account.

1. *Ability.* Job performance depends on the ability a supervisor brings to the task. Here "ability" means both the inherent capabilities of the person and specialized knowledge. In accident prevention, the supervisor must have sufficient training in safety knowledge to control the people and the conditions under which they work.
2. *Role Perception.* More important than ability is role perception. The supervisor's perception of his or her role in safety determines the direction in which efforts will be applied.

    In safety, role perception means management has decided what it wants so far as accident control is concerned, and the supervisor knows what management has decided, knows what the duties are.
3. *Effort.* There are two basic factors which determine how much effort a person puts into the job:

    a. The opinion of the value of the rewards for the work.
    b. The connection perceived between effort and those rewards.

That is true of the total job as well as of any one segment of it, such as safety.

The supervisor looks at the work situation and asks, "What will my reward be if I make this effort and achieve a particular goal?" If the value of the reward is considered great enough, effort will be made.

Here "reward" means much more than just financial reward. It includes all the things that motivate people, such as recognition, chance for advancement, and ability to achieve. In fact, most research into supervisory motivation indicates that the two greatest rewards are advancement and responsibility.

In effect, then, the chances are that the supervisor will decide the rewards are great enough if they offer advancement and additional responsibility rather than some of the lesser enticements that management too often selects. In safety, just as in other areas, management's chosen rewards are very often too small and too unimportant.

OUTSIDE HELP

There are sources of help outside your organization, but they should be chosen with great care. To find a source with more knowledge and skill than is available internally may be difficult.

Insurance carriers generally offer safety service. A recent survey of 35 of the largest casualty insurance companies showed that 90 percent of them did offer an OSHA service. Most of the companies stated they would make precompliance inspections. Obviously, neither their inspectors nor anyone else can certify compliance, but they can point out many areas in which a company is in violation.

A HAZARD CONTROL PLAN

Robert Hughes, Paul Caplan and Frank Godbey, writing in *Professional Safety* magazine, offer these considerations:

A hazard control plan may be for a single process or unit operation, or for an entire plant. It also may be for an existing facility or a new one. The essential elements to be considered are basically the same for all. The elements described herein assume a single process, but will be applicable, in a general manner, to any size or combination of operations or processes.

The process types, plant layouts and operations will vary widely even among plants with the same end product. While the specific elements may vary, the general approach to identify and control the hazards will be very similar. The use of loss prevention and control procedures may be of use in the evaluation of hazards and the establishment of control plans.

It is also essential that management demonstrate support and commitment to the plan through a written policy statement and visible involvement. Just as management commitment is essential to success, so is employee involvement. No plan operates effectively without the support and active participation of all employees. There are two essential elements for ensuring employee involvement: an overall climate that motivates them toward involvement and specific programs that offer opportunities for participation with both tangible and intangible rewards.

Exhibit 8.22 shows a typical approach to the development of a hazard control plan. It is necessary for each plant to provide the necessary elements to evaluate. It

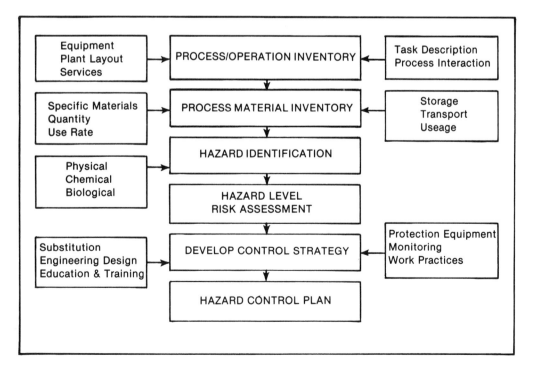

Exhibit 8.22 Typical steps in development of a hazard control plan.

is also important that all aspects of the plan and plant operations be considered and determined by personnel well experienced in both the process and its operation.

While some situations may be amenable to control by a single method, control will usually be best achieved by a combination of several or all the above methods. These overall methods have applicability to both safety and health hazards.

— *Engineering Control.* Control by engineering design encompasses a variety of technical solutions, including elimination or minimization of exposure by material or process modification, substitution of less hazardous material for more hazardous material, isolation of the worker or process, ventilation, and plant layout and design.

— *Monitoring.* A monitoring program is an essential component of a hazard control program in order to maintain the program and to evaluate its effectiveness. There are three basic categories of monitoring that must be addressed.

1. Control Monitoring.
2. Environmental Monitoring.
3. Medical Monitoring.

— *Good Work Practices.* The effectiveness of a control depends in part on the interaction between the control and the worker. The task at hand must be performed in a manner that reduces the potential for emission or eliminates the potential for the worker's intrusion into the area of the hazard.

For example, the effectiveness of a ventilation control can be negated if the worker leans over or into the contaminant generation area such as in a bag emptying and disposal operation or orients his work improperly with respect to the ventilation system. Similarly, bypassing the safety switches on a punch press causes potential for serious injury.

— *Education and Training.* Knowledge of the process, controls, materials used, and procedures are of significant importance in maintaining a continuing safe operation. Appropriate training should be a part of the hazard control plan. Areas to be considered are process operation, control operation and emergency procedures.

— *Maintenance.* The success of any control approach requires the continuing effective operation of both the process and the control. Without proper attention, the operation will deteriorate with time. It is important that a scheduled maintenance program be established as part of a hazard control plan and be strictly enforced.

— *Protective Equipment.* A hazard control plan should first exhaust all means of controlling the hazard by use of engineering design. It is preferable to remove or control the hazard at its source. However, protective equipment can be considered as an alternate or adjunct to engineering means where such means are not available or are not able to completely control the hazard. Protective equipment may also be considered where the consequences of failure of the controls are serious.

Exhibit 8.23 shows how the hazard sequence is controlled, as described by Dawson, Poynter and Stevens from Great Britain, and Exhibit 8.24 describes their strategies for hazard control.

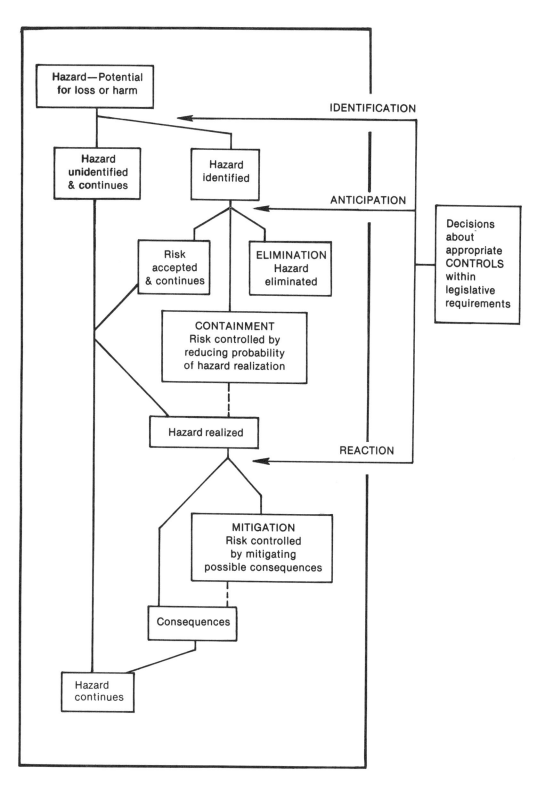

Exhibit 8.23 Controlling the hazard sequence.

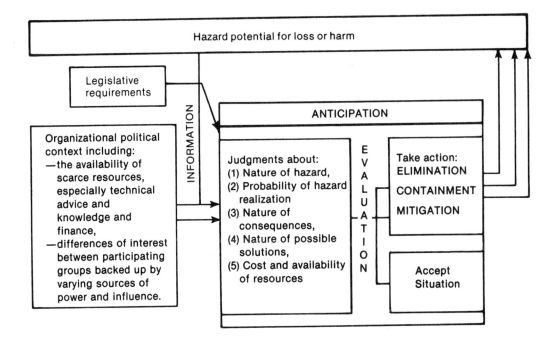

Exhibit 8.24 Formulating strategies to control hazards at work.

REFERENCES

Asfahl, C. "A Ten-point Scale for Workplace Hazards," *Professional Safety,* January 1983.

Fine, W. "Mathematical Evaluations for Controlling Hazards," in Widener, J. (Ed.). *Selected Readings in Safety.* Macon: Academy Press, 1973.

Hudlow, C. D. "Establishing a Safety Program Priority System," *Professional Safety,* December 1976.

Hughes, R., P. Caplan and F. Godby. "Consideration in the Development of a Hazard Control Plan," *Professional Safety,* December 1985.

National Safety Council, *Accident Prevention Manual for Industrial Operations,* National Safety Council, Chicago, 1974.

Partlow, H. and D. Schillinger. "Chemical Industry Loss Prevention," *Professional Safety,* July 1974.

Petersen, D. *The OSHA Compliance Manual,* New York: McGraw-Hill, 1975.

Petersen, D. *Analyzing Safety Performance,* Goshen, NY: Aloray, 1984.

Weaver, D. "The TTC Hazard Rating System," *Professional Safety,* July 1982.

Using Ergonomics

Ergonomics has become a 1980s buzzword. Most organizations are using it (or talking about using it). It is indeed a rich body of knowledge for the safety professional.

Often, however, the safety professional dips into only a portion of that body of knowledge. For instance, one large food processor was experiencing excessive numbers of tendonitis and Carpal Tunnel Syndrom claims. They jumped to ergonomics as the solution, redesigning work stations, table heights, chairs and stools. The result: more claims. They redesigned and retrained, exercised the workers, and basically advertised the problem. The result was even more claims. The more they talked about the problem, the larger the problem grew.

In the above situation, they seemed to perceive only the engineering side of ergonomics. They tried throwing money at the problem with redesign, but it was not purely an engineering problem. People had problems for a number of reasons: overwork, overtime, overuse by management. They were tired physically, physiologically, and psychologically. In many cases the time off was for perceived (or invented) pain. And it worked.

Ergonomics is not a catchall, not a magic pill. Engineering can redesign so there will be less motion, but if you continue to abuse the worker psychologically, the problem will not be solved with redesign. You must also look to the climate as perceived by the worker. But ergonomics does have its place in the design of work and of administration of that work to fit the worker's physical, physiological, and psychological make-up. This chapter looks at how.

Ergonomics by its very definition is safety related. While it is not an exact science, it is a rational approach to the problems of designing and constructing things so that the user will be less likely to make errors resulting in accidents. It attempts to make machines more convenient and comfortable, less confusing, less exasperating, and less fatiguing.

The discipline has various names. Now it is called *ergonomics*. Others call it *biomechanics*, *biotechnology, psychophysics, biophysics, human engineering, human factors,* and *engineering psychology.*

Here are some simple definitions:

- Engineering for the population that will use it.
- Designing the system so that machines, human tasks, and the environment are compatible with the capabilities and limitations of people.
- Designing the system to fit the characteristics of people rather than retrofitting people into the system.

In any man-machine system, there are tasks that are better performed by people than by machine—and, conversely, tasks that are better handled by machines.

In general, machines can usually perform more efficiently on those tasks that must be performed routinely and rapidly with a high degree of accuracy in a given environment. People perform better the tasks calling for responsibility and flexibility in addition to tasks that cannot be anticipated. Exhibit 9.1 outlines the uses of each.

Dr. Julien Christensen outlines some of the key concepts:

> In a modern technological society, all systems—whether they deal with services such as communication, health and transportation, with the production of goods— are man-machine systems. Each system represents the integration of sub-systems composed of humans and hardware interacting with environmental features.
>
> The development and operation of systems in a safe, effective manner requires careful attention on the part of management, personnel specialists, design engineers, safety specialists, and, last but far from least, the workers themselves. Sixteen con-

Man Excels In:	Safety-Design Consideration	Machines Excel In
Detection of certain forms of very low energy levels.	Assume that the ordinary user will have general, but minor sensory limitations.	Monitoring (both men and machines).
Sensitivity to an extremely wide variety of stimuli.	Stick to the big three for the most part: vision, hearing, touch.	Performing routine, repetitive, or very precise operations.
Perceiving patterns and making generalizations about them.	Use this concept to assist in "human engineering" the operation of a product.	Responding very quickly to control signals.
Ability to exercise judgment where events cannot be completely defined.	Average IQ = 100; design for user IQ of about 80.	Performing complex and rapid computation with high accuracy.
Improvising and adopting flexible procedures.	But this capability declines with age and with emotional disturbances.	Sensitivity to stimuli beyond the range of human sensitivity (infrared radio waves, etc.)
Ability to react to unexpected low-probability events.	This is true for highly intelligent, emotionally stable persons only.	Doing many different things at one time.
Applying originality in solving problems, i.e., alternate solutions.	Design for user IQ of about 80.	Deductive processes.
Ability to profit from experience and alter course of action.	True for all users to some extent; avoid design changes for sake of change only.	Insensitivity to extraneous factors.
Ability to perform a fine manipulation, especially where misalignment appears unexpectedly.	Applicable to the user of fairly low IQ, a physiological factor. Largely.	Ability to repeat operations very rapidly, continuously and precisely the same way over a long period.
Ability to continue to perform even when overloaded.	Leads to accidents and misuse of products.	Operating in environments which are hostile to man or beyond human tolerance.

Exhibit 9.1 Man vs. machine.

A. Management considerations
 1. Commitment to safety
 2. Open communications
 3. Stable workforce
 4. Good housekeeping
 5. Motivation
B. Personnel considerations
 6. Careful, unbiased selection
    and placement
 7. Training and maintenance of
    proficiency
 8. Opportunities for self-improvement
 9. Opportunities for advancement
C. Hardware considerations
 10. Design for human use
 11. Removal from hazardous environments
 12. Guards
 13. Warnings
 14. Maintenance
D. Environmental considerations
 15. Physical stressors and their alleviation
 16. Psychological stressors and their alleviation

Exhibit 9.2 Sixteen considerations for developing safe systems.

siderations of importance in the development and operation of safe, effective systems are listed in Exhibit 9.2.

The fundamental sub-system in a production system is the workplace. It is here that the individual worker interacts intimately with hardware in an effort to produce a product efficiently and safely.

The design of a particular work-place is determined not only by the characteristics of the individual human and hardware components but also by the nature of their interactions with each other and with the environment in which these components operate. (See Exhibit 9.3).

These interactions are determined to no small extent by the design of the interfaces or "points of contact" among workers, equipment and the environment. The proper design of these interfaces is central to effective human factors engineering or ergonomics.

Other equally important determinants of effectiveness are training and motivation. Careful consideration of these three factors; viz., human factors engineering (including biomechanics), training and motivation in the design and support of workplaces will do much to assure safe sub-systems which contribute positively to the effectiveness of production systems.

Some Characteristics of People

A.   A general model: input—processing—output

The human can be considered as a component of the workplace sub-system. He receives inputs in the form of physical energy through his senses (vision, hearing, etc.) which are processed by his nervous system. If the input has, or can be made to have, meaning to him, we say that he has received information. He may use the information directly, he may store it for a short time (such as a telephone number

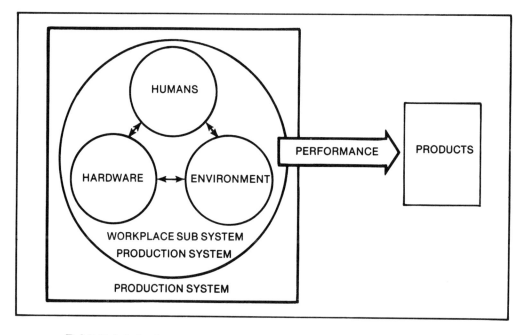

Exhibit 9.3 An interactive model of the workplace sub-system.

that he expects to use only once), he may store it for an indefinite period (such as his home address), or he may combine it with previously stored information immediately or later to generate new information.

The results of this processing usually lead to a decision by the human. If the decision requires some action, he then interacts with the environment through his "effectors"—his voice, his hands, etc. A model that can be used to describe this process is the "input—processing—output" model. As mentioned previously, the proper design of the interfaces, or points of contact, between man and his environment on both the input and output sides is extremely important because it directly affects reliability. This facet of the design engineer's responsibility has been termed "human factors engineering."

B.   Input: The Sensory Sub-System

Before examining the nature of these interfaces, let us briefly examine some general characteristics of man as a sensing-processing-effecting component. Each of man's senses is sensitive to a limited range of a particular type of energy. For example, in the electromagnetic spectrum, man's eyes can sense directly only the range from approximately 400 to 700 millimicrons. He cannot, for example, directly detect energies in the radar range. The ears of a normal young adult can sense physical energies between approximately 20 and 20,000 Hertz. He cannot directly sense radio waves.

Even appropriate energies must be present in certain amounts or people cannot detect them. These minima for the various senses are termed absolute thresholds and demonstrate the exquisite sensitivity of the senses. For example, the dark-adapted eye can detect .000001 millilamberts of brightness (absolute threshold) after 10 to 12 hours in the dark. At the upper limit it can accept as much as 16000 millilamberts, if light-adapted. This is a ratio of approximately 1 to 10 billion!

Energies that are qualitatively similar must differ by a certain amount in order to be distinguished. These amounts are termed differential thresholds. These differences are fairly constant over a rather wide middle range; however, the constancy breaks down at the extremes. The safety engineer and the design engineer should acquaint themselves with absolute and differential thresholds because signal detection and differentiation depends directly on them.

C. Processing: The Central Decision-Making Sub-System

What is generally termed a person's judgment is based to a considerable extent on these thresholds. The number of absolute judgments a person can make (i.e., make without reference to any physical comparative standard) is rather limited. A rule of thumb termed Miller's "Magic Number 7 + 2" provides a convenient way to remember approximately how many absolute judgments can be made along any single dimension such as brightness, weight, etc.

While limited on an absolute basis, people are fantastic comparators. The number of differences that can be detected if the observer can compare the stimulus directly with another of similar quality can run into the tens or hundreds of thousands. This is a very important principle to remember in the design of an inspection station, for example. The human can detect very minute differences if he has a standard present with which he can directly compare the item in question and does not have to compare it with material in his memory bank.

The amount of information that a person can process depends upon many variables—experience, memory, time available, learning ability, motivation and so on. Not the least of these is the way that the information is presented to the person—again human factors engineering.

Current information, properly integrated with information stored in memory ("experiences"), enable people to make decisions. While far from being completely understood, much is known about this complex process that makes people such desirable components in systems. They are preeminent in this area.

D. Output: The Motor Sub-System

Man's motor (muscle) sub-system performs the important functions of stabilization, rotation of segments around the body joints and the transmission of energy to objects outside the body. It is this latter function that is of special importance in human factors engineering.

E. Feedback: (Knowledge of Results)

A person needs to know "how he is doing" to learn effectively; it is also important for continued excellence in performance.

SOME DESIGN PRINCIPLES

If an operator misreads a poorly designed display and operates the wrong control, or the right control in the wrong direction, safety may be jeopardized and the system effectiveness degraded, if not lost entirely. Many accident reports would classify this is an "unsafe act" or "human error" when it is really a design error. Retraining of the operator would not prevent recurrence of the series of events that led to the accident. Much of the effort in occupational safety has been directed toward training. The approach to the machine has been to guard hazards—many of which are simply products of deficient engineering design.

A four element model has been found useful in the development of safer products and workplaces:

1. Design (design the workplace or product to be safer).
2. Remove (remove the worker from the points of high risk; e.g., handle materials in a hot cell with remote manipulators).
3. Guard (insofar as possible, provide guards against contact with remaining points of danger).
4. Warnings (if any residual danger exists after application of the above three steps, warn against residual danger.)

Displays

While information can be "displayed" to all the senses, in most workplaces the visual and auditory channels are the two generally called upon to receive most of the information. However, the designer should always be alert to the possibilities of using the other senses (touch, heat, cold, pain, smell, taste, orientation and movement) as independent or redundant channels. Redundancy is beneficial in many applications. For example, flashing the word "danger" and having it transmitted via the auditory channel is more effective than using either channel exclusively.

An information display is a device used to gather needed information and to translate such information into inputs that the human brain can perceive.

Two general classes of information displays—pictorial and symbolic—are utilized.

Symbolic displays present the information in a form that has no resemblance to what it is measuring. Some examples are a speedometer, a thermometer, a pressure gauge, and an altimeter. In pictorial displays, the geometrical and spatial relationships are shown as they exist. Maps, pictures, and television are examples of pictorial displays.

The two most common types of symbolic displays are the visual and auditory. Much study has been given to the design characteristics of these types of displays and some general principles have emerged:

Principle of Simplicity.   The purpose for which the display is to be read will dictate its design, but as a general principle, the simplest design is the best.

Principle of Compatibility.   The principle of compatibility holds that the motion of the display should be compatible with (or in the same direction as) the motion of the machine and its control mechanism. For instance, a pointer that moves to the right to show an increase should have its corresponding display output value.

Principle of Arrangement.   As the design of the display is important, so too is its location or arrangement with other displays. A poor arrangement of displays can be the source of error.

Sometimes dials must be arranged in groups on a large control panel. If all the dials must be read at the same time, they should be pointing in the same direction when in the desired range. This will reduce check-reading time and increase accuracy. In Exhibit 9.4 for instance, configuration *A* is better than *B*.

Principle of Coding.   All displays should be coded or labeled so that the operator can tell immediately to what mechanism the display refers, what units are measured, and what the critical range is.

A   B

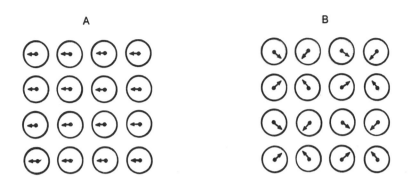

Exhibit 9.4 Principle of arrangement.

Controls

Any display-response that calls for a movement contrary to the established stereotype is bound to produce errors. The designer is calling for errors by asking the operator to change, in this or that unique situation, a behavior pattern that can be described as habit.

General population stereotypes are shown in Exhibit 9.5.

- Handles used for controlling liquids are expected to turn clockwise for off and counterclockwise for on.
- Knobs on electrical equipment are expected to turn clockwise for on or to increase current and counterclockwise for off or decrease in current. *(Note:* This is opposite to the stereotype for liquids.)
- Certain colors are associated with traffic, operation of vehicles, and safety.
- For control of vehicles in which the operator is riding, the operator expects a control motion to the right or clockwise to result in a similar motion of his vehicle, and vice versa.
- Sky-earth impressions carry over into colors and shadings: light shades and bluish colors are related to the sky or up, whereas dark shades and greenish or brownish colors are related to the ground or down.
- Things which are further away are expected to look smaller.
- Coolness is associated with blue and blue-green colors, warmness with yellows and reds.
- Very loud sounds or sounds repeated in rapid succession, and visual displays which move rapidly or are very bright, imply urgency and excitement.
- Very large objects or dark objects imply "heaviness." Small objects or light-colored ones appear light in weight. Large, heavy objects are expected to be "at the bottom." Small, light objects are expected to be "at the top."
- People expect normal speech sounds to be in front of them and at approximately head height.
- Seat heights are expected to be at a certain level when a person sits down.

Exhibit 9.5 General population stereotype reactions.

Research bears out these principles:

Principle of Compatibility.   Control movement should be designed to be compatible with the display and machine movement.

Principle of Arrangement.   This principle provides for the grouping of elements or components according to their function (those having related functions are grouped together) and for grouping in terms of how critical they are in carrrying out a set of operations. It also suggests that each item be in its "optimum" location in terms of some criterion of usage (convenience, accuracy, speed, strength to be applied, etc.).

Also, items used in sequence should be in close physical relationship with each other, and items most frequently used should be placed closer than items less frequently used.

Principle of Coding.   Whenever possible, all controls should be coded in some way. A good coding system uses shape, texture, location, color, and operation. Also all controls and displays should be labeled. Labeling is crucial where the operators change often or the equipment is shared.

Human Abilities

The anthropometric, or human body size, data needed consist of various heights, lengths, and breadths which are used to establish the minimum clearances and spatial accommodations and the functional arm, leg, and body movements which are made by the worker during the performance of his task.

We also need data relating to the range, strength, speed, and accuracy of human body movement. Knowledge of human strength in single maximum exertions and over prolonged periods of time is important for the design of equipment in order to ensure maximum safety in operation. The specific kinds of strength data needed vary from task to task, but might include the maximum weights that can be lifted or carried, or the maximum forces that can be exerted in operating a wheel, lever, or knob during single applications and over extended periods.

Some of the key principles are included in the appendix.

Decision-Making

Dr. Christensen offers these thoughts on design principles for decision-making.

If the informational requirements for a specific workplace sub-system have been carefully determined, if displays have been designed so as to allow the operator to obtain that information quickly and with minimum chance of error, if controls have been designed so as to allow the operator to carry out his intentions quickly and accurately, and if the displays and controls are properly placed in the workplace, errors that might lead to accidents will be reduced and reliability will be increased. Reliability will be further enhanced by appropriate training and motivation.

In addition, however, knowledge of how the operator makes decisions is helpful. Decisions are made under conditions of certainty, under conditions of risk and under conditions of uncertainty. A decision to perform an act that is certain to lead to an accident suggests a psychopathological condition; there is very little that the designer can do about such accidents.

Decisions under conditions of risk imply that the operator knows the probability of an accident associated with each of several alternatives. He may then decide that to save time or effort he is willing to take some action (e.g., remove a guard that inpedes his actions) that he knows carries with it some probability of an accident—he is willing to "take a chance." Certainly, management has a responsibility to inform their workers of all risks and to set up conditions of work and reimbursement that carry with them no reward for taking unnecessary risks.

Decisions under conditions of uncertainty imply that the operator knows neither the nature nor number of alternatives that might lead to an accident. (And, presumably, neither does management, for if they do, they have an obligation to convey such information to the operator.) Under such conditions, the operator is quite apt to set up a hypothesis regarding the possible occurrence of an accident should he take a certain action. If he is injured the first time, he generally will not repeat that action. If he is not injured the first time, he may be tempted to increase the risk on future trials, especially if he derives any benefit from such increase. If reward increases with each increase in risk, an accident lies somewhere in the future. I once knew an individual who in driving home from work, increased his speed around a certain corner one mile per hour each evening! When he got to 54 miles per hour the car slid into the ditch. Fortunately, the major damage was to his billfold. The driver wisely decided to terminate his hypothesis testing at that point.

If there are attractive alternatives in the performance of a job and if the probabilities of an accident associated with these alternatives are unknown, management should conduct tests to establish them. Such information is helpful also in the calculation of the reliabiltiy of the workplace.

Guards and Warnings

On the design principles for guards and warnings, Dr. Christensen suggests:

If complete safety cannot be built into equipment or if the operator cannot be physically removed from potentially dangerous points of contact, the designer then has responsibility to design appropriate guards for the equipment. Since operators have shown remarkable ingenuity in defeating guards, considerable skill on the part of the designer is required to assure that they fulfill their purpose and cannot be defeated. Guards should be permanent and non-removable if at all possible. The operator should be able to use the equipment as easily and effectively with the guards in place as without them. Training programs should emphasize the importance of guards and how effectively they reduce hazards.

If a particular hazard cannot be "designed out" of a piece of equipment, if the operator cannot be removed physically from the potentially dangerous point of contact and if effective guards cannot be installed, then warnings are necessary to help counteract any residual hazard.

Warnings are of two types—dynamic and static. Dynamic warnings consist of such things as warning lights, audible sounds, etc. Guidelines for the design of warning labels, which were derived from a number of sources, are shown in Exhibit 9.6.

Maintainability

Workplace reliability is a function also of the quality of maintenance performed on the equipment in the workplace.

Studies have shown that diagnosis of the cause(s) of malfunctions absorbs the majority of the time of the maintenance person—up to 70 percent, with the remaining 30 percent spent

---

Rules for warning labels

- Conspicuous; attention-getting; urgent; emphatic!
- Clear (comprehensible) and complete for all potential users, regardless of degree of literacy or native language.
- Tell what to expect, why it is a hazard, and what to do if injured.
- Include hazards of use and foreseeable misuse.
- Durable and legible; affixed to product as well as to case or container.
- Pictures, symbols and diagrams OK if universally understood and adequate.
- Conform to standards but consider them as minimum requirements.
- Test effectiveness, using representative sample of potential users.
- Keep warnings current — reflect design changes and user experience.
- Use redundancy.

Exhibit 9.6 Rules for designing warning labels.

almost equally between correction and verification—verification that corrective action was effective.

There are several reasons for what seems to be the inordinate amount of time spent in diagnosis. One is accessibility. Designers of workplaces generally give first priority to the performance requirements of the operator, almost entirely neglecting those of the maintenance personnel. This may not always be the most cost-effective solution. One large company had its machine and shop tools designed with the "plumbing" on the outside instead of hidden inside. Managers reported an immediate improvement in the quality of maintenance and a significant reduction in cost.

One manufacturer of commercial airliners found that approximately 50 percent of the components diagnosed as faulty had nothing wrong with them! This can be a serious matter since removal and replacement of non-faulty components will, in themselves, tend to induce failures, particularly since the careful handling of delicate equipment does not seem to be a universal trait among maintenance people.

Unnecessary removal, due to faulty diagnosis, can frequently be traced to poor preparation of maintenance manuals. One study showed the comprehension level required to understand materials in U.S. Army maintenance manuals to be that of college sophomores, whereas the maintenance people actually read at the level of ninth or tenth graders, in other words, the reading ability of the user. Studies have shown that simpler text, supported by clear, well-labelled diagrams and pictures is of significant benefit to maintenance persons. This is an easy, relatively inexpensive way to increase workplace reliability.

For very complex equipment, studies have shown that simulators are often an aid to diagnosis. The maintenance person enters the symptoms into the simulator and uses that for diagnostic analysis. This saves not only time but also wear and tear on the actual equipment.

Since several industries are finding that maintenance costs constitute on the order of 30 percent of their operating budget, this is an area that is ripe for consideration. Results directly impact reliability and profit.

In summary, the following questions should be considered in planning for human factors:

What is the operator expected to do and what kind of data is required to carry out the task?
Does the task carry an important physical load?
Does the task carry an important mental load?
Are motivation, alertness, and power of concentration strongly involved?
Has the work environment an important influence on the worker (speed, rest pause, etc.)?

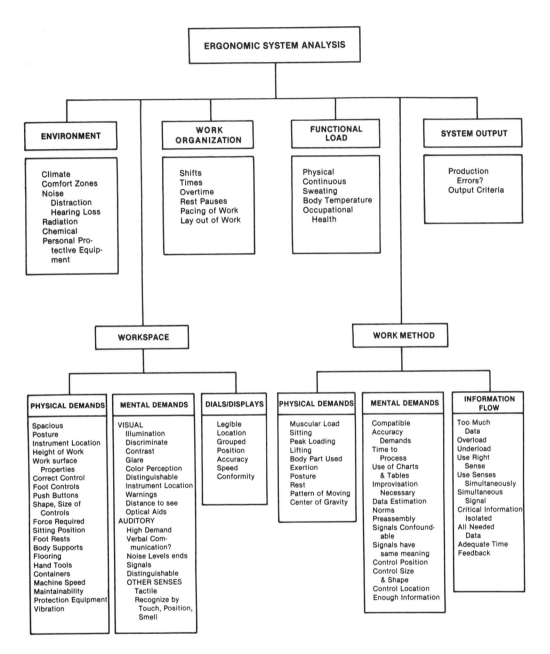

Exhibit 9.7 Ergonomic System Analysis.

Is it desirable to replace the operator, partly or entirely by a machine performance, or the reverse?

Does the task require a learning period of less than one week, less than one month, more than one month?

Is the job so easy, insignificant, or disagreeable that workers feel it unworthy of them?

May the task convey fear or repulsion?

Besides the above questions, Exhibit 9.7, Ergonomic System Analysis from *Human Error Reduction* by the author, attempts to categorize six areas the safety professional might routinely look into in building a safe work environment.

REFERENCE

Christensen, J. "The Human Element in Safe Man-machine Systems, *Professional Safety*, March 1981.

Chapter Ten

Changing the Management System

This chapter looks at the management sub-system—the "safety program" if you prefer.

In this book we'll use the term "safety system" rather than "safety program." The latter connotes something that starts at one point, lasts for a period of time, and then ends. A system is an integral part of an ongoing process: it does not stop at a point in time, it continues as long as the organization continues, and it is a part of management.

In an earlier chapter it was suggested that accidents are caused by a combination of two things; human error and a defective management system. This chapter looks at the latter.

This chapter suggests how we can analyze the management system and improve it. Some of the approaches from the simplistic to the extremely complicated, are:

— Simple checklists.
— Simple yes-no type audits.
— Complicated quantified audits.
— Systems safety approaches.
— TOR analyses.
— Perception surveys.

We'll look at some examples of these approaches.

And finally, we'll look at how to construct and implement an audit and at the current controversy about packaged audits. The appendix contains examples of actual audits used.

The audit concept has been around for a long time. In the 1950s we began thinking about the audit (the predetermination of what must be done, of who should do what, and how we can measure these predetermined actions.).

This, of course, all starts with the determination of the contents of the safety system. There is a lot of research to help as discussed earlier and as covered in detail the companion book, *Safety Management—A Human Approach, Second Edition.*

CHECKLISTS

We first started analyzing the safety system with a simple checklist approach. Back in the early 1950s we developed the loss control analysis guide shown in Appendix A to help assess safety system effectiveness and later the appraisal for multi-plant operations, also shown in Appendix A. These simplistic checklists were followed by the audit.

THE AUDIT

The first checklist in Appendix A is in effect a yes-no type of audit. These were followed by the quantified audit.

R. Diekemper and D. Spartz developed an audit system described in an article in *Professional Safety*:

> Any program, activity, or function which is worthy of continued existence must be designed to permit an objective performance-measurement.
>
> Because of weaknesses in many of the traditional methods of measuring or evaluating safety results, performance, activity, etc., the safety professional continues to search for other measuring techniques.
>
> The various measurements of bottom line results in safety, i.e., frequency and severity rates, monetary costs, injury indexes, etc., do not measure the level and quality of safety activity.
>
> The safety audits which have been devised and used by many safety professionals to measure safety activity have certain limitations when attempts are made to convert audit results to weighted values. Even more frustrating is the problem of conveying the audit results to the management group.
>
> The difficulty in converting the meaning of audit results and the personal knowledge gained by the auditor into concrete terms to inform management is a real challenge.
>
> This article presents a systematic technique for the quantitative and qualitative measurement of safety activity. (Note the word "activity" rather than "results," or "performance.")
>
> The measurement technique (shown in Exhibit 10.1) could be used for a single or multi-plant operation.
>
> The mechanics are simple, and the process is easily understood by plant management. The format consists of three parts: (1) the activity standards, (2) a rating form, and (3) the summary sheet to compute the final score.

Other audit approaches are included in the Appendix.

PROFILING

Profiling is an adaptation of auditing, usually somewhat more elaborate and often accompanied by charts. In profiling, a standard of corporate safety performance in a number of categories considered to be important is developed and are then compared with that standard, the end product being a profile showing how a company compares with the standard in a number of categories.

In his book entitled *The Industrial Environment,* J. Fletcher, a Canadian safety professional, presents a method of rating or grading an organization in the areas of injury prevention, damage control, and total loss control. In connection with injury prevention, Fletcher provides a rating outline for such items as loss-control policy, guarding, inspecting, design and purchasing, audio-visual aids, committees and rules, training, investigation, records and analysis, costs, medical examinations, and personal protective equipment. In the area of total loss control, the book presents a way of profiling fire prevention and control, security, health and hygiene, pollution, and product integrity. Fletcher provides a system of rates from 0 to 5, or from unsatisfactory to excellent, and gives the criteria for each rate.

ACTIVITY STANDARDS

A. ORGANIZATION & ADMINISTRATION

Activity	Poor	Fair	Good	Excellent
1. Statement of policy, responsibilities assigned.	No statement of Loss Control policy. Responsibility and accountability not assigned.	A general understanding of Loss Control, responsibilities and accountability, but not written.	Loss Control Policy and responsibilities written and distributed to supervisors.	In addition to "Good" Loss Control policy is reviewed annually and is posted. Responsibility and accountability is emphasized in supervisory performance evaluations.
2. Safe operating procedures (SOP's).	No written SOP's.	Written SOP's for some, but not all hazardous operations.	Written SOP's for all hazardous operations.	All hazardous operations covered by a procedure, posted at the job location, with an annual documented review to determine adequacy.
3. Employee selection and placement.	Only pre-employment physical examination given.	In addition, an aptitude test is administered to new employees.	In addition to "Fair" new employees' past safety record is considered in their employment.	In addition to "Good" when employees are considered for promotion, their safety attitude and record is considered.
4. Emergency and disaster control plans.	No plan or procedures.	Verbal understanding on emergency procedures.	Written plan outlining the minimum requirements.	All types of emergencies covered with written procedures. Responsibilities are defined with back-up personnel provisions.
5. Direct management involvement.	No measurable activity.	Follow-up on accident problems.	In addition to "Fair," management reviews all injury and property damage reports and holds supervision accountable for verifying firm corrective measures.	In addition to "Good" reviews all investigation reports. Loss Control problems are treated as other operational problems in staff meeting.
6. Plant safety rules	No written rules.	Plant safety rules have been developed and posted.	Plant safety rules are incorporated in the plant work rules.	In addition, plant work rules are firmly enforced and updated at least annually.

Exhibit 10.1 Sample measurement technique.

ACTIVITY STANDARDS

Activity	Poor	Fair	Good	Excellent
B. INDUSTRIAL HAZARD CONTROL				
1. Housekeeping- storage of materials, etc.	Housekeeping is generally poor. Raw materials, items being processed and finished materials are poorly stored.	Housekeeping is fair. Some attempts to adequately store materials are being made.	Housekeeping and storage of materials are orderly. Heavy and bulky objects well stored out of aisles, etc.	Housekeeping and storage of materials are ideally controlled.
2. Machine guarding.	Little attempt is made to control hazardous points on machinery.	Partial, but inadequate or ineffective, attempts at control are in evidence.	There is evidence of control which meets applicable Federal and State requirements, but improvements may still be made.	Machine hazards are effectively controlled to the extent that injury is unlikely. Safety of operator is given prime consideration at time of process design.
3. General area guarding.	Little attempt is made to control such hazards as unprotected floor openings, slippery or defective floors, stairway surfaces, inadequate illuminations, etc.	Partial, but inadequate or ineffective maintenance.	There is evidence of control which meets applicable Federal and State requirements—but further improvement may still be made.	These hazards are effectively controlled to the extent that injury is unlikely.
4. Maintenance of equipment, guards, handtools, etc.	No systematic program of maintaining guards, handtools, controls and other safety features of equipment, etc.	Partial, but inadequate or ineffective maintenance.	Maintenance program for equipment and safety featues is adequate. Electrical handtools are tested and inspected before issuance, and on a routine basis.	In addition to "Good" a preventive maintenance system is programmed for hazardous equipment and devices. Safety reports files and safety department consulted when abnormal conditions are found.
5. Material handling —hand and mechanized.	Little attempt is made to minimize possibility of injury from the handling of materials.	Partial but inadequate or ineffective attempts at control are in evidence.	Loads are limited as to size and shape for handling by hand, and mechanization is provided for heavy or bulky loads.	In addition to controls for both hand and mechanized handling. adequate measures prevail to prevent conflict between other workers and material being moved.

Exhibit 10.1 Continued.

ACTIVITY STANDARDS

Activity	Poor	Fair	Good	Excellent
6. Personal protective equipment — adequacy and use.	Proper equipment not provided or is not adequate for specific hazards.	Partial but inadequate or ineffective provision, distribution and use of personal protective equipment.	Proper equipment is provided. Equipment identified for special hazards, distribution of equipment is controlled by supervisor. Employee is required to use protective equipment.	Equipment provided complies with standards. Close control maintained by supervision. Use of safety equipment recognized as an employment requirement. Injury record bears this out.

C. FIRE CONTROL AND INDUSTRIAL HYGIENE

Activity	Poor	Fair	Good	Excellent
1. Chemical hazard control references.	No knowledge or use of reference data.	Data available and used by foremen when needed.	In addition to "Fair" additional standards have been requested when necessary.	Data posted and followed where needed. Additional standards have been promulgated, reviewed with employees involved and posted.
2. Flammable and explosive materials control.	Storage facilities do not meet fire regulations. Containers do not carry name of contents. Approved dispensing equipment not used. Excessive quantities permitted in manufacturing areas.	Some storage facilities meet minimum fire regulations. Most containers carry name of contents. Some approved dispensing equipment in use.	Storage facilities meet minimum fire regulations. Most containers carry name of contents. Approved equipment generally is used. Supply at work area is limited to one day requirement. Containers are kept in approved storage cabinets.	In addition to "Good" storage facilities exceed the minimum fire regulations and containers are always labeled. A storage policy is in evidence relative to the control of the handling, storage and use of flammable materials.
3. Ventilation— fumes, smoke and dust control.	Ventilation rates are below industrial hygiene standards in areas where there is an industrial hygiene exposure.	Ventilation rates in exposure areas meet minimum standards.	In addition to "Fair" ventilation rates are periodically measured, recorded and maintained at approved levels.	In addition to "Good" equipment is properly selected and maintained close to maximum efficiency.
4. Skin contamination control.	Little attempt at control or elimination of skin irritation exposures.	Partial, but incomplete program for protecting workers. First-aid reports on skin problems are followed up on an individual basis for determination of cause.	The majority of workmen instructed concerning skin-irritating materials. Workmen provided with approved personal protective or devices. Use of	All workmen informed about skin-irritating materials. Workmen in all cases provided with approved personal protective equipment or devices. Use of pro-

Exhibit 10.1 Continued.

ACTIVITY STANDARDS

Activity	Poor	Fair	Good	Excellent
			this equipment is enforced.	per equipment enforced and facilities available for maintenance. Workers are encouraged to wash skin frequently. Injury record indicates good control.
5. Fire control measures.	Do not meet minimum insurance or municipal requirements.	Meets minimum requirements.	In addition to "Fair" additional fire hoses and/or extinguishers are provided. Welding permits issued. Extinguishers on all welding carts.	In addition to "Good" a fire crew is organized and trained in emergency procedures and in the use of fire fighting equipment.
6. Waste-trash collection and disposal, air/water pollution.	Control measures are inadequate.	Some controls exist for disposal of harmful wastes or trash. Controls exist but are ineffective in methods or procedures of collection and disposal. Further study is necessary.	Most waste disposal problems have been identified and control programs instituted. There is room for further improvement.	Waste disposal hazards are effectively controlled. Air/water pollution potential is minimal.

D. SUPERVISORY PARTICIPATION, MOTIVATION AND TRAINING

Activity	Poor	Fair	Good	Excellent
1. Line supervisor safety training.	All supervisors have not received basic safety training.	All shop supervisors have received some safety training.	All supervisors participate in division safety training session, a minimum of twice a year.	In addition, specialized sessions conducted on specific problems.
2. Indoctrination of new employees.	No program covering the health and safety job requirements.	Verbal only.	A written handout to assist in indoctrination.	A formal indoctrination program to orientate new employees is in effect.
3. Job hazard analysis.	No written program.	Job hazard analysis program being implemented on some jobs.	JHA conducted on majority of operations.	In addition, job hazard analyses performed on a regular basis and safety procedures written and posted for all operations.

Exhibit 10.1 Continued.

ACTIVITY STANDARDS

Activity	Poor	Fair	Good	Excellent
4. Training for specialized operations (Fork trucks, grinding, press brakes, punch presses, solvent handling, etc.)	Inadequate training given for specialized operations.	An occasional training program given for specialized operations.	Safety training is given for all specialized operations on a regular basis and retraining given periodically to review correct procedures.	In addition to "Good" an evaluation is performed annually to determine training needs.
5. Internal self-inspection.	No written program to identify and evaluate hazardous practices and/or conditions.	Plant relies on outside sources, i.e., insurance safety engineer and assumes each supervisor inspects his area.	A written program outlining inspection guidelines, responsibilites, frequency and follow up is in effect.	Inspection program is measured by results, i.e., reduction in accidents and costs. Inspection results are followed up by top management.
6. Safety promotion and publicity.	Bulletin boards and posters are considered the primary means for safety promotion.	Additional safety displays, demonstrations, films, are used infrequently.	Safety displays and demonstrations are used on a regular basis.	Special display cabinets, windows, etc. are provided. Displays are used regularly and are keyed to special themes.
7. Employee/supervisor safety contact and communication.	Little or no attempt made by supervisor to discuss safety with employees.	Infrequent safety discussion between supervisor and employees.	Supervisors regularly cover safety when reviewing work practices with individual employees.	In addition to items covered under "Good" supervisors make good use of the shop safety plan and regularly review job safety requirements with each worker. They contact at least one employee daily to discuss safe job performance.

E. ACCIDENT INVESTIGATION, STATISTICS AND REPORTING PROCEDURES

Activity	Poor	Fair	Good	Excellent
1. Accident investigation by line personnel.	No accident investigation made by line supervision.	Line supervision makes investigations of only medical injuries.	Line supervision trained and makes complete and effective investigations of all accidents; the cause is determined; corrective measures initiated immediately with a completion date firmly established.	In addition to items covered under "Good" investigation is made of every accident within 24 hours of occurrence. Reports are reviewed by the department manager and plant manager.

Exhibit 10.1 Continued.

ACTIVITY STANDARDS

Activity	Poor	Fair	Good	Excellent
2. Accident cause and injury location analysis and statistics.	No analysis of disabling and medical cases to identify prevalent causes of accidents and location where they occur.	Effective analysis by both cause and location maintained on medical and first-aid cases.	In addition to effective accident analysis, results are used to pinpoint accident causes so accident prevention objectives can be established.	Accident causes and injuries are graphically illustrated to develop the trends and evaluate performance. Management is kept informed on status.
3. Investigation of property damage.	No program.	Verbal requirement or general practice to inquire about property damage accidents.	Written requirement that all property damage accidents of $50 and more will be investigated.	In addition, management requires a vigorous investigation effort on all property damage accidents.
4. Proper reporting of accidents and contact with carrier.	Accident reporting procedures are inadequate.	Accidents are correctly reported on a timely basis.	In addition to "Fair" accident records are maintained for analysis purposes.	In addition to "Good" there is a close liaison with the insurance carrier.

RATING FORM

	Poor	Fair	Good	Excellent	Comments
A. ORGANIZATION & ADMINISTRATION					
1. Statement of policy, responsibilities assigned.	0	5	15	20	
2. Safe operating procedures (SOP's).	0	2	15	17	
3. Employee selection and placement.	0	2	10	12	
4. Emergency and disaster control planning.	0	5	15	18	
5. Direct management involvement.	0	10	20	25	
6. Plant safety rules.	0	2	5	8	
Total value of circled numbers	____	+ ____	+ ____	+ ____	X .20 Rating __
B. INDUSTRIAL HAZARD CONTROL					
1. Housekeeping—storage of materials, etc.	0	4	8	10	
2. Machine guarding.	0	5	16	20	
3. General area guarding.	0	5	16	20	
4. Maintenace of equipment guards, hand tools, etc.	0	5	16	20	
5. Material handling—hand and mechanized.	0	3	8	10	
6. Personal protective equipment—adequacy and use.	0	4	16	20	
Total value of circled numbers	____	+ ____	+ ____	+ ____	x .20 Rating __
C. FIRE CONTROL & INDUSTRIAL HYGIENE					
1. Chemical hazard control references.	0	6	17	20	
2. Flammable and explosive materials control.	0	6	17	20	

Exhibit 10.1 Continued.

3. Ventilation—fumes, smoke and dust control.	0	2	8	10
4. Skin contamination control.	0	3	10	15
5. Fire control measures.	0	2	8	10
6. Waste—trash collection and disposal, air/water pollution.	0	7	20	25

Total value of circled numbers _____ + _____ + _____ + _____x .20 Rating _

D. SUPERVISORY PARTICIPATION, MOTIVATION & TRAINING

1. Line supervisor safety training.	0	10	22	25
2. Indoctrination of new employees.	0	1	5	10
3. Job hazard analysis.	0	2	8	10
4. Training for specialized operations.	0	2	7	10
5. Internal self-inspection.	0	5	14	15
6. Safety promotion and publicity.	0	1	4	5
7. Employee/supervisor contact and communication.	0	5	20	25

Total value of circled numbers _____ + _____ + _____ + _____x .20 Rating _

E. ACCIDENT INVESTIGATION, STATISTICS & REPORTING PROCEDURES

1. Accident investigation by line supervisor.	0	10	32	40
2. Accident cause and injury location analysis and statistics.	0	3	8	10
3. Investigation of property damage.	0	10	32	40
4. Proper reporting of accidents and contact with carrier.	0	3	8	10

Total value of circled numbers _____ + _____ + _____ + _____x .20 Rating _

SUMMARY

The numerical values below are the weighted ratings calculated on rating sheets. The total becomes the overall score for the location.

A. Organization & Administration _____

B. Industrial Hazard Control _____

C. Fire Control & Industrial Hygiene _____

D. Supervisory Participation, Motivation & Training _____

E. Accident Investigation Statistics & Reporting Procedures _____

TOTAL RATING _____

Exhibit 10.1 Continued.

The British Approach

James Tye, director general of the British Safety Council, offers a similar approach in his book, *Management Introduction to Total Loss Control*. Here he explains his profiling system in which the profiler rates an organization in 30 different areas and rates the performance on a percentage scale (see his Master Evaluation and Development Grid in Exhibit 10.2). Each of the 30 areas is then also broken down into key elements. For example, the key area of management involvement has an evaluation and development grid as shown in Exhibit 10.3.

The South African Approach

Profiling has become the way of life in South Africa. A profiling approach is utilized by the National Occupational Safety Association (the counterpart of our National Safety Council) for grading every industry. The rating report form is shown in Exhibit 10.4. Bunny Matthysen, general manager of NOSA, has described the system as follows:

Management by Objectives the NOSA Way

The industrialist in South Africa has had much of the preparatory work done for him when it comes to Management by Objectives in the field of Accident Prevention.

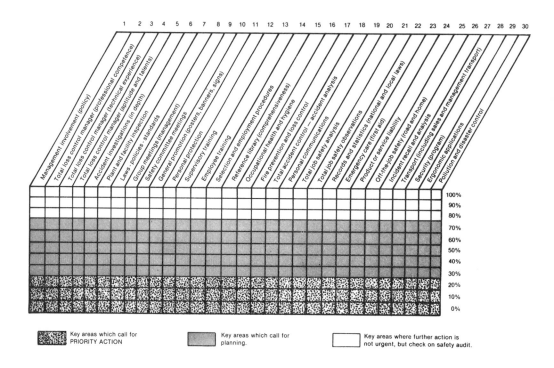

Exhibit 10.2 Master evaluation and development grid.

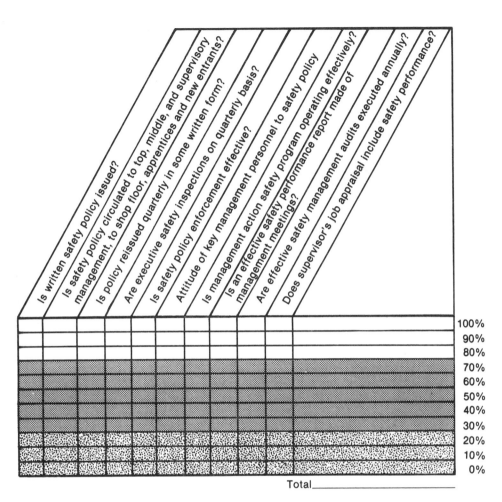

Divide the total percentages by number of questions (10) to obtain
average result to carry over to master evaluation and development chart.

Exhibit 10.3 Evaluation and development grid: key area—management
involvement.

Using the best expertise available, NOSA's technical staff, many years ago, arrived
at the major objectives required to institute a successful safety program. In broad
terms this covered: Industrial housekeeping, electrical, mechanical and personal safe-
guarding; fire prevention and control; accident recording and investigation; and
safety organization. Each major objective was then broken down into subsidiary
objectives giving a total of 50 items, which management should strive to obtain.

Some of the objectives were considered more important than others. So in order to
differentiate them it was decided to quantify the items. The more important ones
were given heavier mark allocations. For example, machine guarding carries 150
marks, whereas clean premises has 40 marks allocated. The total allocation for the
50 items is 2,000 marks.

CONFIDENTIAL   NATIONAL OCCUPATIONAL SAFETY ASSOCIATION   Form 4.13.1
SAFETY EFFORT SURVEY/RATING

SURVEY/RATING OF MESSRS.

NOTE: Items marked "X" require management's attention and should be read in conjunction with the accompanying report. Please refer to the booklet "Management by Objectives" for advice on effective management practice in accident prevention.

	Max.	Actual	Action
1.00 PREMISES AND HOUSEKEEPING			
1.10 PREMISES			
1.11 Buildings and floors — clean and in good state of repair	40		
1.12 Good lighting (natural and artificial)	20		
1.13 Ventilation	30		
1.20 HOUSEKEEPING			
1.21 Aisles and storage demarcated	30		
1.22 Good stacking and storage practices.	50		
1.23 Factory and yard — clean of superfluous material	60		
1.24 Scrap and refuse bins — removal and disposal	30		
1.25 Colour coding — machines, pipelines — other	40		
SECTION RATING	300	%	
2:00 ELECTRICAL, MECHANICAL AND PERSONAL SAFEGUARDING			
2.10 MECHANICAL EQUIPMENT			
2.11 Machine guarding	150		
2.12 Lock-out system and usage	40		
2.13 Labelling of shut-off valves, switches, isolators	30		
2.14 Ladders, stairs, walkways, platforms	40		
2.15 Lifting gear and records	40		
2.16 Compressed gases; pressure vessels and records	30		
2.20 ELECTRICAL EQUIPMENT			
2.21 Portable electrical equipment—monthly check and records	40		
2.22 Earth leakage relays — permanent and portable	30		
2.23 General electrical installation	50		
2.30 HAND TOOLS — All types: condition, storage and use, e.g. hammers, chisels	50		
2.40 PROTECTIVE EQUIPMENT (Issued: use)			
2.41 Head protectors	20		
2.42 Eye protectors	20		
2.43 Foot protectors	20		
2.44 Protective clothing, including hand protectors	20		
2.45 Respiratory equipment	20		
2.46 Maintenance	20		
2.50 NOTICES — Electrical, mechanical, protective equipment, etc.	30		
SECTION RATING	650	%	
3:00 FIRE PROTECTION AND PREVENTION			
3.01 Correct types of extinguishers	40		
3.02 Areas demarcated and clear, extinguishers accessible	20		
3.03 Locations marked	20		
3.04 Maintenance of equipment	30		
3.05 Storage flammable material	30		
3.06 Signs to exits: alarm system	30		
3.07 Fire fighting drill and instructions on fire extinguishers	80		
SECTION RATING	250	%	

EFFORT RATING VALUES

NO. OF STARS	RATING %
5* Excellent	91-100
4* Very good	75-90
3* Good	61-74
2* Average	51-60
1* Fair	40-50

	Max.	Actual	Action
4:00 ACCIDENT RECORDING AND INVESTIGATION			
4.10 RECORDS			
4.11 Adequate accident recording (Register and dressing book)	30		
4.12 Internal accident report form signed by supervisory staff	30		
4.13 Adequate accident statistics kept in accessible place and NOSA informed	30		
4.20 INVESTIGATION of accidents and remedial measures taken to prevent recurrence	30		
SECTION RATING	150	%	
5.00 SAFETY ORGANIZATION			
5.10 SAFETY PERSONNEL			
5.11 One person made responsible for safety co-ordination by management, in writing, i.e. safety officer, permanent/part-time	30		
5.12 Appointment and acceptance of appointment in terms of Factory Regulation C.7.2(a) and (b) or Mines and Works Regulation 2.9.2	30		
5.13 European Safety Committee	80		
5.14 Non-European Safety Committee or any other similar system	40		
5.15 First aider and equipment	20		
5.16 First aid training	30		
5.20 SAFETY PROPAGANDA			
5.21 Poster programme, bulletins, newsletters, use of safety films and internal safety competitions, etc.	130		
5.22 Notice board indicating injury experience	20		
5.23 Suggestion scheme	20		
5.30 INDUCTION TRAINING AND JOB INSTRUCTION, continuous training e.g. poster appreciation, lectures, rule book. NOSA safety training courses	50		
5.40 PLANT INSPECTION system of reporting to management on safety conditions	50		
5.50 WRITTEN SAFE OPERATING PRACTICES and procedure issued, displayed and explained to the illiterate	50		
5.60 ANY ITEM NOT DETAILED			
5.61 COMPANY POLICY (Also Total Loss Control Programme)	100		
5.62 Bonus points awarded or penalty points deducted			
SECTION RATING	650	%	
OVERALL RATING		%	

Exhibit 10.4 Sample safety audit.

Under the guidance of the NOSA technical staff, the optimum requirements for each factory can be determined, using the criterion: HOW MUCH MORE could management reasonably be expected to do within the specific plant; taking COGNIZANCE of the materials, the methods, the men and the money available to the plant.

When the safety state of a plant is to be audited by a NOSA technical staff member, the plant starts with a clean sheet and its full quota of 2,000 marks. When deviations from the set objective are observed they are discussed with management and a note is made thereof. Once all 50 items have been investigated in depth the final mark allocation is made by the NOSA man.

From this unique system has evolved NOSA's Star Grading Scheme, whereby recognition for attaining the set objectives is given. To obtain a Five Star Grading the firm must have a mark allocation of over 90 percent. But attaining the objectives must also result in a reduction in the injury frequency rate. Therefore another criterion is added to the requirements of a Five Star Grading—the injury frequency rate must not be greater than five disabling injuries per 1 million worker-hours exposure. At the other end of the scale, in order to attain a One Star Grading, the mark allocation must be between 40-50 percent and the injury frequency rate must not exceed 25.

The NOSA report form no. 4.13.1 depicted on the following page is sent to management after each safety audit where X marks the spot for major problem areas.

Using the NOSA MBO system management can achieve fantastic results in advertising their objectives of conserving our greatest asset—skilled manpower.

Other auditing and profiling approaches are shown in Exhibits 10.5, 10.6 and 10.7.

THE PACKAGED AUDIT CONTROVERSY

As the audit concept became popular, the idea of the packaged audit also became popular. Perhaps starting in Canada with the book *Total Loss Control* by Jack Fletcher, the concept evolved throughout the world.

Examples of the packaged audit are the NOSA system of South Africa, the British Safety Council System and the ILCI five-star program. Many organizations have been well served by these systems, reporting substantial accident reduction. For instance, a Canadian company shows considerable reduction while utilizing the five-star program over the last five years. Even so, there are some questions that need to be looked at.

The whole concept of the packaged audit is that there are certain defined things that must be included in a safety system to get a high rating or the biggest number of stars. For instance, Exhibit 10.3 shows 30 areas needed and Exhibit 10.5 shows five categories deemed important. How does this thinking jibe with the research? Not too well:

— A NIOSH study in 1978 identified seven crucial areas needed for safety performance. Most are not included in the above packaged programs.
— A Michigan State study had similar results.
— Foster Rhinefort's doctoral dissertation at Texas A & M suggested there was no one right set of elements.

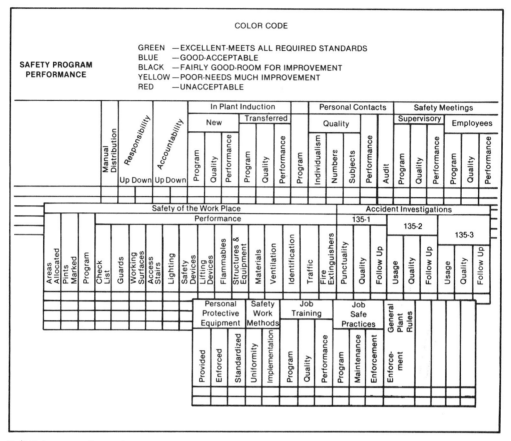

Exhibit 10.5 Sample audit approach. The significant contents of a large 18" x 24" chart have been condensed to this one-page format in order to present the entire picture of this measurement system.

- A 1967 National Safety Council study suggested many of the "required" elements of packaged programs were quite suspect in terms of effectiveness.
- A recent study at the University of Nebraska also had similar results.
- A major study recently done by the Association of American Railroads conclusively showed that the elements in most packaged programs had no correlation with bottom line results.

Clearly the research questions the validity of the concept of packaged safety audits. It does not question the value of the audit concept—only of packaged audits.

In the light of today's management thinking and research, the packaged audit concept has become extremely suspect. When packaged audits became popular apparently no one bothered to ask many questions. In most systems a number of elements were defined; and in most, all elements equally weighted. Thus the right books in the corporate safety library counted as much as whether or not supervisors were held accountable for doing anything about safety. Most of us never thought to question this and most of us also did not question what components were included. Thus some systems had five important components, some 19, some 21, some 30. It all depended on who made up the system.

Safety Performance Ratings

(Ratings are based on measurements of supervisory work performance for the previous period.)

MARCH and APRIL

"We must deeply believe that it is worth our time and effort to prevent even one serious injury.

I trust that each of you will join me in pledging your leadership ability to gain personal commitments to safety as a way of life from those over whom you have supervisory direction and influence."

Sincerely,

ROBERT F. SMITH
President

DEPARTMENT	1 Investigation	2 Facility Inspection Including Housekeeping	3 Rules and Regulations	4 Protective Equipment	5 Group Meetings	6 Proper Job Instruction and Safety Tipping	7 Job Analysis	8 Job Observation	9 Family Safety	10 Safety Performance Rating This Month	11 Safety Performance Rating Year to Date	12 Honor Roll Status
Department A1												
Department A2												
DIVISION A												
Department B1	96	90	85	85	84	78	92	100	88	88	87	
Department B2	93	93	90	85	86	87	93	100	83	90	89	*
Department B3	94	90	85	83	76	83	87	100	90	88	86	*
Department B4	74	90	85	83	84	90	90	100	90	87	86	*
DIVISION B	85	91	85	84	83	85	91	100	87	88	86	*
Department C1	69	90	80	78	75	81	85	100	78	81	75	
Department C2	89	90	76	77	79	94	88	100	75	85	84	
Department C3	90	92	80	79	69	66	86	100	80	82	81	
Department C4	76	90	76	75	58	60	81	100		77	73	
DIVISION C	80	91	78	77	71	76	85	100	73	81	78	
Department D1	88	90	75	75	65	61	85	100	67	78	79	
Department D2	80	90	75	79	58	43	75	100	83	76	73	
Department D3	79	90	82	90	79	78	88	100	72	84	81	
Department D4	68	90	86	79	67	64	67	100	60	76	74	
Department D5	92	90	86	81	80	74	86	100	77	85	80	
DIVISION D	83	90	81	81	79	63	83	100	75	81	79	
ENTIRE PLANT												

TSA — 1084

"ACCIDENT PREVENTION MAKES ACCIDENT ATTENTION UNNECESSARY"

In addition, apparently there was little effort to correlate the packaged audit results to the accident record. Thus safety people were buying in to an unproven concept. When some correlational studies were run, the results were surprising:

— One Canadian oil company location consciously chose to lower their audit score and found their frequency rate significantly improved.
— One chain of U.S. department stores found no correlation between the audit scores and Workers Compensation losses. They found a negative correlation between the audit scores and public liability loss payout.

THE SELF-BUILT AUDIT

The self-built audit could well be the answer—to construct an audit that accurately measures the performance of the safety system. Examples are included in the Appendix.

The process of audit construction consists of:

1. Defining the safety system elements.
2. Defining the relative importance of them (weighting).
3. Defining the questions to find out what is happening.

Some painful lessons have been learned in the above process. In one organization, an audit that looked great ended up with no correlation to results.

The audit *must* be tested against reality. In one organization the accountability section was heavily weighted, but not enough attention was paid to the quality of performance. The result: a paper program only.

Esso Resources Canada Ltd. is one of many organizations to have moved away from the packaged outside audit to the internally constructed audit that puts major emphasis on employee interviews rather than checking paperwork. Their audit uses these key strategies:

- Conduct interviews with employees at all levels to:

  — Test understanding of and commitment to the direction set by area management with respect to safety
  — Measure the level of participation in reaching those objectives
  — Test individual understanding of the required performance and what ownership they have in the results

- Inspect the workplace with emphasis on the following:

  — Observing work practices/planning processes
  — Condition, placement of equipment
  — Condition, placement of protective equipment
  — Orderliness, cleanliness of workplace
  — Identification of hazards, rule and regulation infractions

- Evaluate incident investigation, reporting, analysis:

  — Who investigates and how is it done
  — Use of proactive reporting (near miss, etc.)
  — Reporting processes
  — Analysis and feedback

MANGEMENT CONTROL PERFORMANCE RATINGS							
March 1969 / Department	Investigation	Inspection	Protective Equipment	Rules Practices	Rating this Month	Rating Year To Date	Performance Trend
Transportation						71	
Shipping						85	
Machine Shop						77	
Fabrication						61	
Mechanical						87	
Conditioning						53	
By-Products						79	
Averages						73	

GENERAL COMMENTS

Trends:

Company Wide:

By Department:

Suggested Management Action:

Exhibit 10.7 Sample audit approach.

- Evaluate follow-up to previous audits, inspections or identified hazards
- Evaluate training process:
  - Does it address the need
- Evaluate the communication process:
  - What messages are being sent to the organization (direct and indirect)
  - How are the messages understood
  - What mechanisms are used to communicate in the work place before, during and after a job
  - What is the environment for information exchange

Beyond these there is great flexibility in the audit process.

In Search of Excellence suggests the best management approaches are those that allow, even force, autonomy and entrepreneurship. How does this fare with the audit concept? It doesn't, unless:

1. Audits are built for flexibility. Check performance against the plan; did they do what they said they would do; or
2. An SBO system is used.

Implementing the Audit

There are many choices in audit implementation. For instance:

— By an individual or a team.
— By an insider or an outsider.
— By a staff person or by line managers.
— Announced or Unannounced.
— Quantified or Verbal.
— By checking on or by coaching in approach.
— By looking only or by questioning also.
— By paper looking or by employee interviewing.
— By using questionnaires also.
— By choosing what to look at and what not to look at.
— By determining whom to talk to, and whom not to talk to.

Each makes a major difference. Each is a decision that must be made based upon climate and culture, beliefs, etc. (and the climate desired for the future).

There are no rights or no wrongs. It is what is right for the organization at this point in time.

The Audit as a Motivational Tool

As with any measure, the audit can be a superb motivational tool. For instance, it is common in many organizations to use the audit scores as a primary measure in the performance appraisals of plant managers. Here it is one of the primary motivational tools. Esso Resources Canada Ltd. has also developed the audit technique into an excellent motivational tool, which they label their "Eye-Site" program. Each day each location manager is required to audit the facility for which he or she is responsible and score the facility in terms of a judgment on hazards found, work practices observed, housekeeping, rule compliance, and communication, that would result in recordable injuries, lost time, etc. This scoring is visibly displayed each day to the entire work force, communicating and providing some incentives to them (see Exhibit 10.8).

NON-AUDIT APPROACHES TO ASSESSING SAFETY PERFORMANCE

There are a number of other approaches to assessing the safety system. Technique of Operations Review (TOR) was discussed earlier as a supervisory technique. The results of TOR over time also give an excellent picture of the strengths and weaknesses of the safety system. Similarly, Systems Safety, if properly used, can do the same.

ESSO RESOURCES CANADA LIMITED
JOB SITE EVALUATION

(Esso)

"EYE-SITE"

NMT = NOT MORE THAN

	RI Freq. 1.5 LTA Freq. 0.5	RI Freq. 3.0 LTA Freq. 1.0	RI Freq. 6.0 LTA Freq. 2.0	RI Freq. 9.0 LTA Freq. 3.0
"A" Hazard - LTA or fatality	None	None	NMT 1/20 emp all < 1 day old	NMT 1/25 emp all < 1 day old
"B" Hazard - RI but less than LTA	NMT 1/10 emp all < 1 day old	NMT 1/7 emp all < 2 days old	NMT 1/5 emp all < 2 days old	NMT 1/3 emp all < 2 days old
"C" Hazard - FA or other minor loss	NMT 1/10 emp all < 3 days old	NMT 1/7 emp all < 3 days old	NMT 1/5 emp all < 5 days old	NMT 1/3 emp all < 5 days old
Work practices - use of equipment - procedures, planning	95%	90%	80%	70%
Housekeeping - debris, tools - equipment in good repair	—cleaned daily not allowed to collect -repaired < 1 day	—cleaned daily -repaired < 1 day	—cleaned daily - repaired > 2 days	—cleaned daily repaired > 4 days
Rule compliance - government legislation - local rules/ permits	100% -no violations	95% -no government violations	90% -no government violations	80% -minor government violation
Communication - management objectives understood	-objectives & direction understood & accepted at all levels	90% understand & accept	70% understand & accept	50% understand & accept

Exhibit 10.8 Esso jobsite evaluations.

The same approach can be used for safety system analysis of physical conditions. MORT basically does this, looking beyond the hardware into the management system that is related to safety results. Exhibits 10.9 through 10.12 from MORT show how the MORT fault tree deals with both human and management factors. Exhibit 10.9 describes the fault tree code so that the four trees can be read. Exhibit 10.10 is the master tree, somewhat abbreviated for our purposes. The complete tree is effectively a wall chart because of its size. Exhibit 10.11 takes one section of the master and builds an individual chart to describe the subsystem.

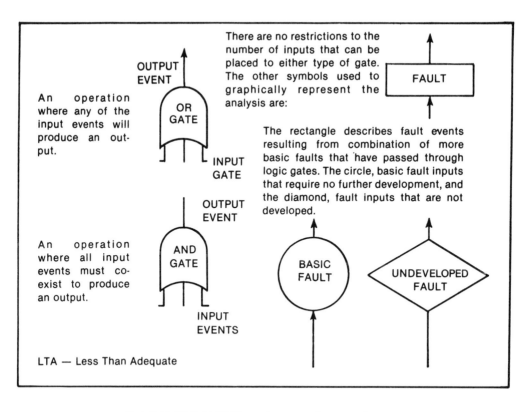

An operation where any of the input events will produce an output.

OUTPUT EVENT

OR GATE

INPUT GATE

There are no restrictions to the number of inputs that can be placed to either type of gate. The other symbols used to graphically represent the analysis are:

FAULT

The rectangle describes fault events resulting from combination of more basic faults that have passed through logic gates. The circle, basic fault inputs that require no further development, and the diamond, fault inputs that are not developed.

An operation where all input events must co-exist to produce an output.

OUTPUT EVENT

AND GATE

INPUT EVENTS

BASIC FAULT

UNDEVELOPED FAULT

LTA — Less Than Adequate

Exhibit 10.9 MORT systems safety code.

SURVEYS

One excellent way of assessing the safety system is through the use of a survey. Some organizations are currently marketing safety climate surveys, which might provide a dipstick of current thinking. One excellent tool was recently developed in the railroad industry.

There seems little question that the way the worker sees the company safety program strongly influences not only his behavior on the job but his ability to learn from and to respond to safety materials. Workers in different companies characterize their safety programs quite consistently in any given company. This was brought out in a study a number of years ago by Social Research, Inc., for Employers Insurance of Wausau. They found these kind of general company types:

The overzealous company. This type of company requires that a great amount of safety equipment be worn. Machines are so guarded as to make them difficult to get near and work with, and the tone of the safety program is heavy. Such a company is likely to impose harsh punishment (a three-day layoff, for example) for some minor infraction of safety procedures. There seem to be endless meetings, films, manuals, and preachments about safety, often not involving the worker directly in any personal way. In such companies response to safety education materials is not lively, and workers feel overexposed to it.

The rewarding company. This type of company might offer prizes for safety records or for entering into safety slogan competitions. While the prizes are relatively small, there is a sense of competition for them. Employees in such companies feel that safety programs are important and that the company generally feels responsible about safety matters.

The lively company. This type of company has a safety program which stimulates competition among the various plants, offers plaques, has boards to record the number of hours passed without accident, or posts a continuing safety record at the plant entrance. These are companies that teach the workers to identify with safety goals and the employees are proud of their record. Safety in these companies becomes one of the lively aspects of the job and is more than avoiding risk or accident; it is a concrete symbol and goal.

The negligent company. This type of company seems to have programs only after the fact. It gets busy about safety only after a major accident happens. The workers here feel that the company does not really care, that they pass out safety equipment because it is current custom and to protect themselves, and that they pass out ssfety information material in the same way.

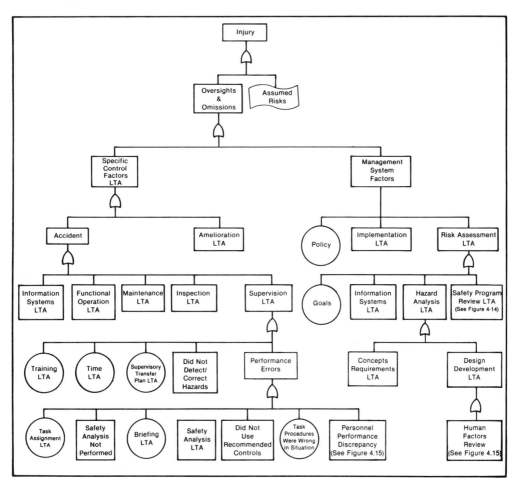

Exhibit 10.10 Abbreviated MORT fault tree.

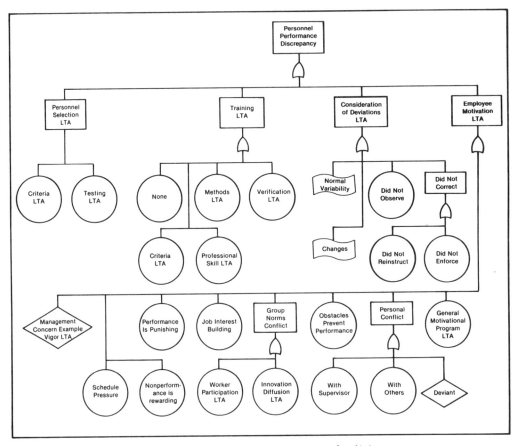

Exhibit 10.11 MORT human error fault tree.

The safety program climate is but one aspect of what we might consider; another is corporate climate in total. This is a major influence on the behavior of both managers and employees. According to B. Scanlon of Oklahoma University, one point of concern is that the employees' perception of the organization's climate and philosophy may be different from that which is intended. Two possible reasons may contribute to this difference in desired versus actual perception. First, perhaps not enough effort has been expended in communicating the guiding philosophy down the line; the second factor may be a discrepancy between that which is professed and actual occurrences. The individual's closest point of contact with the organization is his immediate superior. If the supervisor's actions do not reflect the organizational philosophy, a perception discrepancy occurs.

R. Likert, one of the most famous of the behavioral scientists, discusses climate to a large degree when he describes his "system 4" kind of company. He has isolated three variables which are representative of his total concept of participative management. These are

1. The use of supportive relationships by the manager.
2. The use of group decision-making and group methods of supervision.
3. The manager's performance goals.

The manager's supportive relationship is shown by the degree of:

1. Confidence and trust.
2. Interest in the subordinate's future.
3. Understanding of and the desire to help overcome problems.
4. Training and helping the subordinate to perform better.
5. Teaching subordinates how to solve problems rather than giving the answer.
6. Giving support by making available the require physical resources.
7. Communicating information that the subordinate must know to do the job as well as information needed to identify more with the operation.
8. Seeking out and attempting to use ideas and opinions.
9. Approachability.
10. Crediting and recognizing accomplishments.

Likert measures the relationship of the above to productivity. He states that there is strong evidence to suggest that the organization which exhibits a high degree of supportive relationships, utilizes the principles of group decision-making, and has supervision where there are high performance aspirations has significantly higher levels of achievement. Attitude toward the company, job, superior, and level of motivation are of key importance. The end-results variables refer to tangible items such as volume of sales and production, lower costs, and higher quality.

Participation is one of the main ingredients in gaining employee commitment on an overall basis. It can lead to less need for the use of formal authority, power, discipline, threat, and pressure as means of getting job performance. Thus participation and its resultant commitment become a positive substitute for pure authority. Commitment may be much harder to achieve initially, but in the long run it may prove much more effective.

Likert's description of participation, however, is really a discussion of climate, and an assessment of climate is fundamental to safety program success. Climate builds employee attitudes and acceptance of the program.

Analyzing the Climate

To analyze the corporate climate, it must be divided into two distinct areas—climate of the safety program and corporate climate. Both aspects of climate are difficult to assess—easy to feel but often difficult to get a handle on. Regarding safety program climate, it may be best to ask people throughout the organization, notably line managers and workers. One vehicle with which to do this is a questionnaire.

A major study recently completed almost proves the worthlessness of packaged audits and the worth of climate and perception surveys. In 1979 the Association of Amercan Railroads (AAR) embarked upon a major study in the railroad industry to determine several things. One of the things examined closely was which safety program elements correlate closely with safety success. The examination was performed by the National Space Technology Laboratory (NSTL) with analysis by the Computer Sciences Corporation. They identified twelve subject areas to study:

1. Safety program content.
2. Equipment and facilities resources.
3. Monetary resources.

4. Reviews, audits and inspections.
5. Procedures development, review and modifications.
6. Corrective actions.
7. Accident reporting and analysis.
8. Safety training.
9. Motivational procedures.
10. Hazard control technology.
11. Safety authority.
12. Program documentation.

The NSTL group's hypothesis was that high scores in these areas would correlate generally with lower accident and injury rates. Instead, they found little correlation with these "procedural-engineered" factors. Their report of findings stated, in part:

> It was an unexpected result of this study that so little correlation was found to exist between actual safety performance and safety survey scores. The overall survey score has almost no correlation with train accident rates and cost indicators and is somewhat counter-indicative with respect to personal injury rates. The only two categories which correlated consistently and properly with accident rates were monetary resources and hazards control technology. Two categories, equipment and facilities resources and reviews, audits and inspections, had counter-intuitive correlations.

This conclusion coincided with the findings of a different AAR study group which found that responses to most pilot survey questions based on the presence or absence of these twelve criteria did not distinguish any significant differences between the two railroads studied.

Taken together, the results achieved by the study groups seemed to be saying:

— The effectiveness of safety programs cannot be measured by the more traditional "Procedural-Engineered" criteria popularly thought to be factors in successful programs.
— A better measure of safety program effectiveness is the response from the entire organization to questions about the quality of the management systems which have an effect on human behavior relating to safety.
— The most successful safety programs are those which recognize and deal effectively with worker and supervisor behavior and attitudes which affect safety.

Further Refinement of the Survey Process

The conclusion of the first phase of the project left many unanswered questions. Clearly, it had been proven that many of the things commonly thought to influence safety were having little or no effect. In it's report, the study group cautioned that results could only be considered preliminary and recommended that they be used to develop an instrument and process which could be initiated by companies wishing to evaluate the overall effectiveness of their safety efforts.

Criteria were established by the AAR study group to guide development and evaluation of a second "verification" survey which they proposed be conducted on two additional railroads.

This second survey should:

— Evaluate the organization's perception of management systems which affect safety performance.
— Ask the same questions of managers and employees at five different levels in the organization.
— Be easily and economically administered, analyzed, and evaluated without using a mainframe computer.
— Allow identification and comparisons of specific departments and divisions while maintaining the respondent's anonymity.
— Provide managers with data in a format which will allow definitive comparisons and decision-making.

Survey Verification Study

With the approval of the AAR steering committee and the Federal Railroad Administration, the study group recruited two additional railroads on which to test the revised survey plan. The top safety officers of these roads and their staffs worked with the study group in developing and planning a survey to fulfill the requirements of the established criteria.

A total of 1,815 completed questionnaires were received from the two railroads, allowing a 95+ percent confidence level with respect to inferences concerning a whole railroad and a statistical validity of 90+ percent at the region and division levels.

NUMBER ON EACH RAILROAD PARTICIPATING IN SURVEY

LEVEL	RAILROAD III	RAILROAD IV
Executive	14	1
Top Management	21	6
Middle Management	184	16
1st Line Supervisor	372	23
Employee	878	165
Unspecified	128	17
Totals	1588	227

Description of Analysis Program

Questions which had elicited a statistically significant response on the pilot survey were assigned to 20 categories, each of which defined a specific management system affecting safety. Each question was assigned to one or more categories for the purpose of identifying response patterns.

Data in reports generated was converted to easily read bar charts which allowed comparisons between departments, divisions, regions, and railroads for each of the 20 categories. Other graphic presentations of differences in response between levels in the organization from top executives to employees and areas of high and low positive response were also developed. This information was then presented to top management on both railroads to aid in analysis of safety program effectiveness.

Analysis and Use of Survey Data by Management

Comparison of data from the survey with other indicators of safety performance confirmed that units with the highest positive response to survey questions were generally those with the best performance as measured by other indicators of safety performance.

The analyses of survey data provided by the AAR study group proved to be extremely useful to managements on the participating railroads. For instance, on one railroad the survey clearly identified the weaker management systems which were affecting safety performance as:

- "Recognition for Good Safety Performance"—with a low 48 percent favorable response from hourly-rated employees.
- "Inspections"—with only 51 percent favorable response.
- "Supervisor Training"—49 percent favorable response.
- "Quality of Supervision"—50 percent favorable response.

(See Exhibit 10.12).

This cluster clearly indicated a major problem in perception of supervisory performance and suggested that management needed to immediately train supervisors in all aspects of safety performance, with an emphasis on observations of people and conditions and on positive reinforcement techniques.

On this same railroad, the widest difference in perception of program effectiveness between hourly-rated employees and management was in these same four categories. Not only did they have a serious problem, but they were largely unaware the problem even existed.

One of the regions on one study railroad was consistently lower in positive hourly-rated response in nearly every category than the other regions. Management in this region clearly had a severe credibility problem with employees in nearly every phase of their safety efforts.

On one railroad, one region's scores on "recognition for good safety performance" were significantly lower than those for any other region. When hourly-rated employee perception is at the 40 percent level, it means that six out of ten workers believe they are not receiving the kind of recognition due them for the job they are doing.

On another railroad, two categories were significantly worse than all of the others, clearly indicating the point of attack for improvement.

The survey even documented a problem with injury reporting on one of the railroads where the survey results did not correlate with the accident record.

Ongoing use of the data included:

- One of the railroads is using the survey results to develop performance objectives for individual managers and safety committees. The top executive has decided to personally review the findings with safety committees not meeting their previous year objectives.
- Safety officers are making use of the data to call managers' attention to possible problems requiring corrective action. Their involvement in the survey process provided them with a more balanced picture of the management activities which affect safety performance.
- Because the survey provides a "picture" of the attitudes and beliefs of the organization at a specific point in time, the database generated can be compared with data from future surveys to determine the effects of new safety initiatives. One of the railroads is currently re-surveying in selected areas to determine the effects of the training program conducted as a part of the study.
- Other railroads using the survey to measure safety program effectiveness will have access to a large and complete database with which to compare their data. The AAR study

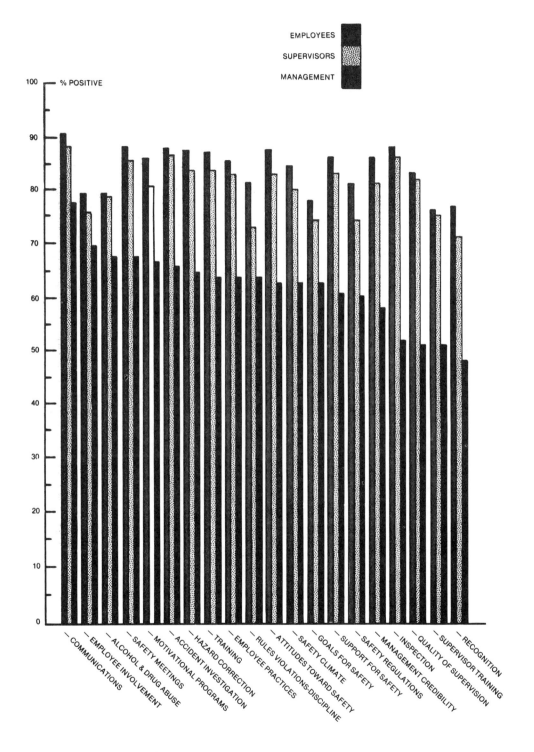

Exhibit 10.12 Safety survey response.

group recommended making a "survey package" of software and instructions available to assist member roads in this regard.

Conclusions

As a result of this nine year study, these conclusions were reached:

— The effectiveness of safety efforts cannot be measured by traditional (procedural-engineering) criteria.
— The effectiveness of safety efforts can be measured with surveys of employee (hourly to executive) perceptions.
— A perception survey can effectively identify the strengths and weaknesses of elements of a safety system.
— A perception survey can effectively identify major discrepencies in perception of program elements between hourly rated employees and levels of management.
— A perception survey can effectively identify improvements in and deterioration of safety system elements if administered periodically.

The above conclusions, based on the data, carry with them considerable importance to safety management thinking. In effect, the data strongly suggests that those twelve elements discribed as a "procedural engineering" or a systems approach are not closely related to safety performance or results.

The twelve elements are not a bad description of the elements of many safety programs in use in industry today. This data *does not* question the audit approach to safety performance improvement. It *does* question the validity of any audit approach made up of arbitrarily decided elements such as these twelve. It does question the validity of any so-called "packaged audit."

It supports rather strongly the belief that a perception survey (sometimes called a "culture survey"), properly constructed, is a better measure of safety performance and a much better predictor of safety results.

THE ALTERNATIVES TO AUDITS

Audits have some severe drawbacks:

— They stifle autonomy.
— They stifle creativity.
— They force uniformity.
— They tend to be based on one person's opinion as to what is "right and wrong."
— They are subjective.

They also have some real advantages:

— They get attention.
— They force performance.

How can we get the advantages without the drawbacks? There is one way to do this: to audit against agreed-to criteria instead of company dictated (or outsider dictated) standards. This allows total flexibility in programs as long as the system meets certain criteria.

What are the criteria? You'll have to decide for yourself based upon what you believe causes injuries and what constitutes an effective safety system (check the earlier chapters).

Then, based upon what you believe, identify your criteria against which to judge the safety system.

These are my criteria (they may not be yours):

— Does your safety system force supervisory performance?
— Does your safety system involve middle management?
— Does your safety system get visible management commitment?
— Does your safety system get total employee participation?
— Is your safety system flexible?
— Is your safety system seen as positive?

REFERENCES

Diekemper, R. and D. Spartz. "A Quantitative and Qualitative Measure of Industrial Safety Activities," in *Self-Evaluation of Occupational Safety & Health Programs,* NIOSH, 1979.

Fletcher, J. *The Industrial Environment,* National Profile Limited, Willowdale, Ontario, 1972.

Johnson, W. *MORT—The Management Oversight Risk Tree,* U.S. Government, Washington, DC, 1973.

Likert, R. *The Human Organization,* New York: McGraw-Hill, 1976.

Petersen, D. *Analyzing Safety Performance,* Goshen, NY: Aloray, 1984.

Tye, J. *Management Introduction to Total Loss Control,* British Safety Council, London, 1970.

Part IV

Reactive System Elements

Chapter Eleven

Records

Part IV deals with what can and must be done after the fact of the accident. For the most part proactive system elements are more important, but once the event has occurred (proaction has failed) the opportunity ought not be wasted to learn from the incident(s) both individually and collectively.

This chapter looks at the recordkeeping function, both the investigation process and the paper flow that exists for legal, compensation, and data analysis purposes.

First, a short look at what is required. The insurance carrier (or department) requires at a minimum some type of first report of injury to trigger their process to ensure that the medical bills and compensation are paid. The data on this report, while essential, accomplishes nothing in and of itself for safety.

Secondly, the federal government requires additional records. Here are some guidelines from a Department of Labor pamphlet:

> The OSH Act of 1970: The Occupational Safety and Health Act requires employers to prepare and maintain records of occupational injuries and illnesses. The Act also requires the Secretary of Labor to develop and maintain an effective program for the collection, compilation, and analysis of statistics of work-related injuries and illness.
>
> The OSH Act provides a basic description of which cases are to be recorded. Sections 8(c)(2) and 24(a) of the Occupational Safety and Health Act provide the basic definition of the types of cases to be recorded:
>
> > ... Work-related deaths, injuries and illnesses other than minor injuries requiring only first aid treatment and which do not involve medical treatment, loss of consciousness, restriction of work or motion, or transfer to another job.
>
> > 1. An injury or illness is considered work related if it occurs in the work environment (defined as any area on the employer's premises, e.g., worksite, company cafeteria, or company parking lot). (See Exhibits 11.1 and 11.2). The work environment surrounds the worker wherever he or she goes—in official travel, in dispersed operations, or along regular routes (e.g., sales representative, pipeline worker, vending machine repairer, telephone line worker).
> > 2. All work-related fatalities must be recorded.
> > 3. All diagnosed work-related illnesses must be recorded.
> > 4. All work-related injuries requiring medical treatment or involving loss of consciousness, restriction of work or motion, or transfer to another job must be recorded.

Recordable and nonrecordable injuries are distinguished by the treatment provided; i.e., if the injury required medical treatment, it is recordable; if *only* first aid was required, it is *not* recordable. However, medical treatment is only one of several criteria for determining recordability. Regardless of treatment, if the injury involved loss of consciousness, restriction of work or motion, transfer to another job, or termination of employment, the injury is recordable.

Recordkeeping forms must be retained for 5 years after the end of the calendar year in which they relate. Additionally, logs (OSHA No. 200) must be maintained for the same period; i.e.,

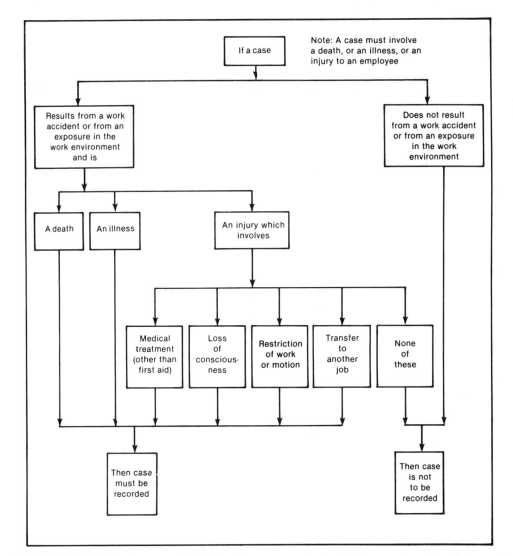

Exhibit 11.1 Guide to recordability of cases under the Occupational Safety and Health Act.

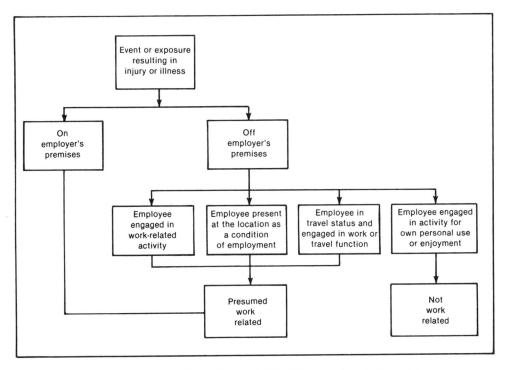

Exhibit 11.2 Guidelines for establishing work relationship

changes in extent of and outcome of cases must be made directly to the original case entries, even though the entries were dated in prior years.

All recordkeeping forms (current and retained) must be available at the establishment for inspection and copying by representatives of the Department of Labor; the Department of Health, Education, and Welfare; or States accorded jurisdiction under the act.

The log, form OSHA No. 200, shall, upon request, be made available by the employer to any employee, former employee, and to their representatives for examination and copying in a reasonable manner and at reasonable times. Access to the log shall be for any establishment in which the employee is or has been employed and covers all logs required to be maintained or retained by the employer.

These records in and of themselves also accomplish nothing for safety.

ACCIDENT INVESTIGATION

This was discussed briefly in Chapter 4 as one of the functions of the first line supervisor. This primary investigation to determine causes is supplemented in many organizations by further investigation by second and upper level managers, by safety staff, by review boards, and in some cases, by executive teams.

The investigation document(s) becomes the source document for most further analysis. Therefore, the accident investigation form (what information is asked for) becomes crucial. Historically the accident investigation process has been based on the old Heinrich domino theory of causation which asks for the identification of unsafe act and unsafe condition. This information is almost worthless as an aid to prevention. Investigation forms that go beyond this are crucial. One such form, based upon the theory of multiple causation, was shown in Exhibit 3.5. Others will be discussed in this chapter and in the next chapter.

The input document will dictate the value of the analysis results. Historically input documents even in the more sophisticated computerized systems have been structured around the old ANSI Standard Z16.2 which suggests the investigation identify:

— Nature of injury.
— Part of body affected.
— Source of injury.
— Accident type.
— Hazards condition.
— Agency of accident.
— Agency of accident part.
— Unsafe act.

One of the problems with the utilization of the standard is that it is very much based upon our old domino theory of accident causation and thus deals primarily at the symptomatic level. Computer programs based on this standard then tend to be summaries and analyses of large numbers of symptoms. Following are examples of how the standard is utilized.

1. A circular-saw operator reached over the running saw to pick up a piece of scrap. His hand touched the blade, which was not covered, and his thumb was severely lacerated.

ANALYSIS

Question	Analysis Category	Answer	Code
a	Nature of injury	Laceration	170
b	Part of body	Thumb	340
c	Source of injury	Circular saw	3750
d	Accident type	Struck against	012
e	Hazardous condition	Unguarded	510
f	Agency of accident	Circular saw	3750
g	Agency of accident path	Blade	3099
h	Unsafe act	Cleaning, adjusting moving machine	052

2. A forklift truck went out of control when one wheel struck a piece of stock lumber which projected into the aisle. It ran out of the aisle and struck a machine operator, breaking his leg between the ankle and knee.

ANALYSIS

Question	Analysis Category	Answer	Code
a	Nature of injury	Fracture	210
b	Part of body	Lower leg	515
c	Source of injury	Forklift truck	5635
d	Accident type	Struck by	029
e	Hazardous condition	Improperly placed lumber	420
f	Agency of accident	Lumber	5700
g	Agency of accident path	None	9999
h	Unsafe act	Unsafe placement of material	657

3. A warehouse employee jumped from the loading platform to the ground instead of using the steps. As he landed, he sprained his ankle.

ANALYSIS

Question	Analysis Category	Answer	Code
a	Nature of injury	Sprain	310
b	Part of body	Ankle	520
c	Source of injury	Ground	5810
d	Accident type	Fall from elevation	031
e	Hazardous condition	None indicated	990
f	Agency of accident	None indicated	9800
g	Agency of accident part	None indicated	9800
h	Unsafe act	Jumping from elevation	503

4. A laborer working in a trench was suffocated under a mass of earth when the unshored wall of the trench caved in.

ANALYSIS

Question	Analysis Category	Answer	Code
a	Nature of injury	Asphyxia	110
b	Part of body	Respiratory system	850
c	Source of injury	Earth	4300
d	Accident type	Caught under	064
e	Hazardous condition	Lack of shoring	530
f	Agency of accident	Trench	1630
g	Agency of accident path	None	9999
h	Unsafe act	Not indicated	999

As can be seen from the above, using the Z16.2 analysis system does not lead very far into corrective mechanisms that really help. The answers generated are guard the saw, be careful in cleaning, place lumber properly, don't jump, shore trenches. Clearly systems are needed that look at the real causes of accidents (management systems and the causes of human error) if past

USING THE COMPUTER

In recent years most large organizations have begun to utilize the computer in their injury recordkeeping. Most companies use an approach based primarily on the American National Standards Institute's Standard Z16.2. Most insurance carriers also have computerized injury records that provide their insureds with a loss run of some type to assist them in their loss-control efforts. Insurance companies also usually gear up their programs on the basis of the Z16.2 standard. This standard includes a general but comprehensive list of items within numbered sequences that can be used in the development of a record analysis system. Utilizing this standard as a starting point, companies then develop their own systems of analyzing their accidents and injuries in categories such as nature of injury, accident type, part of body injured, hazardous condition, source of injury, and unsafe act.

A large number of items are usually listed in each of the categories, allowing the reporter to choose items that come closest to describing what happened in connection with the accident or incident being reported on. Obviously, then, one factor determining the accuracy of input is the initial identification of items which will be on this list. The code shown in Exhibit 11.3 is an example of a basic input document; it was adapted from the report of a supervisor working for a large contractor. Exhibit 11.4 shows the input document of a large insurance company.

Weaknesses in Systems

There are some inherent weaknesses in these systems in terms of current thinking about accident causation. The entire Z16.2 approach to analysis is based on getting a detailed description of the circumstance surrounding an accident, rather than on identifying the contributing causes. It is in actuality more a reflection of the domino theory of accident causation than a reflection of the multiple-causation theory. Thus, in utilizing the Z16.2 approaches many computerized analysis programs end up giving a great deal of easily accessible but not very useful information and little insight into why accidents are occurring. Exhibit 11.5 shows a printout utilizing these approaches.

As a result of our Z16.2 type of input, the computer has provided an analysis that shows a frequency of eight accidents caused by the employee being struck by a tipping, or rolling object, for a cost of $11,726. What are the real causes of these eight injuries? We have no idea. Where did they occur? We cannot tell from this analysis. What should we do about them? There is no way of knowing. A Z16.2 analysis does not deal with contributing causes and thus does not help direct efforts. Perhaps the best information from this type of analysis is a breakdown of where accidents are happening (location, department, etc.), which is not a function of Z16.2.

Input Documents

Thus, while computers are extremely valuable in assembling and analyzing data and information, the amount of assistance they provide is entirely dependent on what and on how it is put in. Exhibit 11.7 shows a printout that is a summary sheet. While most firms and insurance companies are able to provide such summary sheets, it is more common for them to provide simple loss runs, as shown in Exhibit 11.6.

This type of printout is, of course, even less useful in terms of directing safety efforts. It does, however, provide the kind of information that management often likes, for it shows who was injured, tells a little about how they were injured, and gives some dollar information. Two

1. Report number _____
2. Date of injury _____
3. Job location _____

4. INJURED EMPLOYEE (Please print)					TIME OF INJURY		
LAST NAME	FIRST INITIAL	BADGE NO.	JOB NO.	HOUR	MIN.	AM/PM

5. CRAFT			6. AGE GROUP			SEX
A. Birmkr.	F. IW	K. PF	A. 18-20	E. 36-40	I. 56-60	
B. Corp.	G. Lab.	L. Surv.	B. 21-25	F. 41-45	J. 61-65	M
C. Cm. Msn.	H. MW	M. Trnstr.	C. 26-30	G. 46-50	K. 66-70	
D. Elec.	I. OE	N. Other	D. 31-35	H. 51-55	L. 71 +	F
E. Insul.	J. Paint.					

7. HOW IT HAPPENED: _____

8. | DR | CASE | LOST TIME | FATAL | WITNESS _____ EMPLOYEE _____

9. NATURE OF INJURY
A. Abrasion, scratch
B. Amputation
C. Burn
D. Concussion
E. Conjunctivity (eye irritation)
F. Confusion, crushing, bruise
G. Cut, laceration, puncture
H. Dermatitis, rash
I. Electric shock, electrocution
J. Flashburn (eye)
K. Foreign body
L. Fracture
M. Heatstroke, exhaustion
N. Hernia, rupture
O. Inflammation, irritation
P. Respiratory problems
Q. Sprains, strains, dislocations
R. Multiple injuries
S. Injury or disease (NOC)
T. Nonoccupational claim (record purposes only)

10. PART OF BODY
A. Head (not face, eyes, internal ears)
B. Ears internal (including hearing)
C. Eyes
D. Face (nose, mouth, lips, teeth, etc.)
E. Neck
F. Arm(s) (including wrist)
G. Hand(s) and or finger(s)
H. Abdomen (including internal organs)
I. Back (including muscles, spine, spinal cord)
J. Chest (including ribs, internal organs)
K. Hips (inc. pelvis, buttocks, internal organs)

L. Shoulders
M. Leg(s) (including ankle)
N. Foot and/or toe(s)
O. Multiple body parts
P. Circulatory system (heart, blood, vessels)
Q. Digestive system
R. Respiratory system
S. Body part (NOC)

11. SOURCE OF INJURY
A. Burning or welding operation
B. Hand tools
C. Power hand tools (except grinder)
D. Grinder (hand or pedestal)
E. Material, equipment being handled, worked
F. Body motion (not overexertion)
G. Ladder
H. Scaffold
I. Object improperly placed, located
J. Work surface, floor, ground, stair
K. Falling, flying, moving object
L. Building, structure, installed equipment, etc.
M. Rigging operation
N. Construction equipment, vehicle
O. Electrical equipment
P. Irritants (chemicals, dusts, fumes, etc.)
Q. Insects, plants, (wasp sting, poison oak, etc.)
R. Fire
S. Wind
T. Source of injury (NOC)

12. ACCIDENT TYPE
A. Struck against
B. Struck by
C. Fall from elevation

D. Fall on same level
E. Caught in, under, between
F. Overexertion (lifting, throwing, pulling, pushing)
G. Contact with electric current
H. Contact with temperature extremes
I. Contact with temperature extremes
J. Accident type (NOC)

13. UNSAFE CONDITION
A. No unsafe condition
B. Defect of source of injury (dull, rough, sharp, slippery, broken, etc.)
C. Improper or inadequate clothing
D. Unsafe method or procedure
E. Unsafe placement or storage
F. Inadequate guarding
G. Inadequate illumination
H. Inadequate ventilation
I. Unsafe condition (NOC)

14. UNSAFE ACT
A. No unsafe act
B. Working on moving, energized, pressurized equipment
C. Failure to wear protective equipment (hard hat, safety glasses, rubber goods, foot guards, etc.)
D. Operating without authority
E. Failure to secure or warn
F. Horseplay
G. Improper use of equipment
H. Taking an unsafe position or posture
I. Inattention to footing or surroundings
J. Making safety devices inoperative
K. Undue haste
L. Driving error by equipment/vehicle operator
M. Unsafe loading, placing, mixing, etc.
N. Using unsafe equipment
O. Unsafe act (NOC)

Exhibit 11.3 Supervisor report and injury analysis.

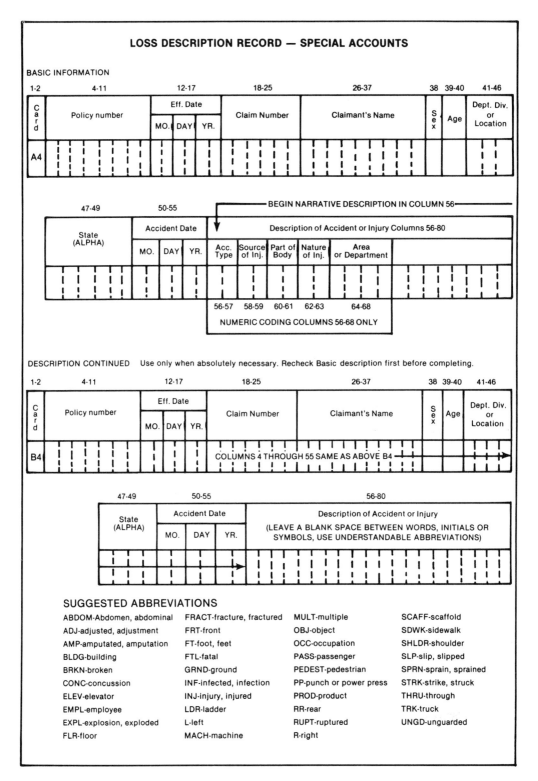

Exhibit 11.4 Input document, insurance company (front)

ACCIDENT TYPE (56-57)

01 Arrest or Slander, False
02 Bodily Reaction
03 Caught In, Under or Between
04 Contact with Caustics, Noxious, Toxic or Radiation
05 Contact with Electric Current
06 Contact with Temperature Extremes
07 Fall-Stairway or Elevation
08 Fall-Same Level
09 Over-exertion (Lifting, Carrying, Pushing, Pulling)
10 Product Installation or Service
11 Rubbed or Abraided
12 Struck Against Objects
13 Struck by Objects
14 Transportation, Public
15 Vehicle, Motorized Equipment
98 Other, Not Classified Above
99 Unclassified, Insufficient Data

SOURCE OF INJURY (58-59)

01 Bodily Motion (not over-exertion)
02 Boilers and Pressure Vessels
03 Boxes, Barrels, Containers, Packages
04 Buildings and Structures
05 Carts, Hand Trunks
06 Chemicals, Chemical Compounds
06 Chemicals, Chemical Compounds (solids, liquids, gases)
07 Civil disturbances, Riots
08 Clothing, Shoes
09 Conveyors, Hoists
10 Elevators, Escalators
11 Environmental (atmosphere, cold, heat, steam)
12 Drugs, Medicines
13 Dusts
14 Electrical Apparatus
15 Flame, Smoke, Fire
16 Food Products
17 Furniture, Fixtures, Furnishings
18 Glass
19 Hand Tools
20 Heating Equipment
21 Ladders
22 Liquids (water, other NOC)
23 Machines
24 Mechanical Power Transmission (belts, chains, gears etc.)
25 Noise
26 Particles (Foreign Objects, various)
27 Sidewalk, Street, Grounds, Misc.
28 Soaps, Detergents, Cleaning Compounds NOC
29 Stairway
30 Transportation, Public
31 Utilities, Overhead and Underground
32 Vehicle, Motorized Equipment
33 Weather
34 Working Surfaces (Floors, Scaffolds, etc.)
98 Other, Not Classified Above
99 Unclassified, Insufficient Data

PART OF BODY (60-61)

01 Head
02 Brain
03 Ear(s)
04 Eye(s)
05 Face
06 Multiple Parts
07 Other, NOC
10 Neck
20 Upper Extremities
21 Shoulder
22 Arm
23 Wrist
24 Hand
25 Finger(s)
26 Multiple Parts
27 Other, NOC
27 Other, NOC
30 Trunk
31 Abdomen
32 Back
33 Chest
34 Groin
35 Multiple Parts
36 Other, NOC

40 Lower Extremities
41 Leg
42 Knee
43 Ankle
44 Foot
45 Toe(s)
46 Multiple Parts
47 Other, NOC
50 Multiple Parts

60 Body Systems
61 Circulatory
62 Digestive
63 Excretory
64 Muscle
65 Nervous
66 Respiratory
66 Respiratory
67 Other, NOC

98 Body Parts NOC

99 Unclassified, Insufficient Data

NATURE OF INJURY (62-63)

01 Amputation, Enucleation
02 Bruise, Contusion, Crushing
03 Burn or Scald
04 Concussion, Brain, Cerebral
05 Cut, Laceration, Puncture
06 Dermatitis-Rash, Skin or Tissue Inflammation
07 Fatal
08 Fracture
09 Hearing Loss
10 Hernia
11 Multiple Injuries
12 Pneumoconiosis
13 Poisoning, Systematic (inhalation, ingestion and absorption)
14 Occupational Disease, NOC
15 Strain, Sprain
98 Other, Not Classified Above
99 Unclassified, Insufficient Data

AREA OR DEPARTMENT (64-68)

01 Cafeteria, Lunchroom
02 Elevators, Escalators
03 Exit & Entrance Ways
04 Exterior Parking Lot
05 Exterior Other
06 Maintenance
07 Office
08 Production (mfg. Only)
09 Rest Rooms
10 Shipping and Receiving
11 Stairways
12 Stockroom
13 Supermarkets-Other (Identify Specific Store Area)
14 Bakery
15 Check Out
16 Delicatessen
17 Drug
18 Frozen Food
19 Grocery
20 Meat
21 Produce
22 Sales Area, NOC
98 Other, Not Classified Above
99 Unclassified, Insufficient Data

Exhibit 11.4 Input document, insurance company *(back)*

Cause Code	Cause of Accident	Frequency		Claim Cost		Premium	Loss Ratio
		New	Old	New Claims	Up to Date		
	10 Workmen's Compensation						
10 01	Empl struck-injured by hand tool or machine in use		1		46		
10 02	Empl struck-injured by falling or flying object		6		584		
10 03	Empl struck by tipping sliding or rolling object		8		11,726		
10 11	Empl strain-injury lifting		2		202		
10 13	Empl strain-injury pushing or pulling		2		325		
10 17	Empl cut-scrape by hand tool-utensil-not powered	1	1	15	15		
10 19	Empl cut-scrape by object being lifted-handled		1		52		
10 21	Empl fell or slipped on same level		3		1,618		
10 22	Empl fell or slipped from different level	1	7	2,831	22,184		
10 23	Empl fell or slipped slipped-did not fall		1		42		
10 31	Striking against or step on object being handled		1		65		
10 32	Striking against or step on stepping on sharp object		2		78		
10 42	Caught in or between mechanical apparatus		1		513		
10 43	Caught in or between object handled-other object		5		325		
10 80	Empl injury foreign body in eye		8		602		
10 91	Empl injury vehicle accident-misc.		9		14,294		
	TOTAL	2	58	2,846	52,671	81,341	64.8%
	OPEN		4		32,412		
	30 Automobile liability						
30 25	Auto liab-intersection our unit straight across		5		136		
30 31	Auto liab-non intersection sideswipe collision	1	8	15,000	30,859		
30 32	Auto liab-non int hit parked or stnding veh		3		188		
30 40	Auto liab-rear end our unit hit other vehicle	1	4	350	4,559		
30 41	Auto liab-rear end other veh hit our unit		5		1,093		
30 42	Auto liab-backing backing or rolling back		6		706		
30 61	Auto liab-fixed object fixed object-overhead		7		2,792		
30 67	Auto liab tractor trailer unit jackknifed		1		13		
30 96	Auto liab object from truck	1	2		1,289		
30 99	Auto liab miscellaneous-unclassified	1	1	82	82		
	TOTAL	4	43	15,432	41,717	106,602	39.1%

CLAIM CAUSE ANALYSIS	PAGE 1	01-01	07-10
		CONTROL DATE	ANALYSIS DATE

Exhibit 11.5 Computer printout showing analysis.

other typical analyses are shown in Exhibits 11.7 and 11.8. Exhibit 11.7 is a breakdown by job classification. This information is perhaps a little more usable, allowing a look at types of jobs in which unusual frequencies or severities of accidents are being experienced.

Exhibit 11.8 breaks down the information by seriousness of accident or, rather, by cost of accident. It becomes obvious that there are some real problems connected with computer printouts dealing with dollars. In any claim involving a large amount, there is always a long delay before anyone has much of an idea as to the exact amount of the claim. In Exhibit 11.8, for instance, most of the information is based on "open" claims, which means that the dollar figures are only estimates of what the real costs will be. These estimates are often a far cry from the actual dollar figures and for the most part they are almost unusable in safety efforts.

Obviously, it is fairly easy to generate almost any kind of document if the information is already in the computer. And, as is also obvious, if the wrong information is in the computer, it is impossible to get usable information out. The systems which contain only Z16.2 information, and which therefore are unable to determine trends of causes, are a perfect example of this.

Another good example is the approach chosen years ago by some large insurance carriers to lighten the load of information to be coded and put into the computer. Only claims expected to cost over a certain amount are looked at and coded for insertion into the computer; all

	Cause Code	Employee Driver Claimant	Cause of Accident Injury Type	Present Valuation	O F
OPEN CLAIMS					
12-05-70M	10 91	Emp Bettes T A Driver	Empl Injury Vehicle Accident - Misc. Trailer Jackknifed-truck ran into embankment	3,500	*0
12-05-70T	10 91	Emp Bettes T A Driver	Empl Injury Vehicle Accident-Mis. Trailer Jackknifed-Truck ran into embankment	6,150	*0
68-18-70	50 15	Dvr Jones N. Univ Mach Co-Clmt	Transit Cargo Damage by loading or unloading Dropped generator while unloading	1,000 250	*0 D
12-11-70	50 02	Dvr Tugwell D US Goven-Clm	Transit Cargo Collision of our unit Dmg to plastic glass canopy on tractor tanks	450 250	*0 D
12-27-70	50 16	Dvr Loviette M T Bechtel Corp-Clm	Transit Cargo Improper packing Wooden skid broke-damaged compressor	555 250	*0 0
			TOTAL OPEN CLAIMS	11,655	
CLAIMS FINALIZED SINCE LAST REPORT					
06-27-70M	10 91	Emp McGee R L Truck driver	Empl Injury Vehicle accident-misc. Vehicle struck-back-neck injuries	187	F
06-27-70T	10 31	Emp McGee R L Truck Driver	Empl Injury Vehicle accident-misc. Vehicle struck-back-neck injuries	1,107	F
11-23-70	10 91	Emp Johnson H O Truck Driver	Empl injury Vehicle accident-misc. Slipped off truck-wrenched back-cut shin	0	F
11-29-70	10 03	Emp Thomas E J Truck Driver	Empl struck by tipping, sliding or rolling object Hit on chin by cheater pipe-cut lower lip	158	F
03-23-70	30 99	Dvr Spurlock Unknown clmt	Auto Liab Miscellaneous-unclassified Facts unknown-auto property dmg	82	F
09-08-70	30 31	Dvr Colley T A Johnson R Jr-Clmt	Auto Liab-non intersection sideswipe colllision Clmt attempted to pass-hit drivers vehicle	0	F
09-09-70	30 32	Dvr Bell H Mem Funeral home-clmt	Auto Liab-non int hit parked or stnding vehicle Ambulance knocked into side of dvr veh	0	F
12-09-70	30 25	Dvr Little R E Jr Pel Trkg Co-clmt	Auto liab-intersectionour unit straight across Dr crossing intersection-struck by clmt	136	F
06-25-70	50 02	Dvr Pawkett G Simmons Steel- clmt	Transit cargo collision of our unit Hit tree limb with load-dmg cab on crane	9,761 250	F D
11-18-70	50 02	Dvr Colley T A Western com-clmt	Transit Cargo collision of our unit Clmt alleges damages to top of oil well unit	3,730 250	F D
12-05-70	50 01	Dvr Barnes K D	Trasnit Cargo collision of carriers unit	2,500	F
12-11-70	50 02	Dvr Mason R D US Goven-clm	Transit Cargo collision of our unit Dmg to plastic glass canopy on tractor tank	132 250	F D
			Unlisted final claims	17,793	
			Total open and final	12,088	
			Total deductibles	2,250	

CLAIM COST CONTROL	Page 1	01-01-	07-10-
		CONTROL DATE	ANALYSIS DATE

Exhibit 11.6 Loss run.

		Actual Values					Computed values based on Proportional Distribution of unknown			
SUMMARY OF ANALYSIS BY JOB CLASSIFICATION	¢ ¢ ¢ ¢ ¢ ¢ ¢ ¢									
ANALYSIS PERIOD 10-01-69 TO 10-01-70										
FOR COVERAGE 10 AND 11 WORKMEN'S COMPENSATION							PAGE NO. 022			
	No. Claims	% Total Claims	% Total Value	Value	Average Value	No. Claims	% Total Claims	% Total Value	Value	Average Value
	0	.60	$0		$0					
Office manager (Const.)	1	.60	$133	.27	$133	1	.60	$133	.27	$133
Dist. Welder	2	1.19	$43	.09	$21	2	1.19	$43	.09	$21
Welder	2	1.19	$61	.12	$30	2	1.19	$61	.12	$30
Welder 4 year	1	.60	$10	.02	$10	1	.60	$10	.02	$10
Station Attendant	1	.60	$209	.42	$209	1	.60	$209	.42	$209
Main Operator	8	4.76	$10,637	21.29	$1,329	8	4.76	$10,637	21.29	$1,329
Asst Main Oper	5	2.98	$609	1.22	$121	5	2.98	$609	1.22	$121
Equip. oper.	5	2.98	$253	.51	$50	5	2.98	$253	.51	$50
Field Booster St. Op.	3	1.79	$143	.29	$47	3	1.79	$143	.29	$47
Main. man-repairman	43	25.60	$7,020	14.05	$163	43	25.60	$7,020	14.05	$163
Maint. man	31	18.45	$12,065	24.15	$389	31	18.45	$12,065	24.15	$389
Repairman, Roustarout, Util. help	5	2.98	$60	.12	$12	5	2.98	$60	.12	$12
Laborer	31	18.45	$3,213	6.43	$103	31	18.45	$3,213	6.43	$103
Cor con tech 1	1	.60	$11,000	22.02	$11,000	1	.60	$11,000	22.02	$11,000
Utility man 1	3	1.79	$182	.36	$60	3	1.79	$182	.36	$60
Handler	1	.60	$10	.02	$10	1	.60	$10	.02	$10
Welder 3 year	1	.60	$14	.03	$14	1	.60	$14	.03	$14
Machinist Sr.	1	.60	$2,440	4.88	$2,440	1	.60	$2,440	4.88	$2,440

Exhibit 11.7 Analysis by job classification.

CLAIM ANALYSIS BY SIZE RANGE   EXCLUDES DEDUCTIBLE

Experience By Size Range	Experience Period One Number Claims	% Of Total	01-01-69 01-01-70 Amount Claims	% Of Total	Experience Period Two Number Claims	% Of Totals	01-01-70 01-01-71 Amount Claims	% Of Total	Experience Period Three Amount Claims	% Of Total	01-01-71 07-10-71 Claims	Total
Zero Value Claims	724	10.6%		.0%	686	11.8%		.0%	133 2-OPEN	6.2%		.0%
1 TO 100	4,708 11-OPEN	69.3%	144,209 350-OPEN	5.0%	3,920 48-OPEN	67.6%	125,182 1,620-OPEN	5.4%	1,626 1,184-OPEN	76.0%	51,466 33,780-OPEN	7.5%
101 TO 250	523 2-OPEN	7.7%	82,815 383-OPEN	2.8%	462 5-OPEN	7.9%	72,287 1,016-OPEN	3.1%	138 30-OPEN	6.4%	21,502 4,827-OPEN	3.1%
251 TO 500	270 3-OPEN	3.9%	94,510 1,070-OPEN	3.2%	222 9-OPEN	3.8%	77,486 3,533-OPEN	3.3%	56 3-OPEN	2.6%	19,398 1,264-OPEN	2.8%
501 TO 1,000	186 4-OPEN	2.7%	131,537 3,120-OPEN	4.5%	95,103 8-OPEN	4.1%	19 6,038-OPEN	.8%	14,261 6-OPEN	2.0%	4,813-OPEN	
1,001 TO 2,500	181 22-OPEN	2.6%	286,257 44,674-OPEN	9.9%	205 65-OPEN	3.5%	342,202 130,304-OPEN	14.7%	118 101-OPEN	5.5%	235,196 208,844-OPEN	34.4%
2,501 TO 5,000	41 27-OPEN	1.1%	274,135 88,378-OPEN	9.5%	81 49-OPEN	1.4%	270,596 163,362-OPEN	11.6%	34 33-OPEN	1.5%	102,740 98,962-OPEN	15.0%
5,001 TO 10,000	60 28-OPEN	.8%	440,097 213,502-OPEN	15.2%	43 37-OPEN	.7%	315,810 274,826-OPEN	13.6%	5 5-OPEN	.2%	44,971 44,971-OPEN	6.5%
10,001 TO 25,000	41 31-OPEN	.6%	647,442 497,086-OPEN	22.5%	30 26-OPEN	.5%	472,033 417,720-OPEN	20.4%	8 8-OPEN	.3%	93,468 93,468-OPEN	13.6%
25,001 TO 50,000	9 9-OPEN	.1%	313,519 313,519-OPEN	10.9%	6 6-OPEN	.1%	219,732 219,732-OPEN	9.4%		.0%		.0%
OVER 50,001	5 5-OPEN	.0%	463,109 463,109-OPEN	16.0%	2 2-OPEN	.0%	323,809 323,809-OPEN	13.9%	1 1-OPEN	.0%	100,600 100,600-OPEN	14.7%
Incurred Total	6,798 142-OPEN	100.0%	2,877,630 1,625,191-OPEN	100.0%	5,791 255-OPEN	100.0%	2,314,240 1,541,960-OPEN	100.0%	2,138 1,373-OPEN	100.0%	683,602 591,529-OPEN	100.0%

Exhibit 11.8 Analysis by cost ranges.

others are ignored. While this is perhaps fine from the standpoint of time saved in the company's home office, it certainly has an adverse effect on the usability of the information that comes out. One carrier that adopted this approach initially has, on the basis of the information in its computer system, been publishing data for a number of years on accident "causes" by type of industry, needed controls, etc. The inference of these data sheets, or technical guides, is that the information is based upon a comprehensive analysis of all accidents. It is not; only a few selected accidents are analyzed—those few that a worker in the claims department thought, at first glance, would cost over a preselected dollar figure.

The companies usually do not divulge their cutoff point. Thus there is no way to know what is being analyzed. This, however, is probably not the biggest problem with such information.

The biggest problem is the validity and accuracy of the input. Accident reports received in the claims departments of insurance companies must be coded for insertion into the computer system. The only one place that this coding can be accurately done is at the place of business where the accident happened. If coding is not done there (and it usually is not), it must be done at the insurance company. In some way, someone at the insurance company must translate the description given on the accident report into the company's code. This person must interpret what is written in the report and somehow make it fit an available description that has a code number. This is the beginning of an almost infinite number of possibilities for interesting errors. The person who is doing the actual coding holds the most important key to controlling error. If this person is familiar with the company in which the report originated, the chances of error are lessened, especially if he or she has a technical background. In most cases, however, the coding is done by an inexperienced person in the claims department, usually with no quality control check. Thus at this point it often becomes impossible to ensure accuracy. Some carriers have attempted to remedy this situation by having field safety engineers do the coding, but this has seldom worked well. One company that prides itself on coding integrity because it utilizes field personnel to code found in a recent year that over 40 percent of all claims that were to be coded came in uncoded from the field and had to be coded by a clerk at the home office, using his or her best judgment. And no attempt was made to determine the accuracy of the 60 percent of the claims that were coded in the field by checking back with the customer.

If coding is done at the company generating the source document (first report of injury), there is some chance for input accuracy. If coding is done by the insurance company, there is almost no chance. Thus when embarking on a computer analysis system and working with the insurance carrier, retain the coding element; otherwise, the end result will be well-analyzed but unusable information. If there is an answer to the input-accuracy problem, it lies in (1) getting a source document that will provide answers that are detailed enough to enable coding the information and (2) having the coding under some kind of tight quality control.

The U.S. Department of the Interior System

Exhibit 11.9 shows an input document that is different from the traditional ones. This United States Department of Interior document was devised and implemented by Bill Pope. It has some rather special advantages over traditional input documents. First, it is based on the belief that accidents are symptoms of management problems and operational errors, and it asks the supervisor to look for and record these things, rather than merely the circumstances surrounding accidents. Second, coding is done by the supervisor rather than by someone far removed from what actually happened.

The key to this approach is the analysis made on the back of the form in terms of the identification of the management problem; the personnel problem; or the problem of property, equipment, and the environment. The codes used for these sections are shown in Exhibit 11.10.

A flow diagram of the Department of the Interior's safety management information system is shown in Exhibit 11.11. This is only one of such systems; it is presented in this chapter as an example of a system which is not based entirely on Z16.2 and which attempts to deal with the input-accuracy problem.

Computers have a very real place in our future safety management efforts. However, some philosophical questions (what information to keep) and some system questions (how to get accurate information) will have to be answered. If this can be accomplished, the computer will become a valuable safety tool.

A somewhat similar system was described in a recent issue of *Professional Safety* by Guy Fragala:

> When a computerized system is developed for creation of a data base, what is referred to as a dictionary must be formulated. This dictionary contains the variables to be included in the system, and the specific names of the elements contained in the individual variable categories. The variables which have been included in the dictionary for this data base being discussed are shown in Exhibit 11.12.

	Field Report No.
UNITED STATES DEPARTMENT OF INTERIOR (For Safety Management Use Only) SUPERVISOR'S REPORT OF ACCIDENT	DATE OF THIS REPORT

Refer to DI-134A for Instructions, Definitions and Standard Coding Details

Section A. Identity *(Supervisor to Complete)*

(1) Organizational Unit *(Area, Region, etc.)*

Last No. Here

(2) Reporting Station *(Name & Address)*

Last No. Here

(3) State in which Accident Occurred

(4) Mo. _____ Day _____ Year _____

(5) Nearest Hour of Accident

_____ a.m. _____ p.m.

(6) Name *(Employee, visitor or other involved)*

Last _____ First _____

Soc. Sec. No.

Use separate form for each employee involved

(7) Employment Status *(Circle one)*
1 Permanent*   5 Contractor   9 Public *(other)*
2 Y.O.C.   6 Vista   0 Other *(specify)*
3 Temporary   7 Job Corpsman
4 Emergency   Public *(visitor)* _____
*Include Job Corps Staff

(8A) CSC Occupational Code   (8B) Work Environment

Last No. Here

(9) Result of Accident *(circle one)*

01 PI *(personal injury)*   07 PF *(fire)*
02 PI with PF *(fire)*   08 PM *(motor vehicle)*
03 PI with PM *(motor veh.)*   09 PB *(boat)*
04 PI with PB *(boat)*   10 PA *(aircraft)*
05 PI with PA *(aircraft)*   11 PO *(all other property*
06 PI with PO *(all other)*   *damage)*

(10A) Identification of Property Damage *(if any). (Give name, model, number, size, make, type, etc.)*

Employee operated:

"Other" operated:

(10B) Make of Govt. Vehicle _____

(11) Property Ownership *(Circle one)*

                          5 Employee-owned on O.B.
0 No. prop. involved   6 Inter-agency (GSA) motor
1 Interior owned   pool
2 Other Federal   7 Interior Leased
3 Contractor   8 Privately owned
4 Concession   9 Other *(Explain on No. 18)*

(12A) Age:
        a. Of Employee *(year of birth)* _____
(12B) Year of Mfg. _____

(13) Is Tort Claim Expected? Circle one: Yes or No
If "Yes", has Tort Claim
Officer been contacted? Circle one: Yes or No

Section B. Medical *(Supervisor to Complete)*

(14) Nature of Injury

(15) Nature of Injury

(16A) Severity of Injury *(Circle one)*
0 No injury involved
1 First aid attention only
2 Medical attention only
3 Disabling injury *(fatal)*
4 Disabling injury *(temporary)*
5 Disabling injury *(permanent)*

(16B) Z16 Compliance
_____ Chargeable
_____ Not Chargeable

(16C) BEC Forms Prepared
Circle: Yes   No

a. Leave date: Mo. _____ Day _____ Year _____
b. Return date: Mo. _____ Day _____ Year _____
c. Death date: Mo. _____ Day _____ Year _____

d. No. days lost

Section C. Story of Accident *(By Supervisor or Employer)*

(18) Narrative: Include: Who, What, When, Where, and How
Facts are important—fault finding not.

NOTE: Condense Story Here

Section D. Supervisory Opinion *(Supervisor to Complete)*

(19) I think accident might have been prevented if:

Employee Fault Finding Adds Nothing to Management Improvement.

(20) I suggest the following policy or procedure change by mgmt. to help prevent similar accidents:

Signature and Title of Reporting Supervisor

Exhibit 11.9 Source document from the Department of the Interior.

ANALYSIS OF MANAGEMENT PROBLEM

FORM DI-134(*back*) If the title you need is not in DI-134A, tell your safety officer so that it may be added. Information on this side represents a "team" effort by safety and other functional managers to assist the Line Officer to discover underlying causes of accidents and to plan for their ultimate correction. All items require an entry. Follow bureau instructions for analyzing the problems on this side.

Section E. Line Management Problem (*Operations, Design, Construction, Maintenance, Plant Mgmt., etc.*)

(21)	Type of Accident (*Event*)	(22)	What was used, done, contacted (*Source*)
(23A)	Human error (*first selection*)	(23B)	Human Error (*Second selection*)
(24A)	Condition Defect (*first selection*)	(24B)	Condition Defect (*second selection*)
(25)	Review of the Mgmt. problem cited in sections D and E		

Sig. _____
Reviewing Mgmt. Official

SECTIONS F, G, H, I, and J to be completed by responsible mgmt. analysis identifying problems in the system to be resolved that will reduce accident loss. Keep remarks brief. Use coded information. Leave no blanks. In each section, consultations with supervisor and appropriate mgmt. official is desired.

Section F. Personnel Problem (*Consultation with Supervisor and Personnel Official is Desired*)

(26)	Supervisory Control and Training	(27)	Fitness for duty evaluation
(28)	Opinion of the Mgmt. problem related to personnel services		

Sig. _____
Reviewing Personnel Official

Section G. Property/Equipment/Environmental Problem (*Consultation with Engineering and Property Official Desired*)

(29)	Maintenance and Environmental Control	(30)	Fitness-for-use Evaluation
(31)	Opinion of the Mgmt. problem related to property services		

To be repaired: Yes or No Est. Cost $_____
To be replaced: Circle: Yes or
                                No

Sig. _____
Reviewing Engineering or property official

Section H. Finance Problem (*Consultation with Administrative and Finance Official Desired*)

(32) Amount of Property Loss	(33) Amount of Tort Claim Award
To Govt. Prop. $_____ To "Other" Prop. $_____	To Government $_____ To "Other" Party $_____
(34) Opinion of Mgmt. problem related to financial services	

Sig. _____
Reviewing Finance Official

Section I. Legal Problem (*Consultation with Tort Claim or Legal Official Desired*)

(35)	Opinion of public safety problem	(36) Possibility of recovering from a 3rd party? (*CheckOne*) ___Yes ___No *If "No," explain why in (37)*
(37)	Opinion of the cause, not related to a government employee or operation	

Sig. _____
Reviewing Tort Claim or Legal Official

Section J. Corrective Action (*Taken to make less probable the recurrence of this accident*)

(38) Local corrective action taken or planned:
                      When: Now _____ Fiscal Year _____

Sig. _____   _____
Mgmt. Official Taking Action   Title

(39) Recommended Bureau or Department Action to Assist in solving identified problems:*

*A Bureau response is expected if request for action is made here.

Signature of Reviewing Safety Officer	Date	Signature of Reviewing Authority	Date	Initial of Bureau Safety Officer	Date

Exhibit 11.9 Continued.

OPINION OF THE MANAGEMENT PROBLEM RELATED TO PERSONNEL SERVICES

Definition: Identifies a specific personnel service that may be related to a cause of accident.

Many human errors are products of how management plans for, receives, develops, and utilizes employees. Work errors can occur because such services need revision to upgrade their utility to the line supervisor.

If there is any reason to connect the personnel services listed below to a cause of accident under investigation, it should be checked off to provide a bank of information that will guide the personnel function.

Code	Personnel Service	Comment
00	No cause factor related to a personnel service	
01	Task being performed unrelated to that for which he was hired.	Regardless of "other duties" assigned
02	Questionable physical ability for work done	See item (27), performance
03	Recruitment problem	Unfamiliar with locale assigned
04	Labor-management dispute involved	Involved or behind accident
05	Compensation too low for demands of more skill	Imbalance of local wage scale producing accidents.
06	Employee relations problem	Underlying cause of accident
07	Lack of or insufficient training	See item (27)
08	Injury compensation trouble	Problem with handling claims
09	MV licensing problem	SF 46 not issued, not in possession
10	Physical handicap	Placement problem, policy
11	Problem related to classification or qualification	Hazardwoud work identifcation
12	Employee or student conduct involved	Disciplinary action arising out of accident
13	Health services needed locally, or insufficient for needs.	First aid facilities needed, enough to support program.
14	Disability retirement involved	Should be considered for this employee
15	Physical examination problem	Employee's phys. cond. to be checked out
16	Dual employment	Cause related to 2nd job, overwork, etc.
17	Check National Driver Register*	Countrywide history of poor driving*
18	Check SMIS driver accident records*	Interior history of poor driving*
19	Local staffing problem	Need more/better qual. employees
20	Hazard duty pay may be considered*	Research to determine the need in SMIS*
21	Communications problem	Between levels of management or between supervisor and employees
22		
23		

*If these codes are entered - be sure to repeat request in Item (39). The Safety Management Information System (SMIS) will be asked for a special printout. It will be sent, through channels, to the originating field station.

Exhibit 11.10 U.S. Department of Interior code.

OPINION OF THE MANAGEMENT PROBLEM RELATED TO PROPERTY EQUIPMENT/
ENVIRONMENTAL SERVICES

Definition: Identifies a specific property service related to a cuse of accident.

Many condition defects are products of how things are purchased, stored, maintained and used. Defects of building structures, machines, materials, tools and the like will produce accidents. Their prompt removal or repair improves the supervisor's chances to operate free from error.

If there is any reason to connect the property services listed below to the causes of the accident under investigation, please check it off in the report.

Code	Propery service	Comment
00	No cause factor related to any property service can be found.	A property official may be most qualified for this decision.
01	Design, specifications, operation changes needed	Unsafe, guards missing, etc.
02	Storage, handling and issue	Bin, bulk, coal and fuel, etc.
03	Building operat., maintenance, and grounds	Substandard bldg. mgmt.
04	Equipment operation, tools, etc.	Repair and maintenance problems
05	Substandard protection program	Fire, personal clothing, mechanical safeguards, etc.
06	Motor Pool Management (GSA) problem	Vehicle condition when loaned
07	Motor vehicle management (interior)	Official use, defense of suits
08	Purchasing/specifications/safety	Compliance with safety codes
09	Negotiated contracts	Safety clause—"hold safe"
10	Policy, regulations not enforced	
11	Periodic test, inspection needed	Hoists, boilers, elevators, etc.
12	Property disposal problem	Disposal problems—housekeeping
13	Acquisition of excess property problem	Method of acquisition, safety of
14	Space management problem	
15	Board of survey action	Liability for damage problem
16	A non Federal maintenance problem	Icy conditions, holes in sidewalk, etc. not under Federal maintenance system. Correction due elsewhere.
17	Check SMIS—may need lift device*	
18	Check SMIS—how many other cases*	
19		See item (37) cause unrelated to Govt. employee/operation.
20		
21	Inadequate fire protection facilities	Lack of water, sprinklers, fire escapes, fire dept. alarm, etc.
22	Inadequate fire prevention activities	Fire inspections, fire drills, fire plan, etc.
23	Need moving equipment	Dolly, cart, etc. to reduce handling
24	A contractor's safety problem	Problem not under control of bureau

Exhibit 11.10 Continued.

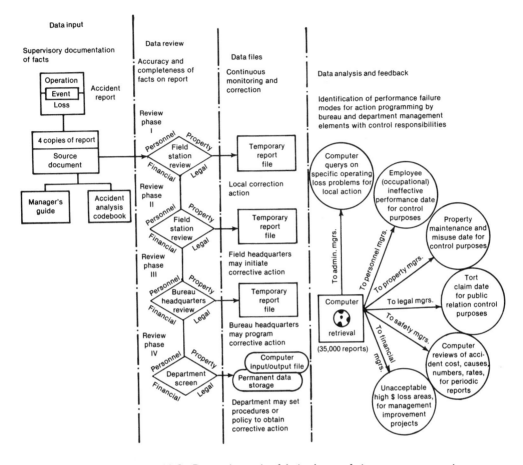

Exhibit 11.11 U.S. Department of Interior safety management information system.

It is expected that many injury data base systems in existence would contain most of these variables. There is, however, a unique concept contained in this particular data base. This unique feature is a set of four variables which comprises a new injury sequence.

This sequence is modeled after that of H. W. Heinrich who presented his domino theory, or accident sequence, in the early years of accident prevention. This new sequence is consistent with that of Heinrich's, up until the unsafe act and unsafe condition. In Heinrich's sequence, unsafe acts and unsafe conditions were said to result from a personal fault or fault of the individual.

If properly investigated, each injury occurrence will readily display a sequence of events which are adaptable to this new injury sequence. A diagram of this new injury sequence is shown in Exhibit 11.13.

Application of this new sequence is demonstrated using the hypothetical incident below:

A machine operator reached into the point of operation of a punch press to remove a piece of work which did not automatically eject. He had never been properly instructed regarding the procedure for removing a piece not ejected. The

1. Date of Injury Occurrence
2. File Number
3. Sex of injured
4. Age of injured
5. Marital Status
6. Occupation
7. Length of Service
8. Department Category
9. Specific Department
10. Location of injury
11. Time of injury Occurrence
12. Day of the Week of injury
13. Accident injury
14. Injury Type
15. Treatment Obtained
16. Whether or not injury is loss time
17. Time Lost
18. Has a compensation claim been filed.
19. Unsafe Act
20. Unsafe Condition
21. System Fault
22. Date Report Filed
23. Person Filing Report
24. Date of Entry into the System
25. Person inputting report into system.

Exhibit 11.12 Database variables.

machine repeated and his hand became caught in the point of operation. The result, his hand was crushed.

Analyzing the above incident yields the following sequence components:

Injury—Crushed hand

Accident—Caught in, on or between

Unsafe Acts—Reached into the point of operation

Unsafe Conditions—Machine repeated; Operator able to reach into point of operation.

System Faults—Lack of Proper Instruction; Poor preventive maintenance; Poor machine design, allowing operator into point of operation.

Along with the other data base variables, the above information would be entered into the computerized system. The data base dictionary created contains coded classification for categories included in the sequence elements. Once information has been entered into the system, the various modes of data retrieval available to the user make possible a great deal of useful information.

A basic form of data retrieval is the cross referencing of two variables. For example, an organization could retrieve data on system faults and cross reference this information with any of the other variables in the system.

A specific example is cross referencing individual departments with system fault types. This analysis might indicate that a particular department is experiencing 60% of their injuries due to lack of proper instruction. The obvious corrective action would be to strengthen that department's training efforts.

Another example might be to cross reference system fault types with injury locations. This analysis might indicate that in a particular location malfunctions with machinery are occurring which are resulting in injuries. As a result of this analysis, recommendations for improved preventive maintenance for particular locations may be made.

As stated previously, the computerized system being discussed was used to create a data base containing three years of injury reports from an institutional setting. From the data base created, a number of cross reference matrices were generated for analytical purposes. With the variables selected, 84 cross reference tables are possible; 51 of these were generated and investigated. An illustration of the set of cross references matrices generated for the analytical work conducted as shown in Exhibit 11.14.

COST CONTAINMENT

While not actually a part of the recordkeeping process, there are some things that can be done that will help a great deal in lowering the cost of the incident that has occurred.

The need for cost containment is shown by the following from Joseph Califano (1985):

The statistics regarding health care costs are shocking:

- This year, for the first time in our history, Americans began spending more than $1 billion a day on health care.
- Health care costs rose from $41.7 billion in 1965 to $355 billion in 1983—an increase of 751 per cent.
- Health costs jumped from $13.9 billion in 1965 to $150 billion in 1983—an increase of 979 per cent.
- Physicians' fees increased from $8.5 billion in 1965 to $68.1 billion—an increase of 700 per cent.
- During that period, the Consumer Price Index rose—but only by 242 per cent

Health care is still the most inflationary sector of the economy. During 1983, the costs of medical care rose at a 10 per cent rate, more than triple the 3.2 per

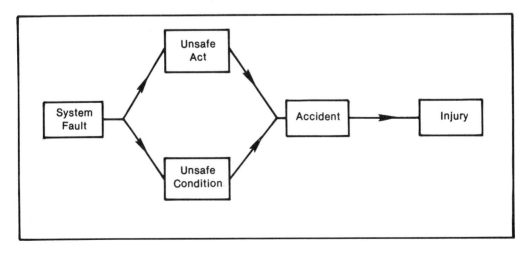

Exhibit 11.13 New injury sequence.

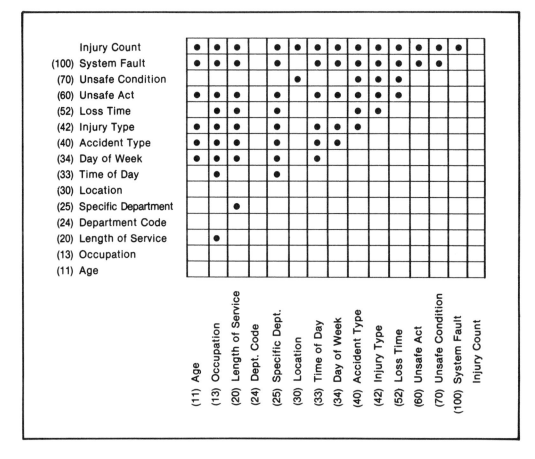

Exhibit 11.14 Cross reference matrices.

cent increase in the overall Consumer Price Index. The daily cost of a hospital room rose 12.2 per cent, to an average of almost $400 per day. The 1983 bill of $355 billion was a levy of almost $1,500 on every man, woman, and child in America.

First, hospitals have generally been reimbursed on a cost, or in the case of for-profit hospitals, a cost-plus basis. Doctors are paid on a fee-for-service basis. Thus, the more hospitals have spent, the more money they have received; the more services the doctors perform, the more money they make.

The Chrysler Story For the past two years I have been serving as head of a special committee on health care of the Chrysler Board of Directors created by Chairman Lee Iacocca. This is the only committee of its kind in American business.

At Chrysler, as we fought for survival, we had to address the cost of health care.

It has not been an easy task. During 1984, Chrysler's health care costs exceeded $400 million, making the Blue Cross/Blue Shield Chrysler's single largest supplier. That's more than $1.1 million each day. This past year Chrysler's total health care bill (which included Chrysler's Medicare payroll tax and a portion of the health insurance premiums of its suppliers) will exceed $550 for each car we sell. That's down somewhat from $600 a car last year—not because inflation in health care costs has abated, but because we are selling more cars. This year Chrysler must sell about 70,000 vehicles just to pay for its health care bills.

Excessive health care costs are eroding American's ability to compete with foreign companies. Mitsubishi Motor Corporation, a Japanese car manufacturer in which Chrysler has an investment, spends only $815 a year for an employee's health care costs while each employee pays approximately $374. Unlike Chrysler, Mitsubishi has no direct cost for retirees or their surviving spouses, because of Japan's national health coverage. Chrysler's comparable cost per active employee is $5,700–400 per cent higher.

That gap may well increase. The Japanese government is moving aggressively to control health care utilization by seeking a law to require a substantial co-payment for employees, beginning at 10 per cent and rising to 20 per cent.

What does Chrysler get for its health care dollar?

A health care industry that is expensive, wasteful, and inefficient. Here are a few examples of what we are discovering as we analyze our own health care plan in depth:

- Among the nation's Medicare recipients, a very common medical procedure is cataract surgery—lens extraction and implant. The procedure takes about 20 minutes, and rarely requires a general anesthetic.

  The average ophthalmologist charge for this procedure in the Detroit area is about $2,000.

  If a doctor performed three of these procedures a day, four days a week, 42 weeks a year, he would earn more than $1 million, for less than 200 hours of actual surgery, and have a 10-week vacation to boot.

  Compare this with the typical charge of $1,500 for serious abdominal surgery lasting four to five hours.
- We asked some doctors to investigate eight Detroit area hospitals with extraordinarily high percentages of non-surgical admissions for low back problems.

  This study showed that two-thirds of the hospitalizations—and 2,264 out of 2,677 of the total hospital days—approximately 85 per cent—were inappropriate.

  With respect to three of the hospitals audited, none of the admissions were found to be appropriate.

  In more than 60 per cent of the cases, patients were subjected to electromyograms—an invasive and expensive procedure that is necessary only if surgery has already been clinically indicated. All the test results were normal.

  Had the inappropriate admissions not occurred. Chrysler would have saved approximately $1 million.

  We have no reason to believe that Chrysler's experience is unique. Similar waste and inefficiency exist in almost every health benefit program in this country. Chrysler's preliminary investigation suggests that as much as 25 per cent of its hospital costs may be due to waste and inefficiency. For Chrysler, elimination of those costs would save almost $50 million in 1984.

THINGS TO DO AHEAD OF TIME

There are a number of things that can be put into place ahead of time that will help immeasurability in containing costs. Here are a few:

- Have one person oversee the Workers' Compensation program.
- Evaluate the facility for "light duty" productive work areas and jobs. This identification becomes crucial later.

— Carefully select a company physician. There are doctors that:

   — Charge a lot or not so much.
   — Are injured employee centered or company centered.
   — Are knowledgable or not about Workers' Compensation.
   — Are knowledgable or not about your plant.
   — Know or don't know about the real world.

Here are some tips on appropriate selection:

   — Consult with other employers in the vicinity as to their recommendations on which physicians are most helpful and most experienced with occupational injuries and illnesses.
   — Communicate with your Workers' Compensation insurance company. They are also able to provide assistance and recommendations for selection of qualified and capable doctors.
   — Select physicians, surgeons and specialists who do Workers' Compensation work willingly and will abide by a fee schedule if there is one established.

The selection process requires:

   — Meeting with and interviewing the doctor.
   — Making sure the doctor tours and understands the facility, process, and culture of the company. In the physician-company relationship, you are the customer and the physician is the salesperson who must meet your needs.
   — Carefully selecting an insurance company/insurance claim adjustment service.

If insured, look carefully at the claim department policies and abilities before deciding on the carrier. This is crucial in cost containment efforts. If your claim representative (whether you are insured or self insured) uses an open check book (never fights) your are heading for trouble. If you use an adjustment service (are self insured) and you don't know what's going on or have no say in what's going on, you're in trouble also.

In short; you must be in control.

THINGS TO DO AFTER

After the event has occurred, here are some actions that can help:

   — Ensure immediate qualified care. If the supervisor is required to take the employee for care, it shows care and it maintains some control. If the employee is brought to the company doctor or clinic, considerable money is saved. Emergency rooms tend to cost four to eight times more.
   — Ensure the accident investigation looks for third party recovery possibilities. If the injury is the result of design defect or negligence of a third-party, the losses can be recovered.
   — Maintain periodic contact with the claim person on the case. Claim people respond to those customers who are interested. If an employee is off three weeks, and no target

Employee's Name	Date of Injury

EMPLOYEE DISABILITY LOG
(Follow the steps below to obtain needed information when
an employee has sustained an **On-the-Job Injury**)

WEEK #1
- Report accident/injury immediately.
- Contact Employee.
  - ...Date of doctor's appointment:
  - ...Doctor's Name:
  - ...Date released to return to work: ...Date of actual return to work:
- Contact Employee's Doctor.
  - ...Date examined:
  - ...When can employee return to regular job/alternate productive work?
  - ...Describe injury:
  - ...Contact insurance claims representative. Exchange information.

#2
- Contact employee. Date contacted:
- Alternate productive work available?
- Return to work target date:

#3
- Contact Employee. Date contacted:
- Contact doctor:
  - ...Return to work date:

#4
- Contact employee. Date contacted:
- Contact insurance claims representative:
  - ...If no target return-to-work date, insurance claims rep./plant management should schedule an independent exam.

#5
- Contact employee. Date contacted:
  - ...Advise of and instruct employee to keep independent exam.
  - ...Date employee last saw doctor:

#6
- Contact employee. Date contacted:
- Has employee been evaluated by our doctor for return to work?

#7
- Results of Exam:
- Contact employee:
  - ...Confirm return to work date by phone and letter to employee and insurer.

#8 — Second Eight Weeks Continue with this Weekly Follow-Up For:
- If employee hasn't returned to work, contact employee. Date contacted:
- Contact insurance claims representative regarding status of claim.
- Return to regular job tasks, or:
- Return to alternate productive work, or:
- Examination with our doctor for return to regular or alternate productive work or:
- At such time that it is obvious that employee absolutely cannot return to work at your facility, work closely with the insurance claims representative and consumer loss control department to evaluate other alternatives:
  e.g. Job placement outside Pillsbury; retraining for job placement.

Employee Injury Follow-Up is a Location Management Responsibility

Exhibit 11.15 Employee disability log.

return-to-work date has been provided, instruct the insurance company to set up, *immediately,* a second opinion examination with a doctor of your choice. Purpose of exam: Obtain a return-to-work target date.
- Use a light duty program.
- Use a disability log (see Exhibit 11.15).

REFERENCES

Califano, J. "Controlling Health Care Costs." *National Safety News,* Chicago, January, 1985.
Fragala, G. "A Modern Approach to Injury Recordkeeping," *Professional Safety,* January, 1983.
Pope, W. C. "In Case of Accidents, Call the Computer," in *Selected Readings in Safety,* Macon, GA: Academy Press, 1973.

Chapter Twelve

Statistical Safety Control (SCC)

Safety professionals have used statistics and statistical concepts such as safety sampling, safe-T-scores, and control charts, for many years but these and other techniques have never become an integral part of most safety systems.

In recent years the concepts of statistical process control (SPC) have become popular in American management. Actually statistical concepts were used widely in the 1940s and 1950s in quality control but were never used much beyond that application until the 1980s. Today SPC is used in all areas of business for problem identification, problem solving, and for almost everything. It is a way of managing. Its use in safety, however, is almost non-existent.

This chapter looks at the uses of SPC in safety. First some background on SPC from James C. Siegel of Ford Motor Company:

> In this regard, some managers have rediscovered a fundamental building block needed in their management systems that can create "meaningful change": The use of statistical methods as a basic tool to improve both quality and productivity on an on-going basis.
>
> In a way, this rediscovery is painful, because American industry discarded statistical methods in the 1950's and they subsequently became one of the primary building blocks of industrial progress in Japan.
>
> The use of statistical methods is not an all-purpose remedy for every corporate problem. But it is a rational, logical, and organized way to create a system that can assure continuing, ongoing improvements in quality and productivity simultaneously.

Today we add safety also.

As Japanese industries grew in size and importance in world markets during the 1970s, it became apparent that the Japanese had developed a cost and product quality advantage for certain products vis-a-vis American companies, many of which were well-entrenched and seemingly invincible. A great deal of interest developed to determine how the Japanese had become so successful.

American businesspeople, business journalists, and educators rushed to Japan in great numbers in order to discover the "secrets" of the "sudden" Japanese success. Through their findings there emerged a pattern of business practices that were consistently present among most successful Japanese companies. Exhibit 12.1 contains a listing of some of the more important factors. Although this list is neither all-inclusive nor in priority order, it does highlight a general business philosophy of long-term commitment toward specific goals (particularly quality goals) and the elimination of waste.

The importance of the use of statistical techniques did not emerge as a significant factor early in these studies because of its fundamental inclusion in many of the other practices. For instance, the success of Japanese quality circles is due in large part to their methodical use of

269

- Management Commitment to Quality
- "Just in Time" Inventory System
- Schedule Stability
- Automation
- Tight Plant Layouts
- Human Resource Management
- Extensive Training
- Quality Circles
- Statistical Techniques

Exhibit 12.1 Japanese business practices.

effective problem-solving tools, particularly statistical techniques. It has been only recently that the pervasive role of the statistical approach in Japan has been understood and appreciated.

The American statistician who introduced these techniques to the Japanese was Dr. W. Edwards Deming. The crucial factor in Dr. Deming's success in Japan was that the executives and managers who attended his lectures responded to his statistical concepts. They built their manufacturing systems on a foundation of statistical quality control and began the cycle of continual improvement that has brought them to their present dominant position in many world markets.

ROLE OF THE STATISTICAL APPROACH

The statistical approach, as advocated by Dr. Deming, fits well with the changing management control concept in many American companies, from a concept of "detection" to one of "prevention."

Exhibit 12.2 is a schematic representing the control method typically used in many American companies. This method, which might be termed "detection," relies primarily upon some type of after-the-fact inspection to separate acceptable product from that which is unacceptable. Based on information about unacceptable product, the process (or system) is adjusted. For a manufacturing process, this method of process (or quality) control is often performed by a quality control organization. However, this schematic could apply to non-manufacturing situations as well, from banking to sales.

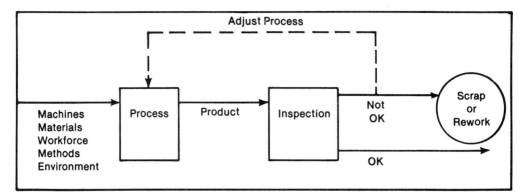

Exhibit 12.2 "Detection" control concept.

The drawback associated with "detection" is that unacceptable product must be produced before people can determine how to adjust the process. Obviously, this wastes resources, for it costs just as much to produce an unacceptable product as an acceptable one, yet the unacceptable product must still be either reworked or scrapped. The ideal situation would involve being able to monitor (and if necessary, adjust) the process on a periodic, "real-time" basis in order to minimize the possibility of producing unacceptable products. This approach might be called "prevention."

Exhibit 12.3 is a schematic representation of the "prevention" approach. In some situations it still might be necessary to have end-item inspection. But the key difference here is *periodic, selective* measurement just as product is produced, measurement of either the product itself or the process. In manufacturing situations, for example, this might be the diameter of a hole or the temperature of a plating bath. Based on these "real-time" measurements the process can be monitored and adjusted as necessary.

With that short discussion on SPC uses in quality, it becomes obvious that there is an application in safety. If we can (and we can) monitor before the fact of the accident, and adjust the process (or the behavior) before someone is hurt, we are much ahead in every way.

Statistical techniques are among the best tools for evaluating these selective measurements. They provide a method for logically and systematically evaluating information. Specifically, they help determine process stability, the ability to consistently meet consumer requirements, and the causes of accidents and other problems (should they arise).

It is important to understand that selective in-process measurement is *not* a substitute for inspection—its function is not to separate unacceptable from acceptable products. Rather, it is process (or system) control, for the purpose of *immediate* evaluation, interpretation, and appropriate process action.

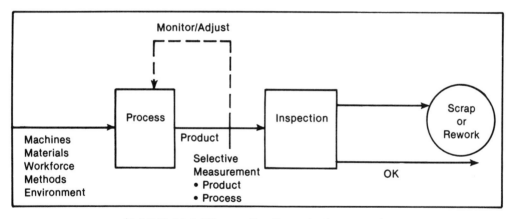

Exhibit 12.3 "Prevention" control concept.

THE SPC TOOLS

Here are the basic tools to use in statistical process control:

— *Pareto Chart*—A bar graph of identified causes shown in descending order of magnitude or frequency. For example, in the evaluation of paint defects on a motor vehicle, the first bar might represent the percentage due to dirt-in-paint, the second bar due to runs/sags, the third bar due to chips/scratches, etc. (See Exhibit 12.4.)

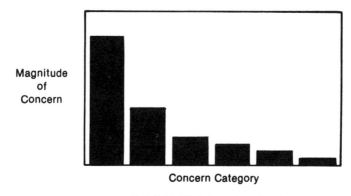

Exhibit 12.4 Pareto chart.

— *Fishbone Chart*—A cause and effect diagram for analyzing problems and the factors that
contribute to them. For example, the problem might be variation in the output of a
fabric weaving process; branches of the diagram could relate to categories of possible contributing factors (machines, materials, methods, environment, workforce). (See Exhibit
12.5.)

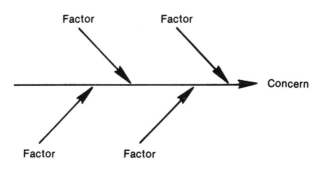

Exhibit 12.5 Fishbone chart.

— *Histogram*—A bar graph displaying a frequency distribution. For instance, this might be
used to categorize pistons according to diameter after the final machining and plating
operations, with the X-axis being divided into diameter measurement increments. (See
Exhibit 12.6.)

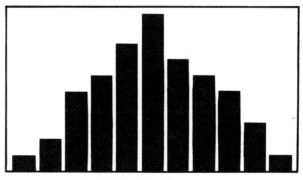

Exhibit 12.6 Histogram.

— *Scatter Diagram*—A graph displaying the correlation of two characteristics. For example, a scatter diagram might be used to compare the tensile strength of a wire versus its diameter. (See Exhibit 12.7.)

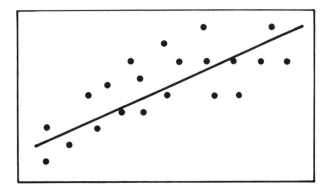

Exhibit 12.7 Scatter diagram.

— *Control Chart*—A method of monitoring the output of a process or system, through the sample measurement of a selected characteristic and the analysis of its performance over time. (See Exhibit 12.8.)

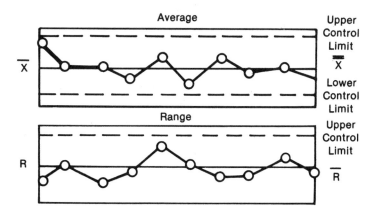

Exhibit 12.8 Control chart.

Of the analytical tools described above, the control chart is a particularly useful and effective technique. However, because it is generally not well understood by many American managers, the control chart is explained in more detail below.

— *Statistical Control Charts*—In monitoring, the control chart is the working tool of statistical control. On it the observed accident rates are plotted against time, with the overall accident rate or mean for the entire period. Finally, upper and lower control limits are computed such that the probability of an accident rate exceeding the limits by chance alone is very the probability of an accident rate exceeding the limits by chance alone is very small. The control chart is based upon the *bell-shaped curve,* which is applicable to many things, such as height and weight of people, frequency of accidents, or number of drinks consumed in a given period of time.

As an example, using the number of drinks consumed last night by a group at a lengthy cocktail party (see Exhibit 12.9), the average was five drinks per person. A few people had only one drink each (pretty rare), a few people had as many as ten drinks (also rare), and most people had between three and seven drinks.

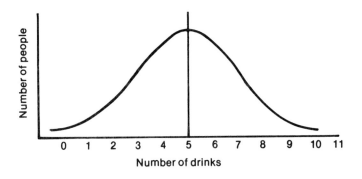

Exhibit 12.9 Normal distribution curve.

Now apply the chart in Exhibit 12.9 to the accident frequency rate of the plant (see Exhibit 12.10). The average in this case is a frequency rate of 3.0.

It is important to put limits on the chart—limits to stay within the future—that is, an *upper control limit* and a *lower control limit* (see Exhibit 12.11). Those limits are such that if any month's record falls outside them, it is not strictly because of chance—something is wrong.

If the limits are set so that there is only a 1 percent possibility that a month's record will lie outside those limits by chance, it would look like the shaded area in Exhibit 12.11. The months are plotted along the bottom (see Exhibit 12.12). Each period (biweekly) is plotted.

From a study of statistical control chart data it can be determined whether the system is a relatively constant one. If so, it is a stable situation. Conversely, if an accident rate exceeds the

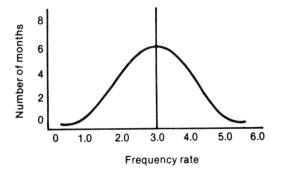

Exhibit 12.10 Normal distribution curve.

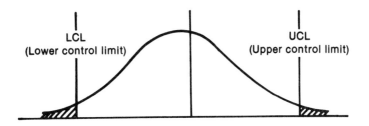

Exhibit 12.11 Control limits.

upper limit, it signals a change for which there is an assignable cause. Similarly, when a point falls below the lower limit, there has been a significant change for the better.

Statistical control techniques offer a means for making the work of accident reduction more effective and efficient. They cannot assign cause, but they can point out where and when to look for causes.

Preparation of Control Charts The basic data needed for an accident control chart are the number of accidents and the number of person-hours worked for successive periods. If these data are not available from company records, they will have to be collected on a weekly, bi-weekly, or monthly basis.

The ratio of accidents to person-hours for any given period will give the rate for that period. Customarily, 20 such rates should be available before a control chart is drawn up, although there are occasions when a smaller number are used. When these rates, together with the mean accident rate for the total period and upper and lower control limits, are plotted, the resultant chart might resemble the one shown in Exhibit 12.13.

One percent control limits were used in this instance. This gives a 0.005 probability of the accident rate falling above the upper limit and a like probability of it falling below the lower limit when the accident rate is stable. If the true accident rate or mean is identical with the estimated rate, the probability of observing a point out of control is 1 percent. (The greater the number of person-hours covered, that is, the greater the length of time over which data are collected, the more closely the estimated rate will approximate the true rate.) Under these conditions only 1 rate in 100 will be out of control. Should the underlying accident rate change, the probability of observing a point out of control is more than 1 percent.

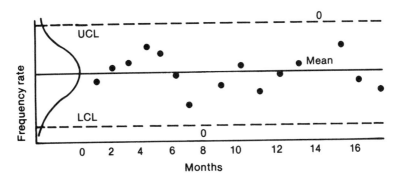

Exhibit 12.12 Control chart.

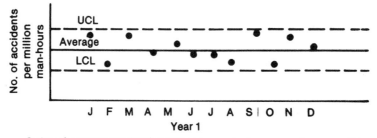

Exhibit 12.13 Control chart for accident frequency, XYZ Manufacturing Company.

One percent control limits have proved satisfactory for industrial accident studies, but should a situation arise in which all data points are found to fall within control limits, 5 percent limits might prove more satisfactory. Smaller differences in accident rates between periods would be detected, but at the price of more false signals, since the probability of a point being out of control when there is no change is 5 percent versus 1 percent for the 1 percent limits.

Control Limits Both 1 and 5 percent limits can be computed very easily with a form similar to the one shown in Exhibit 12.14. The following computations should be carried out:

For 1 percent limits:

1. Enter in column 1 the number of accidents X occurring in each period.
2. Enter in column 2 the number of person-hours worked MH in each period.
3. Divide the sum total of column 1 by the sum total of column 2. This will give

$$A = \frac{X}{MH}$$

Time interval	(1) Number of accidents	(2) Man-hours worked	(3) $\sqrt{\dfrac{A(1-A)}{MH}}$	(4) $\sqrt{\dfrac{A(1-A)}{MH}}$	(5) $2.576\sqrt{\dfrac{A(1-A)}{MH}}$	(6) UCL	(7) LCL
(1)							
(2)							
(3)							
(4)							
(5)							
(6)							
(7)							
(8)							
(9)							
(10)							
(11)							
(12)							
Total							

Exhibit 12.14 Sample form for calculating control limits.

4. Substract A from 1 and multiply by A:

$$A (1-A)$$

5. Divide the product of step 4 by the number of person-hours for each period, and enter in column 3.

$$\frac{A (1-A)}{\text{MH}}$$

6. Find the square root of each entry in column 3, and enter in column 4.

$$\sqrt{\frac{A (1-A)}{\text{MH}}}$$

7. Multiply each entry in column 4 by 2.576, and enter in column 5.

$$2.576 \; \sqrt{\frac{A (1-A)}{\text{MH}}}$$

The upper control limit for each period is A plus the corresponding entry in column 5.

$$\text{UCL} = A + 2.576 \; \sqrt{\frac{A (1-A)}{\text{MH}}}$$

The lower control limit for each period is A minus the corresponding entry in column 5.

$$\text{LCL} = A - 2.576 \; \sqrt{\frac{A (1-A)}{\text{MH}}}$$

For 5 percent limits, substitute 1.96 for 2.576 in step 7. Procedures for computing are otherwise identical.

It takes little more than a glance at year 1 of Exhibit 12.15 to see that the accident rate was very stable throughout the year. No rate falls outside the limits. At this moment in time the safety director would have had every assurance of a well-controlled accident situation but would have had no knowledge of what the future might hold.

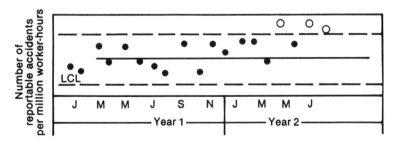

Order of occurrence (each point represents frequency for a month)

Exhibit 12.15 Control chart for accident frequency.

Let us suppose that the safety director continued to use this chart and plotted data points as the months passed. By the middle of year 2 the chart would have taken on quite a different appearance. The rate of April of year 2 signals a change and the situation should be examined closely. In this particular instance the records disclosed that a plant expansion program had been instituted for which no provision had been made in the safety program.

Once a rate has fallen outside the limits, signaling a change to the chart user, there is no way of telling at that time when stability will be reestablished and at what level. If subsequent points fall within the limits, this indicates one or two possibilities: (1) either the cause of the change was determined and corrected, or (2) the cause was transitory and self-correcting. Should subsequent points continue to fall outside the limits, one might infer that the reestablishment of stability has taken place at another level, but this could not be concluded until several points have accumulated. For this reason, practice calls for delaying the computation of a new mean and new control limits until at least ten such points have accumulated.

To distinguish between instability and a new level of stability, one can apply control limits. If all the points falling outside the original limits are contained by these new limits, one can assume a new level of stability, but if they are not so contained, there is continuing instability.

Using SPC in Safety

The above shows one way that SPC concepts can be used in safety. There are other, easier ways to use SPC. While the above indicated a number of analyses that can be used about 80 percent of all safety problems can be handled with three. All three require no knowledge of or use of statistics.

The three SPC tools that will get the most results are:

1. The *Pareto chart* which will help discover what the problem really is.
2. The *fishbone diagram* which will help identify what caused the problem.
3. The *process flow diagram* which will help assess potential future problems.

The first step is to determine what data to analyze. Many choices are available as evidenced by measurement theory discussed in Chapter 6. Exhibit 12.16 shows some of the options.

Here's an example of the use of SPC in safety:

The top executive of the XYZ Corporation made a decision to improve the safety results in his organization. Triggered by a fatality and the realization that his organization was one of the worst in the industry, he did what executives typically do: He put pressure on the safety staff to "turn it around."

Results Data	Sampling Data	Performance Data
Number of accidents	Number of Unsafe	Number of Inspections
— Lost Time	Acts	Made
— Total Recordables	Number of Unsafe	Number of Meetings
— Near Injuries	Conditions	Held
— Severity of Injuries	Number of OSHA	Number of One on One
— Days Lost	Violations	Contacts
— Dollars	Number of Complaints	Number of Positive
		Reinforcements

Exhibit 12.16 Possible data to analyze.

Where do you start? This safety director started with SPC, using first a series of Pareto charts to find out where the biggest problems were:

—Pareto 1.—Where is the problem? A Pareto of the fifteen plants in the corporation showed:

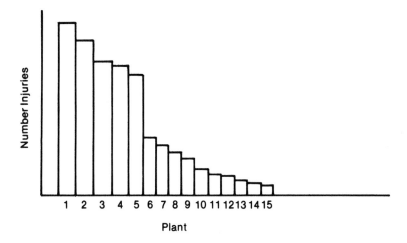

The problem was primarily in five plants. A quick computation showed that bringing the accident record down in those five to the level of the other ten would make the company second best in the industry. An additional halving of that would make them the best.

Pareto 2 answered which type of injury that was predominant in those five plants:

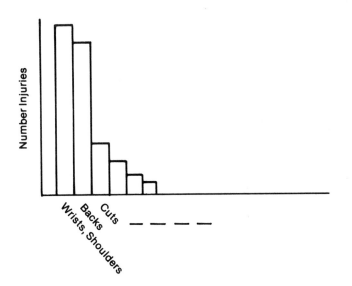

The problem was the soft tissue injury in wrists, shoulders, and backs.

Pareto 3 showed which department caused most of the soft tissue injuries.

From three Pareto charts it was decided that a step change could be made through reducing the soft tissue injuries in two departments of five plants, the packers in the packaging department and the box handlers in shipping.

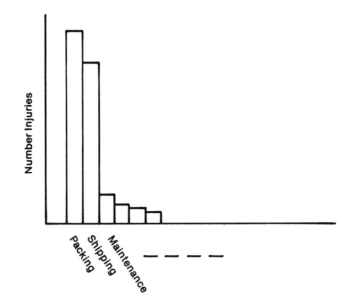

Next a fishbone diagram was used to determine the causes of the soft tissue injuries in each of the above departments. To get the best possible input here the workers themselves were asked to problem-solving using the fishbone approach. The group identified the items in Exhibit 12.17.

From the fishbone, the group went to a process flow diagram to provide further information (see Exhibit 12.18):

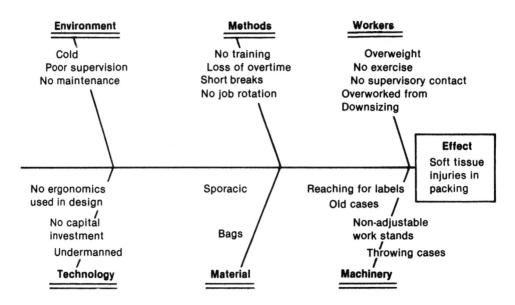

Exhibit 12.17 Fishbone diagram showing problem-solving ideas from workers.

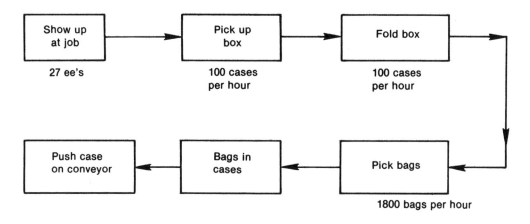

Exhibit 12.18 Process flow diagram to help access future problem areas.

Finally, the group decided which factors to concentrate on:

— Get adjustable packing tables.
— Do pre-shift warm up exercises.
— Use wrist bands.
— Rotate packers hourly.

A substantial improvement was achieved. The approach became a normal way of life in the five plants and was used on other safety, quality, and productivity problems. In one year the five worst plants became the five best plants and the company became the best in its industry. The secret here is the same secret that Dr. Deming brought to Japan in the 1950s: some SPC concepts coupled with employee involvement in the problem solving process.

The above story showing the use of SPC in safety was hypothetical, although drawn from reality. Here is a story that is not hypothetical; it really happened.

Exxon Chemical's Bayton, TX, Butyl Plant published the following in their internal magazine *Chemsphere*. The article was entitled "Premeditated Safety" and was described as "A report on 300 injuries that never happened—the best safety story you've ever read":

> The evidence was plentiful, obvious and grim.
>
> Based on hard statistics, the coming turnaround (overhaul) of the Baytown Butyl Plant was likely to disable four workers and injure 28 others seriously enough to require a government report. Another 287 workers would need first aid.
>
> The forecast: 319 injuries in all.
>
> That prediction was made as Baytown prepared for the butyl unit's first major turnaround in five years.
>
> When the operation was over, some 90 days and 450,000 work hours later, Baytown totaled up the cost in terms of human hurt.
>
> The most serious injury was found to be a cut thumb for which a contract employee needed stitches.
>
> There were no disabling injuries, instead of the expected four.
>
> There was a total of six—not 28—injuries requiring a government report.
>
> There were 13 mishaps calling for first aid, rather than 280. One of these involved a supervisor who, sitting in his office, caught a speck of dust in his eye that was somehow attributable to the turnaround.

Total actual injuries: 19. Anticipated injuries that never occurred: 300.

What kept those 300 injuries from happening, and what made the turnaround more productive than expected?

'Before the project began,' according to Larry Haynes, manager of the butyl plant, 'we recognized that with so many people in so small a space and with more stringent government rules for reporting 'recordable' injuries, we had to do something different if we were to avoid injuries.'

There was another factor: the unit's tradition as a strong safety performer and its position as an exporter of safety ideas to other company units provided an added incentive to get the job done safely. No one wanted that tradition sullied, or any of their colleagues hurt.

Enter Cecil Guidry, a Baytown native whose father and grandfather before him had worked at Exxon's Baytown, Texas, complex.

In 1965, young Guidry came to Exxon Chemical as an operator and process technician and was assigned to the butyl unit. He worked himself up to supervisor and, in the early 1980s, became one of the increasing number of Exxon Chemical employees known as "quality-improvement professionals."

These are people who specialize in finding ways to apply the growing body of quality-improvement technology to all facets of the company's operations, including safety.

Says Haynes: 'We asked Cecil to channel his considerable quality-improvement skills into ensuring a safe turnaround.' Guidry was named safety coordinator for the project.

Consider the scope of the challenge:

- The sheer size of the undertaking. The turnaround required taking apart, examining, fixing or replacing and then putting back together distillation towers, blending tanks, reactors, finishing and packaging areas, and a control room. Every bit of insulation (an important material for manufacturing processes that take place at 150 degrees Fahrenheit) had to be replaced, all 1,400 valves checked and/or replaced, and every bit of piping inspected for leaks.

- The number of people and the amount of space. The project involved 200 company employees and up to 800 contract workers who lacked the level of safety awareness common to Exxon Chemical people. The contract workers were performing non-routine, unfamiliar tasks—a factor likely to increase the risk of accidents. And they were doing it in a space normally occupied by fewer than 200 people. An analysis of prior accident data at the plant, based on a widely used quality tool, indicated that when the number of people at a work site doubles, accidents at that site will likely triple. During the turnaround, the population at the unit increased as much as fourfold.

- The mix of specialists. An army of pipefitters, welders, insulation workers and electricians—15 separate crafts and skills in all, each with its own potential safety hazards—invaded the site.

  There was another underlying factor: the said but true industrial fact of life that turnarounds offer a great potential for injury. A 20-member safety team, consisting of process, mechanical and administrative personnel, was assembled under Guidry's direction. A comprehensive, 12-point safety program grew out of the group's efforts. Its execution of the program identified and removed 20,000 potentially hazardous conditions and prevented injuries that would have cost individuals considerable suffering and the company about $500,000. In addition, the program generated an estimated $4 million in quality and productivity savings.

It was an initiative that obviously paid off. The start-up of the rebuilt plant was the smoothest on record.

The success of the turnaround and Guidry's role in it were featured in a presentation by Exxon Chemical President Gene McBrayer to the Exxon Corporation board of directors on ways quality techniques can bring about a safer work place.

Cecil Guidry stated the following approach is how the above results were achieved:

LEARNING TO LET SPC TECHNIQUES LEAD THE WAY TOWARD CONTINUOUS IMPROVEMENT

- Statistical analysis of safety performance indicated variation is due to common cause system problems not special causes. Solutions aimed at the individual are not the answer.
- The data base for analysis includes: injury, near miss, and safety audit reports.
- Ishikawa's technique of asking "WHY" at least five times or more leads the way to identifying system problem root causes.
- Pareto and other prioritizing techniques help determine where we should concentrate our efforts.
- Cause and effect flow chart analysis of management systems surfaces weaknesses.
- Analysis of time studies, people, objectives, activities, and locations predicts accidents. This important study gives insight to needed system modifications before the accident takes place.
- Planning to remove accidents from the work place.

  - Planning, scheduling, coordinating, etc.
  - Adequate and timely communications
  - Parts supply
  - Tool supply

  Key is providing the worker with everything needed before work begins including "Why the job is important."

- Realizations for the future

  - To accomplish our goal we must remove thousands of hazards at a time. This can only be done through system modifications.
  - Past cures have relied so much on memory that they have not been permanent or irreversible in nature and allow problems to repeat 94 percent of the time. The time and resources needed to develop irreversible cures for safety and quality are keys to our success.

- There is a direct relationship between safety, quality, productivity and costs. The above four cannot be separated. Any problem solution for one must improve all four, or it will not keep us in business. Any solution must help the people accomplish their goals.
- Use of SPC techniques has generated several "success stories" at the Baytown Butyl Plant:

  - FHB Safety Problems Solvers Team improvements.
  - 1987 IRD safety, quality, and productivity improvements.
  - Improvement in mechanical workforce safety record.

Like most other Exxon locations, Baytown Butyl has safety rules, safety committees, permit systems, inspections, audits, SOPs and many other *systems* aimed at safety. Our efforts are aimed at why these systems do not function as they should on their own. Safety systems seem to parallel quality and productivity systems instead of being interwoven. The conflict between these parallel systems prevents the workers from accomplishing their goals in a safe manner. We are redesigning the interfaces between safety, quality, productivity, and costs to make them compatible and additive.

REFERENCE

Siegel, J. *Managing with Statistical Methods,* Society of Automotive Engineers, 1982.

Purcell, A., "Premeditated Safety," *Chemsphere* New York: Exxon Chemical Company, Second Quarter, 1989.

Part V

Other System Elements

Chapter Thirteen

Complying with OSHA

The emphasis in this chapter is on the Occupational Safety and Health Act (OSHA) rather than a broader emphasis on all regulatory compliance. With the constantly changing regulatory scene and the broad issues that regulations cover, any attempt to cover more than some thoughts on OSHA compliance is beyond the scope of this book.

Part V of this book also deals with other, closely allied programs that should be seriously considered by any organization today. These include programs in wellness, substance abuse, stress management (all in Chapter 14), fleet safety (Chapter 15), and product safety (Chapter 16). Each of these are not integral to the safety system, but are important to consider in the totality of loss control systems for the organization. Some people consider OSHA compliance to be a part of safety, others do not. This author is in the second school of thought, which believes that compliance with the OSHA rules and regulations, some of which help safety, some of which are irrelevant to safety, some of which are even counter-productive to safety, is necessary.

WHAT IS OSHA?

The Occupational Safety and Health Act of 1970 was signed by President Nixon on December 29, 1970, and became effective on April 28, 1971.

The law was (and still is) met with a mixed reaction by American industry. Some felt like Exhibit 13.1 shows, others like Exhibit 13.2. Few were delighted at the time (still true).

It covers some 57 million workers in over 4 million businesses. Essentially, it requires every employer to provide a job environment that is "free" from recognized hazards that cause or are "likely to cause" death or serious physical harm. To provide such a job environment, each employer is required to comply with safety and health standards which are specifically spelled out by the law.

In general, OSHA standards consist of rules for avoiding hazards which have been shown by research or experience to be harmful to personal safety and health. Every employer is obliged to know and observe the standards which apply to that business.

The scope of the standards is almost unbelievably broad. Regulations touch on such things as the location and size of exits, methods for guarding floor and wall openings, scaffolding, ventilation, noise and light exposure, air contaminants, use and storage of gases and flammables, protection equipment—such as headgear and goggles, sanitation facilities, medical and first aid facilities, fire protection—such as sprinklers and alarms, compressed air equipment, materials handling, machinery and machinery guards, hand and portable power tools, welding and cutting operations, and electrical equipment—such as outlets, grounding, switches, cords, and cables.

It is safe to say that the regulations are so broad and all-inclusive that there is no way in which a company can be in complete compliance with all aspects of the law at any one time. We can only do our best and, especially, try to find the most important points.

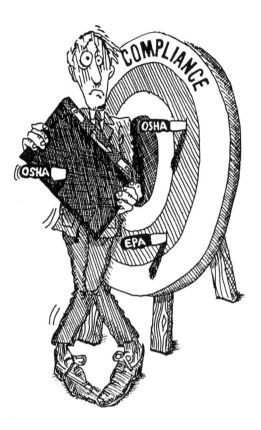

Exhibit 13.1.

ENFORCEMENT

To enforce compliance with its standards, OSHA provides for inspection of premises and penalties of fines or even jail sentences for noncompliance. The law also authorizes Department of Labor (DOL) inspectors to enter any establishment at any reasonable time, and without delay or prior notice. An inspector has authority to inspect whatever he or she wishes, and may question any employer, owner, operator, agent, or employee.

The initial standards were adopted from key safety consensus standards developed by the American National Standards Institute (ANSI), the National Fire Protection Association (NFPA), the American Society of Mechanical Engineers (ASME), and the American Society for Testing and Materials (ASTM). Also included were standards that have long applied to government contractors (such as those of the Walsh-Healy Act, the McNamara-O'Hare Service Contract Act, the Construction Workers Act, and the Longshoremen's and Harbor Worker's Act). Those standards, as OSHA, went into effect upon their publication in the *Federal Register* on May 29, 1971.

Because the standards were adopted hurriedly, without time for careful review, problems arose: some standards were too vague; others were too specific. Many of the national consensus standards, for example, were never intended to be standards at all. They were used by industry as guidelines for good management practice, with little thought of how they could be enforced. One such standard says that employers are "responsible for the safe condition of the tools used by employees," which does little more than leave employers wondering what OSHA will consider "safe." In other cases, standards set very specific requirements. For example, one standard states the exact height for all staircase railings. Does that mean a company must tear out and replace a railing that falls two inches short of OSHA requirements? According to the law, it does.

OSHA's own brief list of most frequently violated standards follows and it can serve as a guide for setting priorities for self-inspection activities. Each item can be found in Part 1910 of OSHA's general industry standards.

The violations are listed in no particular order.

— National Electrical Code: Missing ground plugs, frayed or improperly spliced cords.
— Mechanical power transmission apparatus: Failure to guard belts, gears.
— Portable fire extinguishers: Improperly mounted or inaccessible extinguishers.
— General requirements for all machines: Missing guards to protect against nip points, rotating parts, flying chips or sparks.
— Woodworking machinery: Missing guards on saws.
— Guarding floor and wall openings and holes: No rails around or covers on pits; unprotected drops of four or more feet.

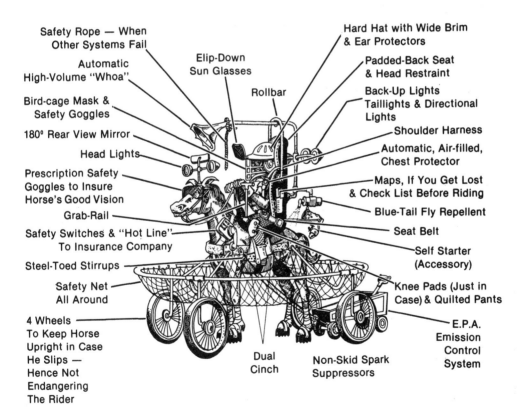

Cartoon (from an unknown American source) by courtesy of Schuberth-Werk Braunschweig, Fed. Rep. of Germany.

Exhibit 13.2 Cowboy after OSHA.

— General requirements—walking and working surfaces: Failure to keep work areas clean, orderly, and sanitary.
— Abrasive wheel machinery: Guards missing or tool rests not adjusted to within 1/8 inch of wheel face.
— Welding, cutting, and brazing: Improper storage and handling of compressed gases.
— Powered industrial trucks: Lift trucks left running and unattended.
— Saw mills: Unguarded saws, equipment, open walkways along conveyors.
— Means of egress, general: Unmarked exists.
— Flammable and combustible liquids: Excessive storage or containers of flammable or combustible material left uncovered.
— Sanitation: Failure to provide a clean, sanitary wash place.
— Spray-finishing using flammable or combustible liquids: Failure to post or observe no-smoking signs in paint spray areas.
— Hand and portable power tools and equipment—general: Missing guards.
— Handling materials—general: No marking of permanent aisles for forklifts.
— General requirements, means of egress: Blocked or inadequate exits.
— Overhead and gantry cranes: No load capacity posted on crane.
— Portable wood ladders: Missing rungs, weak or wobbly legs, missing braces.
— Noise exposure: Noise in excess of 90 decibels on an eight-hour time-weighted average.
— Medical services and first aid: Missing or inadequate first aid kit.
— Eye and face protection: Failure to wear protective goggles where there is strong possibility of danger to the eyes from flying objects.
— Personal protective equipment—general: Missing hardhats where there is a danger of dropped items, striking against objects, or electrical shock.
— Fixed ladders: Rungs too widely spaced, too little clearance to sides or rear of ladder.

A PREPARATION PLAN

Exhibit 13.3 outlines the preparation plan. Following this general plan will, we believe, help you to achieve compliance systematically. How to carry out the plan is detailed in the following pages. Briefly, there are nine basic steps, which are summarized in the Exhibit.

---

1. Become familiar with what the law says and what it requires.
2. Know and put into effect record-keeping obligations under OSHA.
3. Obtain the standards and determine which subparts apply to the standards.
4. Learn how to read the standards.
5. Find the physical violations in the facility.
6. Locate administrative violations of the law.
7. Develop a plan of corrective action.
8. Develop a system through regular management and supervisory methods that ensures keeping in compliance.
9. Document everything done.

Exhibit 13.3 A preparation plan.

The Act requirs that employers:

1. Furnish a place of employment which is free from recognized hazards that could cause or are likely to cause death or serious physical harm (the general duty clause); and
2. Comply with safety and health standards promulgated under the Act.

In turn, each employee must comply with the safety and health standards and all rules, regulations, and orders which are applicable to his or her own actions and conduct.

The law provides for sanctions against the employer in the form of citations and penalties if the employer fails to comply. There is no provision for penalizing an employee for failure to comply.

The Act created three new federal agencies, each of which bears its share of responsibility:

1. *Occupational Safety and Health Administration (OSHA)*, which is headed by an assistant secretary of labor, was created to promulgate and enforce occuptional safety and health standards.
2. *Occupational Safety and Health Review Commission (OSHRC)*, which is an independent establishment of the Executive Branch of the U.S. Government, adjudicates some disputes between an employer or an employee and the Secretary of Labor.
3. *National Institute for Occupational Safety and Health (NIOSH)*, whose function is to develop and establish recommended occupational safety and health standards, to conduct research, experiments, and demonstrations relating to occupational safety and health, and to conduct education programs. NIOSH is part of the National Institutes of Health, under the wing of the Department of Health, Education and Welfare.

INSPECTIONS

Perhaps with the proviso of reasonable expectation that something is amiss, compliance officers may enter and inspect all facilities of any establishment covered by OSHA. The employer or designated representative must then participate in the opening conference, where the compliance officer will set the stage, answer questions, solicit information as to what the company is doing in safety, etc.

After that, both an employer representative and an employee representative may accompany the compliance officer during the inspection. In the absence of an employee representative, the officer will confer with employees picked at random. Any employee may be questioned in private.

OSHA inspectors are not permitted to make a "partial" inspection. If they inspect at all, they must make a complete inspection—and they will not give advance warning. Neither may a company invite an inspector to come and advise whether any condition violates a standard. Once the inspector is on the premises, a complete official inspection is made.

During an inspection, an OSHA representative may point out one or more violations of standards and discuss them with the employees as well as ascertain whether the employer has complied with other promulgated regulations, including posting of the OSHA poster and the required recordkeeping. Any apparent violation of either the standards or the general duty clause will be noted.

The compliance officer is required only to note apparent violation—*not* to present solutions or suggest methods of correction.

After the inspection, the compliance officer will confer with the employer, advising of all conditions and practices which may constitute safety or health violations, any citations that may be issued for the alleged violations, and any penalty that may be proposed for each citation.

An employer has the right to contest a citation, a proposed penalty, notice of failure to correct a violation, the time allotted for abatement of a violation, or any combination of these. Employer rights in the process are shown in Exhibit 13.4.

Knowing recordkeeping obligations. An employer with more than seven employees must do certain things in the way of keeping certain injury records. Put these procedures into effect immediately. See Chapter 12.

It should be noted that one of the major thrusts of OSHA in the mid-1980s has been the accuracy of injury records. Inaccurate records have resulted in fines of millions of dollars.

OBTAINING THE STANDARDS

The general standards are at 29 CFR 1910. Construction work may also be subject to the Safety and Health Regulations, Part 1926 (Construction); and maritime work may also be subject to Parts 1915 through 1918 (Maritime).

Part 1910

A quick glance at the 590 pages of Part 1910 is an eye-opener for the uninitiated. A number of things that become readily apparent dictate how a company might go about getting into and staying in compliance with the law. However, bear in mind the following:

- These pages contain only a small portion of the standards that are law. Other standards that have the force of law are referred to in the *Code*.
- The standards are all-inclusive. It seems safe to say that no one company of any size can ever be in complete compliance with the law at any one time. Almost everything is covered in the standards, either directly or by reference. And achieving compliance with everything today does not preclude violation again tomorrow, for keeping conditions safe is infinitely more difficult—but even more important—than getting conditions safe to begin with.
- Much in the standards does not apply to every operation. Since the standards *are* all-inclusive, all of it cannot possibly apply. Unfortunately, someone must go through the entire lot to find out just which standards do apply.
- There is more to the standards than simply correcting physical conditions. Built into (or hidden in) the standards are numerous references to required inspections, records, signs, training, and so forth, all of which are as much a part of the law as the rest. A company can be fined for such things as having no training program, no licenses for fork-truck drivers, no crane inspection, or no medical authorization for first aid kits.

There seems no way around having someone in the organization find out what the standards actually say. That is no easy task. Reading the standards has all the excitement and drama of reading the telephone directory. However, it must be done. And it seems preferable that it be done by a reader with these two qualifications:

1. Sufficient technical knowledge to understand, which may mean a different reader for each subject.

Before Inspection Begins		
	General Principles	1. Receive courteously, ask for official identification. 2. Refer only to lop mgr. at facility (for designee) who will act as official guide (OG) & accompany inspector at all times. 3. Brief inspector: wants to cooperate, please ask questions only from OG who will get answers (ex as noted below) 4. Ask purpose of particular inspection & what particularly wants to see or know; try to avoid "hunting expedition" 5. Do not volunteer information, only required information should be given (counsel can advise) 6. OG makes notes of everything said or done, however trivial. 7. If disagree with inspection statements, so state clearly, but don't argue.
	Note: For OSHA inspection because of employee complaint, inspector must show written notice before starting	

During Inspection

Areas Open to Inspector

- All areas of plant, warehouse, branch, etc. where work is performed for employer

What Inspector Can See

- Can See
  1. Conditions, structures, machines, apparatus, equipment, materials
  2. OSHA recordkeeping reports

- Cannot See
  Any other records

What Inspector Can Do

1. Make tests using...equipment, OG record date.
2. Question privately any employee
3. Employee who filed complaint can accompany inspector (without pay)
4. Shut down machine or process if imminent danger

What Inspector Can Take Away

1. Environmental samples (air, water, dust, chemicals); inspector seals sample in OG presence, OG keeps one for analysis, one for inspector
2. Photos - (gives or inspector takes) if related to inspection; OG take some photos, or have inspector send copies

After Inspection

1. OG writes summary for management; keep original notes
2. Inspector should state if citation is to be recommended
3. OG can conduct closing discussion

Exhibit 13.4 OSHA inspection process.

2. Enough knowledge of the company's operations to know what applies. And, again, that may mean a different reader for each section.

READING THE STANDARDS

The outlining method used in the standards is difficult to follow and is often tricky. The vast body of federal regulations is broken down into fifty "titles." OSHA regulations are in Title 29, Chapter XVII.

The chapters group together *parts,* which are usually numbered in Arabic numerals running consecutively through a title. (The parts do not begin anew with a different chapter. Chapter I in a title might contain Parts 1 to 199; Chapter II might contain Parts 200 to 299; and so forth. Frequently, parts are "reserved" for future regulations within a title, so that regulations pertaining to the same issue can be grouped together as they are written.) A part consists of a unified body of regulations that apply to a specific function of the issuing agency or are devoted to specific subject matter under control of the issuing agency. For instance, Part 1926 of Title 29 refers to occupational safety and health standards for the construction industry, and it would be referred to as 29 CFR 1926.

In turn, parts are divided into *sections* or *subparts,* each of which, according to the Office of the Federal Register, will "ideally consist of a short, simple presentation of one principal proposition." Sections are usually just a few paragraphs long, but occasionally run pages. Each section begins with a boldface heading. Sections are numbered consecutively within parts; thus the third section in Part 1926 would be 1926.3 and the thirteenth would be 1926.13. (The Office of the Federal Register calls that a "modified decimal system".) Sections also are subdivided, when necessary, with paragraphs.

Recognizing physical violations. Next, look for the physical violations. That usually means some detailed inspections. Since in most cases there are many violations (since the law covers a multitude of things), take it slowly, department by department, or perhaps subject by subject (e.g., ladders first, exits second).

There are two primary ways to find violations: inspecting or excerpting. Under inspection, one person (usually the safety manager) is responsible for finding the violations. Under excerpting, the safety manager (or someone) excerpts which standards apply to each department and someone else (probably a line manager) is responsible for seeing that the department is in compliance.

For OSHA violation purposes, each supervisor should have, as a self-training device, a checklist for the department. Once the supervisor is trained, the form shown in Exhibit 13.5 is far more helpful, for as a violation is spotted, the cause and plan for removal are noted.

The forms are designed for the supervisor and, therefore, do not really refer in detail to OSHA. If safety staff is performing the OSHA inspection (as our exercises assume), you obviously will not need those forms but instead need the standards and a clipboard to write on.

FINDING ADMINISTRATIVE VIOLATIONS

The first approach tends to achieve compliance only on physical conditions. But, as previously mentioned, the Occupational Safety and Health Act (OSHA) standards (and referenced standards) do not speak only to physical conditions, although that is where enforcement emphasis has been and will no doubt remain. The standards also require that management train employees, periodically inspect, keep specific records, and adequately warn through signs.

SUPERVISOR'S INSPECTION FORM		
Name _____ Date _____		
Symptom noted Act/Condition/Problem	Causes Why-What's wrong	Corrections made or suggested By you-By others

Exhibit 13.5 A suggested supervisor's inspection form.

Those requirements are often and too easily overlooked. Next, find the administrative requirements of the law by using references on inspections, tests, and recordkeeping; then find the requirements for signs, markings, and signals, for training, and for extinguishers. Make another list of violations.

Beginning corrective action. You should by now have compiled a rather long list of violations—some physical, some managerial or administrative. To begin corrective action, draw up a plan of attack that spells out what must be done, who will do it, and when it will be completed. Obviously, items on the list will have to be given priorities and then scheduled accordingly. (See Chapters 4 and 8.)

Keeping in compliance. Develop a system through regular management methods that ensures keeping in compliance. Getting in compliance, difficult as it is, is much easier than keeping in compliance.

For instance, the regulations require that the tool rests of abrasive wheels be adjusted to within 1/8 inch of the wheel. An adjustment made on Monday at 8:00 a.m. will no doubt be changed by 8:05, a violation of the law. The trick is to devise plans and systems that *keep* the tool rests at 1/8 inch.

Obviously, the required items are highly specific, and each is an attempt to get at only a narrow section of the total job. We will not try here to get at other, unmentioned narrow bands or sections of the total job, because that could be a never-ending task. Rather, we will look briefly at some general approaches, most of which are fairly common techniques. They are inspections, job safety analysis (JSA), hazard hunts, committees, and outside help.

COMMITTEES

One way to get employee participation is through committees.

If a safety committee is used, the safety professional would do well to examine closely the workings of the committee. For effective operation, these rules are essential:

1. Define clearly the duties and responsibilities of those who are to find OSHA violations.
2. Choose members in view of their duties. Choose those with some knowledge of the standards.
3. Provide any necessary staff assistance.
4. Design procedures for prompt action.
5. Choose the chairperson carefully.

DOCUMENTING

Finally, document everything. You must prove you have a plan, and it pays to document to show good faith. Companies that have made good-faith efforts toward worker safety will receive much lower fines for the same violations than firms with little or no safety record.

The above offers a kind of generic approach to OSHA compliance. However, as each new thrust comes from OSHA, additional efforts will be required. For instance, in the last few years there have been several major new thrusts, each requiring a major change in the safety professional's efforts and the line manager's activities. A new asbestos standard spelled out in considerable detail exactly the compliance requirement of the organization. There are others, and more will come.

HAZARD COMPLIANCE STANDARD

A good example is the Hazard Communication Standard, which spelled out what must be done in detail. Phillip S. Howard explained:

> The following information is bare fact direction towards speedy compliance. Do not lose sight of the main purpose of the HCS. The purpose is to provide employees with information concerning th presence and potential hazards of chemicals in the workplace. OSHA uses the term "exposed employees." Not all employees are exposed to chemicals but because of their assigned work may very well be "affected employees."
>
> An affected employee may be the engineer who frequents the production area that uses chemicals. The engineer may never be actually exposed to any chemicals but may be required to be trained in HCS. Similarly, a production worker may operate totally enclosed equipment with good ventilation, etc. and never be exposed but is definitely an affected employee.

Compliance Flow Chart

Now, let's quickly review the basic requirements of HCS. Please review the employer's HCS Compliance Flow Chart (Exhibit 13.6). In the *Flow Chart* many items not required by HCS are also included to expand your injury prevention and to decrease business liability exposures.

The *Flow Chart* includes the following required items of HCS:

1. Identify all chemicals.
2. Prepare a master list of chemicals for each facility (and maybe each department).
3. Acquire a file of MSDS for each chemical.
4. Place a copy of MSDS file for employee access.
5. Establish a communication system of chemical hazards.
6. Write an HCS program stating how all other requirements will be met.
7. Train affected employees in:
   a. the standard
   b. generic MSDS and its terms
   c. written HCS program of the business
   d. specific hazards of each chemical or category of chemicals
   e. location, detection, and emergency response on each chemical or category of chemicals

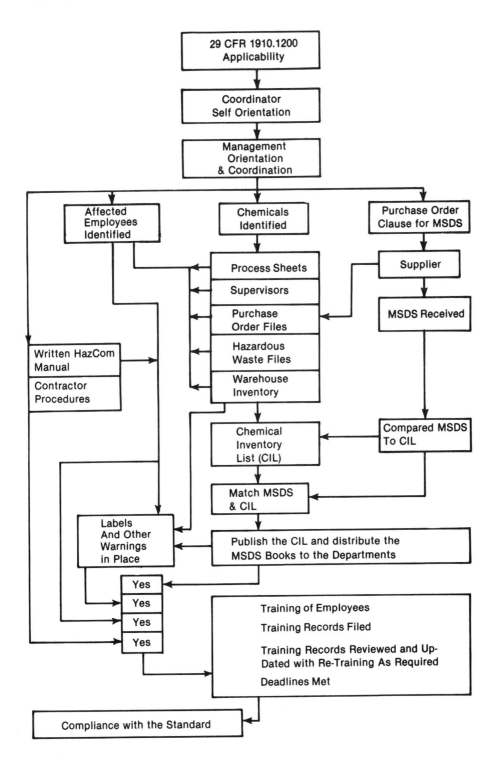

Exhibit 13.6 Employers' HazCom compliance flow chart.

8. Maintain good training records.
9. Maintain control of new chemicals and training of new employees before they are introduced to the workplace.
10. Establish a communications system for contractors entering your premises.

Choose now an efficient and speedy method to compliance. Your personal and corporate health will be better if you organize and streamline your efforts.

FIRST: The *Activity Chart* see Exhibit 13.7) should be followed to ask yourself how this standard applies to your organization. If the strict adherence to the standard says you are not required to comply, consider also the financial aspects of defending or losing an intentional tort suit, a workers' compensation claim and/or specific safety violations. The expense of compliance is very cheap insurance that cannot be otherwise bought.

SECOND: Develop a list of activities that you need to accomplish. Then decide if you are a manufacturer, importer, distributor, and/or an employer within SIC codes 20 through 39. The information for this decision is in the Code of Federal Regulations (CFR), Part 29, subport 1910.1200 (b) (1), (2) and (3) on Scope and Application. It can be found in the Federal Register published on November 25, 1983 on page 53340 and in the middle of column 2. For your quick reference and starting point the *Activity Chart* follows this same information sequence.

In the *Activity Chart*, each subject is generally identified. If you are familiar with the subject or the category within the subject, skip to the next one. If you want more information, then follow the line across for clues of where to look further. You can also use this as a checklist of your own activities and your own progress.

THIRD: Along the way to standard compliance, you will need task assignments and self-imposed deadlines. For the fast track system, use the employer's fast track *Assignment Schedule*. (See Exhibit 13.8). Make realistic target dates and stick to it. Make your plan and then work your plan towards compliance.

As a final effort to give you a quick look at the sequence of needed events, the HazCom Compliance *Flow Chart* has been furnished. It has key points identified with the sequential and parallel activities easily seen.

For clarity, these charts and lists are prepared with the employer's responsibilities in mind. The manufacturer, importer and distributor have these same tasks as an employer, but also have to complete the Material Safety Data Sheet (MSDS) for their customers.

HCS Basic Parts

The three basic parts of the HCS are: Applicability, Hazard Evaluation, and Communications. These are well detailed in the *Activity Chart* (see Exhibit 13.7).

The main intent of the standard is fulfilled with the communications part. The rest is supportive information on how to prepare for the communications with employees. This applies to the sales chain from the manufacturer to the user because each business along the way is an employer also.

The communication details of the HCS are the main intent of the standard. It is intended entirely to prevent accidents, illnesses, and injuries in the workplace. Actual experiences with implementing an HCS program have highlighted many benefits to employees and employers. More than once, an employer has related a reduction of employee relations problems after implementing the HCS in the facility.

HazCom ACTIVITY CHART

Subject	Category	1910.1200 Sub-Part Number	1910.1200 Sub-Part Name	Fed. Reg. Page # 11/25/83	Location On Page	Comments
A P P L I C A B I L I T Y	Hazardous Chemical?	(c)	Definitions	53341	mid-3	See also definitions for Health Hazard and for Physical Hazard
	SIC #20-#39	(b)	Scope	53340	mid-2	
	Manufacturer Importer Distributor	(j)(1)	Date	53346	top-1	Compliance by 11/25/85 for MSDS issued
	Employer	(j)(3)	Date	53346	top-1	Compliance by 5/25/85 for written manual, training, lists.
	Exemptions Total	(b)(5)	Scope	53340	mid-3	Product exempted
	Exemptions Partial	(b)(3) (b)(4)	Scope	53340	mid-2 bottom-2	Lab exemptions Label exemptions
	Preemption of State	(a)(2)	Purpose	53340	top-2	
H A Z A R D	Identify All Chemicals used!	(d)(2) (e)(1)(i)	Hazard Determination Written Program	53342 53343	mid-2 mid-1	Use Supervisors Process Sheets Purchase Orders Etc. to get Chemical Inventory List (CIL)
	MSDS File for Mfg. & Employer	(d)(1)	Hazard Determination	53342	mid-2	Put Standard Clause on every Purchase Order.
	Evaluation by Mfg., Importer & Distributor	(d)(1) (g)(1)	Hazard Determination MSDS	53342 53343	mid-2 top-3	
E V A L U A T I O N	Known Hazardous Chemicals	(d)(3) (d)(4)	Hazard Determination Hazard Determination	53342 53342	bottom-2 top-3	These must be in your list if on site. Chemicals not on these lists must be evaluated per Appendixes A & B, page 53346.
	Evaluation Procedures	(d)(5)	Hazard Determination	53342	mid-3	
	Evaluation Procedures	(d)(5)	Hazard Determination	53342	mid-3	
	MSDS Format	(g)	MSDS	53343	top-3	What must be on an MSDS—guide for Mfg., Import. & Dist. Check for Employer
C O M M U N I C A T I O N S	Written Manual	(e)(1)	Written Hazard Communication Program	53343	mid-1	Must include all the categories on this subject.
	Chemical Inventory List (CIL)	(e)(1)(i)		53343	mid-1	
	Methods of Communication to Employees	(e)(1)(ii)		53343	mid-1	
	Methods to Inform Contractors	(e)(1)(iii)		53343	bottom-1	
	Manual Available to Employees	(e)(3)		53343	bottom-1	
	Labels and Usage Defined	(f)	Labels	53343	bottom-1	
	Other Warning methods	(f)(5)	Labels	53343	mid-2	
	Availability	(f)(5) (g)(a)(9) (g)(a)(10)	Labels	53343 53343	mid-2 top-3	
	Training and Sub-Parts	(h)	Information & Training	53344	bottom-2	
	Trade Secrets	(i)(2)(i)	Trade Secrets	53344	mid-3	
	Keeping Current	(h)	Information & Training	53342	bottom-2	

This is the main intent of the standard. The rest is supportive information on how to prepare for the communications. This applies to the sales chain from manufacturer to user because each organization along the way is an employer also.

Exhibit 13.7 HazCom Activity Chart.

Employers' HazCom
ASSIGNMENT SCHEDULE

Topic	Target Date	Responsibility Of Whom	Actual Date Completed
Coordinator Appointed			
Management Orientation			
MSDS-supplier			
Publish Chemical Inventory List (CIL)			
MSDS Master File			
Written HazCom Manual			
Contractor Arrangements			
MSDS Binder and Distributed			
Labels-Signs			
Process Sheets Other Warnings			
ID of Affected Employees			
Training			(May 25, 1986!)
Training Records			
Training Updated			

Exhibit 13.8 Employers' HazCom Assignment Schedule.

Personally, I strongly support that a well-informed employee is a safer, more productive and better quality producing employee. The employee has less uncertainties of the workplace when any communications are improved.

A Hazard Communications Program is another part of employee communications and employee safety when properly organized and implemented (see Exhibit 13.9).

I. *Identification of all chemicals*
1. Use purchase orders of last three years, engineering designs, process sheets and any other paper source to make an initial master list. If you bring any chemical into the workplace and are not willing to eat it or drink it, put it on the initial list.
2. Make an actual physical inventory of all chemicals used in the facility and compare it against your initial master list.

II. *Make the Master List* of chemicals for the facility and each department. Show why certain chemicals were deleted from this list when applicable, i.e., articles and janitor supplies used in normal consumer manner.

III. *MSDS File*
1. Obtain an MSDS on every chemical from the supplier. Document every request. If no more than 30 days each request then send a letter to OSHA stating your attempts at HCS compliance on that chemical.
2. Place each MSDS in separate file folder and never throw out old MSDS on same chemical.
3. Make a working copy of each MSDS for your HCS Coordinator.

IV. *Employee MSDS File*
Make a copy of each MSDS and make available to all employees whenever they are at work. Departmental employees MSDS files need only contain chemicals used in that department.

V. *Hazard Communication System*
1. Decide on communication procedure for every chemical. Pick one or mix these options:
   a) Container labels
   b) Area signs or placards.
   c) Process sheets (or menus, work orders, job instructions)

VI. *Training must use a sign-in sheet.*

1. Basic Class—
a) Discuss the standard's requirements of MSDS, Labels, Signs, and Training.
b) Tell employee rights AND responsibilities.
c) Present a generic MSDS and how to read it including most new terminology.
d) Present the written program manual for the business and how to use it and where it is kept.
e) Explain the hazard communication system(s) for the facility.
f) List the categories that specific training will include.
g) Give a quiz and score it (numerical or pass/fail).
h) Make known all the chemicals included in the category of the class.
i) Discuss the hazards of the worst chemical in the category and variations of all chemicals in the category.
j) Discuss the location, use, detection, and emergency response including first aid for this category of chemicals.

VI. *Records*
Keep the sign-in sheets, instructor outline, audio-visual(s) used, master list, categories list and department training lists.

VII. *Contractors*
Treat the contractor as one employee plus the following:
1. Give a copy of the MSDS on all chemicals the contractor uses
2. Require the contractor to provide a copy of the MSDS of each chemical brought in.
3. Require the contractor to sign an acknowledgement of receipt of MSDS's and plant rules.

Exhibit 13.9 Details of required items of HCS.

REFERENCES

Howard, P. "Hazard Communications on the Fast Track," *Professional Safety*, April 1986.
Petersen, D. *The OSHA Compliance Manual*, New York: McGraw-Hill, 1978.

Chapter Fourteen

Other Programs

Up to this point, this book has concentrated on the absolutely essential elements of a safety system, both proactive and reactive. This Part looks at other elements and programs that many organizations believe to be essential also, although they tend to be outside of or adjunct to the traditional safety system which tends to focus heavily on personal injury control on the job.

This chapter discusses possible problems for off-job safety, wellness programs, substance abuse programs and stress management programs. The next chapters overview fleet safety and product safety. Each is worthy of a book in and off itself (and such books are available). The purpose here is to ask the reader to consider them.

OFF-JOB SAFETY PROGRAMS

We know that more workers are injured off-job than on. Raymond Kuhlman describes the problem:

A lot of interesting numbers on off-the-job accidents are published in various books and journals. Most are little more than that—interesting.

> *In the United States, in 1984, there were 91,500 accidental deaths and over seven million accidental injuries. The job-related deaths were 1 of 9, and the job-related injuries were 1 of 4. The costs of personal accidents totaled 93.7 billion dollars. Canadian loss ratios were similar.*

These figures are nice to quote in a safety talk or article, but they don't help get a senior manager's decision on how much time or money to budget for an off-the-job safety program.

In order to sell off-the-job safety programs, some people have developed data for their own companies or plants. Personalized data is more meaningful when dealing with local programs.

One major manufacturing company thought the ratios were too high, until it researched its own. It found it had 18 people injured off the job for each person hurt on the job. Its fatality ratio was even higher: 35 to 1!

A large chemicals manufacturer with a very comprehensive occupational safety and health program thought the employees were so well versed in safety that they took it with them off the job. The company was surprised to find that there were 26 injured off the job for each work injury or illness. It started an off-the-job safety program and quickly achieved a 67 percent reduction in off-the-job injuries. The private vehicle fatalities and injuries became about half that of community averages at its plant locations. Also, the company noticed a 50 percent reduction in the occupational injury rates which were previously very low for the industry.

303

Managers often wonder why employees have such high off-the-job accident rates. They say that employees seem to change attitudes and behaviors when they get outside the company environment. That isn't what happens. The employees don't change at all; they just show their true selves. The environment that controlled their accident exposure is what changes.

The need for off-the-job safety is not a need for little bits of information. The need is an ability to develop and conduct a personal safety program. That requires an understanding of the methods to use, and the motivation to do all the work that is necessary. Home mailings, booklets, posters, and group meeting safety talks supplement but do not take the place of basic employee development to meet off-the-job safety needs.

Every plant, office, or other workplace is unique and has to have its own safety program if that program is to be effective. The same is true for each employee's personal life. Family size and ages, home and furnishings, motor vehicle use, recreational interests, and other aspects of a person's environment all differ. So, everyone will organize their personal safety program a bit differently.

There are, however, certain universal guidelines that are common to effective off-the-job safety programs. These are:

Guideline 1

Overall direction should be given by an experienced professional. Many aspects of a good safety program require special knowledge. Line supervisors, trained in fundamentals of safety management, can give employees advice on their problems.

The supervisors have daily contact with employees and can use this to prompt off-the-job safety. They can assist employees with personal problems and programs. They know the employees' general interests and can give appropriate information. They can refer the employees to the safety coordinator as an additional resource when more help is needed.

Guideline 2

Management commitment must be visible and consistent through all levels. Senior, middle and line managers must know, understand, and believe in the off-the-job safety program. Managers need to budget the time and other resources to complete the program activities adequately. They need to perform their tasks enthusiastically, and assess the performance by subordinates.

The managers need to be visible to employees, through such modes as talks at employee and community meetings. They need to reinforce the program through the example set by their own personal home, motor vehicle, and recreation safety programs.

Guideline 3

A well-organized program prompts individual participation and produces the best results. The major activities need to be planned so that the necessary time and materials are available when needed.

At the start, time is needed for senior and middle managers' planning meetings. In turn, time and materials are needed for supervisors, then employees' training sessions. Later, follow-up educational and promotional materials are needed. Some of these will be seasonal and need to be in hand at the start of the respective sport season, or at the start of a period of exposure to a seasonal hazard.

Guideline 4

Employee participation in activities should be spontaneous. Each program activity should include an invitation for employees to participate voluntarily in planning, preparing, presenting, and measuring.

In the beginning, selected employees may have to be asked, personally and privately, to help. A quality program motivates participation. Soon there are waiting lines forming to work on the next activity.

Activities that are meaningful, educational, practical, and helpful draw employee interest and participation. Most important, each employee must be recognized, personally and publicly, for specific participation. The greater efforts deserve the higher recognition. When in doubt, give just a little more recognition than is apparently deserved.

Guideline 5

Off-the-job safety deserves a broad loss control approach. The program suggested to employees should go beyond personal injury prevention. It should also include controlling exposures to toxic substances, fires and explosions, personal property damage, personal and property security, and personal liability. Even past that, it might help personal risk management; helping in the areas like determining insurance needs and deductible limits, or in risk transfer activities such as contracting for home repairs. The program should help employees identify potential risks and assess their loss potential so they put their efforts on the most significant risks.

Guideline 6

The program should be family oriented. The family is the key group for sharing ideas and for changing behavior. The employee and his or her spouse influence each other's actions. Parents influence their children, and vice versa. The potential for improving safety through family involvement is exciting!

A Program Plan

The organization's involvement in off-the-job safety programs should be concentrated on helping people identify and treat the exposures that could be harmful to their families. These exposures will vary a lot. The risks are so broad that no program could treat them all. The risks vary with culture, climate, heredity, economic level, and personal interests. But, the risks for a particular family must be pinpointed before the appropriate control measures can be picked.

The general types of risks employees face as a group must first be identified as the starting point. Once the general risks are known, then the program can help employees define their specific, personal loss exposures.

Project Team

Set up an off-the-job safety project team of managers from all levels and employees. The team members will do the preliminary work to organize the program, will develop or obtain program materials, and will review the progress and problems in program development. The team should be kept small; five to seven people at most. This has been found to be the most efficient working group.

Exposure Inventory

Make an inventory of the off-the-job loss exposures of employees as a group. Any effective safety program starts with an inventory of the exposures, followed by an evaluation of the degree of risk for each exposure. This allows one to give proper priorities to the most serious problems. Off-the-job, this inventory is a listing of the general risks in the following areas where appropriate:

Domestic. Housing and its upkeep, schools, shopping areas, and travel for life's necessities.

Recreation. Facilities and equipment used for sports, hobbies, family outings, community gatherings, and personal leisure.

Vacations. Activities and the travel needed to reach them.

Other Employment. Work done by the spouse, by the employee in a second job, and by other family members.

Future Activities. Projected changes as children grow, and as economic status and interests change with age.

The preliminary inventory is made to see what subjects need to be included in program educational and guidance activities. It is refined as accident, medical treatment, and insurance information comes in through program activities. The preliminary inventory can be done by any, or a combination, of the following:

Surveys of Employees. Questionnaires can be used to get information on interests and activities.

Discussions at group meetings. Employees can be asked what things they like to do and what hazards they have found.

Compensation and Insurance Records. Reports and summaries can be analyzed for agencies of accidents and types of injuries.

Accident Reports. If there is a comprehensive investigation program, reports may include off-the-job injuries.

Supervisor Interviews. Individual supervisors can be asked about what activities their people engage in and other personal interests they have learned of through their daily contacts.

Discussions Among the Project Team. Members can "brainstorm" the general scope of employee activities and the loss exposures common to the local area.

Select Key People

Organize the loss exposure inventory into general areas, such as gardening, home repair, hunting, etc. Pick an employee with the most appropriate knowledge or experience in the subject area of loss exposure to be the key person for that area. The key person will coordinate the development of the program for that subject. He or she will refine the loss exposure identification, assess the risks under varying conditions, and be a resource on safety measures to control the loss exposures. The key person might be responsible for selecting general promotion materials, for outlining pertinent safety talks, for an exhibit at a safety fair, for coordinating an extensive theme campaign, or for organizing a club.

Set Up Safety Circles

Organize small project teams for subject areas. These teams assist the key person. They meet to help define all the possible accident exposures and to develop accident prevention measures to suggest to families.

When possible, include family members in the project teams. A spouse or adolescent child might have interests and knowledge that employees do not have. The safety circles develop ideas and present them for company approval and sponsorship. Circle members assist other employees with the parts of their family safety programs related to the circle's subject area.

Audit Performance

Measure the quantity and quality of work in the program done by project teams, supervisors and employees. Measure the guidance and direction given by key people and the safety staff. Provide systems for the employees to measure their own personal safety programs. Finally, measure the results of the program in changes in accident costs.

Commend Good Performance

Recognize those who have contributed to the program. Verbal commendation is usually all that is necessary, but many organizations like to give tangible items. The most appropriate awards are items that will add to the person's safety program, such as smoke detectors and fire extinguishers. Companies have also found it beneficial to award items such as distinctive jackets or shirts. These, when worn to work or at social gatherings, prompt others to get involved in the program.

While off-job injuries are not a Workers' Compensation problem, they remain a serious problem to any organization. The organization pays through the group insurance and through other hidden costs.

Off-job safety programs often include the entire family. There are many excellent off-job or family safety programs. Michael Lacivita of Commercial Shearing, Inc., describes some elements of his:

The "Hear and See" Program

The basis for the program is a series of relatively large, amusingly illustrated posters with a different topic every month. There is no dearth of subject material. The supervisor expands upon the poster's main message while it is facing his audience. It is then displayed in his department for a month to remind and reinforce that particular topic.

Recognizing that safety encompasses everyone's lives, we decided to promote off-the-job safety as well and bring employees and their families into the safety picture.

Involving entire families in safety consciousness pays dividends, both economic as well as humanitarian. The matter of personal attitudes which can spread to every member can be created and furthered in a positive way by providing information about safety practices to this important audience—important because the majority of accidents occur outside of the workplace.

In 1979 we developed and distributed an informative booklet titled "Did You Know," again well illustrated, which emphasized safety tips applicable to home, school, and office as well as plant. We related safety subjects to the safety instincts of a wide variety of animals appealing to people of every age. Understanding that accidents can be prevented by using good common safety sense, it was natural to relate such sense to the safety instincts of the animal kingdom.

Catchy decals were also developed and quickly began to appear on bikes, lunch boxes, school books, bumpers and around homes and on hard hats, of course. The visibility is high and the chances of retaining the message are fortified by the constant presence of these colorful decals. Judging from the number of requests received for additional copies, and more decals, the campaign did not go unnoticed.

In fact many of our customers in the mining industry have requested our "Be Safe-Not Sorry" decals picturing a sad-faced bassett hound after seeing these decals on our field salesmen's hard hats.

To encourage family involvement, in 1980 we launched an annual safely contest with prizes to youths of families. Here, the idea was to give children free rein to pose a safety problem and provide a solution. Both the number of entries and the ideas depicted were surprising for their thought and quality. It was evident, too, that some entries were serious family projects.

This contest was limited to employees' children under 12 years old. The appeal was to "Help Safety Sam Stop the Accident Bad Guys." Simple enough, but effective in starting conversations and, it is hoped, planting the seed of safety awareness in the young. To help the youngsters get started, examples of bad guys were "Pinch Point Pete," "Thoughtless Theodore," and "Glassless Gary."

The winners gained recognition by having their pictures published in our house organ and their entries used as a topic of our monthly "hear and see" safety poster.

We believe that this safety competition made a lasting impression on everyone, including those employees with older or no children, since entry blanks, available at all time clock locations, were widely posted prior to the contest.

For the most part off-job safety campaigns fall into the category of communications campaigns, perhaps with some gimmickry. Do they work? The companies involved believe they do. Most cite improved accident records both on and off job. In actuality, this writer is not aware of any off-job program that has incorporated in its design any measurement to really know the effect of the campaign.

WELLNESS PROGRAMS

One of the thrusts of the 1980s has been the structuring of company wellness programs. While these are covered in more detail in the companion book *Safety Management—A Human Approach, Second Edition,* a short overview of the elements of such a program is provided here. The health (physical and psychological) of a person is one factor in staying well, being injured, being stressed out, therefore, the growth in wellness programs in industry in the 1980s. There is a strong trend in industry toward adopting wellness concepts and programs, of dealing with the whole worker, whatever that might mean. What wellness is and means might vary somewhat depending upon to whom we are talking.

Traditionally, medicine "cures" us. As individuals, we take little or no responsibility for our lifestyle or habits. We go along with our overweight, overdrinking, smoking habits and shun any medical advice until we're in the throes of a crisis. Then we expect the medical community to bail us out—pills, surgery, whatever it takes to "fix" us and make us well again. Unfortunately, what we attempt to "fix" are only the symptoms, not the root or cause of the problem.

"Well" medicine, on the other hand, takes the opposite approach. The individual is responsible for his or her own lifestyle and habits. Each individual manages his or her physical fitness, nutrition, stress, spiritual anchoring. When a person is ill, that person looks for the cause of the problem instead of at the runny nose or hacking cough. The medical profession works as a partner in resolving the illness, rather than as a "fix-it" handyman.

Some estimates suggest that as much as 70 percent of all our illnesses and over 50 percent of all our medical costs are related to our lifestyle. For instance, look at the factors related to coronary heart disease, the leading cause of death in America:

- Cigarette smoking
- Blood fats
- Family history
- High blood pressure
- Obesity
- Diabetes
- Sedentary habits
- Personality

Each increases our likelihood of heart attack. Most are lifestyle related.

Elements of a Wellness Program

The following elements are a part of a number of current programs:

- An exercise/fitness program.
- A health appraisal program.
- A substance abuse program.
- Smoking cessation programs.
- Nutrition programs.
- Stress control.
- Asymptomatic approaches.
- Employee Assistance Programs.

Some of the above are obvious as to content. Some need explanation.

Health Appraisal

In a health appraisal program the various risk factors of an employee are identified. The employee completes a questionnaire which covers current medical status, past medical history, family history, and social history. Lifestyle questions are included, plus numerical data such as height, weight, blood pressure, and cholesterol levels. The information is then computer analyzed and related to the twelve leading causes of death in this country. The result gives the individual's current health status, how the status relates to the health status of the average healthy person of the same sex and age, and, if needed, points out ways of changing health risks. The individual can then be directed to whatever changes might be indicated in the appraisal: smoking withdrawal, nutritional counselling, safety, alcohol usage, exercise program, or stress management.

Asymptomatic Approaches

Part of a wellness program addresses finding diseases in the asymptomatic person. Obviously it would be expensive and wasteful to look for all diseases during the asymptomatic period. The decision whether to include a disease in any surveillance program is based upon the following criteria to determine whether such a program would be worthwhile:

— The disease must occur frequently enough to be looked for.
— There should be some inexpensive and relatively easily performed test available that would detect the disease during the asymptomatic phase.
— There must be an effective treatment available.
— This treatment introduced during the asymptomatic phase must yield better results than treatment introduced during the symptomatic phase.

For example, cancer of the colon would qualify as a disease that meets the above criteria. It is the second most common cancer in man and the third most common cancer in women. There is an inexpensive test available. There is very effective treatment, and lastly, if treatment is introduced during the asymptomatic phase the cure rate is better than 90 percent. This cure rate falls to approximately 50 percent if introduced during the symptomatic phase. Disease of the gall bladder should *not* qualify. Although it occurs frequently, the test is not inexpensive. There is an effective treatment available but the treatment introduced during the asymptomatic phase yields no better results than treatment introduced during the symptomatic phase. In other words, treatment after the individual becomes sick achieves the same results.

Employee Assistance Programs (EAPs)

Many of the problems which impact health and safety on the job originate outside the work environment. The changing family structure, single parents, and alcohol/drug problems are some of the major problems in our society which affect employers and employees resulting in increasing absenteeism, accidents, increased costs of health care or declining productivity. EAPs are emerging as one of the ways industry can reduce the cost of these problems.

The problems employees present to a comprehensive EAP include the following:

— Alcohol and/or drug problems.
— Marital and family relations.
— Emotional problems.
— Single parenting.
— Financial.
— Legal.
— Work related problems.

An EAP should offer features such as:

— 24 hours/7 days a week service.
— Confidential services.
— Voluntary participation.
— Serves employees and families.
— Free of charge to the employees and their families.

- Covers any problem.
- By appointment.

Substance Abuse Programs

Another area receiving major attention today is substance abuse. There are no statistics on how many industrial injuries result from substance abuse, but it undoubtedly is a major factor. In *Safety Management—A Human Approach, Second Edition* the problem is reviewed in detail, but here's an overview with thoughts as expressed in that book and by author Bruce Wilkenson.

As big a problem as it is, it is only recently that controls for the problem have been developed in industry. The elements of a control program are:

- Preemployment Screening

    Business groups of all types are evaluating different methods of screening drug users at the preemployment level. It stands to reason that drug users will avoid applying for jobs where drug screening is required. Therefore, it simply means that if a company is not screening for drug users, these potential employees will appear on the job site.

    Employers have the right to know of any potential problem that may be encountered when hiring employees. Urine drug screening at the pre-employment level is an excellent loss prevention method. Management has an obligation to provide a safe and healthy work place for all employees and has the right to hire the most qualified applicants for the available job positions.

- Identifying problem employees

    Drug users are sometimes difficult to identify unless some type of urine or blood and plasma screening procedures are incorporated in a company's drug program. However, drug users do establish a pattern of unusual behavioral habits, that, if alert, company representatives can detect.

    Some of the symptoms include:.

    - An increase in quality control problems.
    - Low production output.
    - An increase in automobile liability premiums.
    - Increased absenteeism.
    - Workers' Compensation rates increases.
    - Signs of employee theft surface through inventory shortages.
    - Morale begins to deteriorate.

Supervisors are the first line of defense in combating employee drug abuse. Their observations come from working with the employee at the point of operation. Some of the signs they notice are:

- Some employees taking more breaks than others.
- Employees going to the restroom or their lockers in groups.
- A sudden change in individual personalities; short tempers where patience once existed.
- An increase in employee reprimands.

— A noticeable rise in employee complaints.
— Missing tools and equipment.
— Individual loss of work quality or production output.
— Employees sleeping on the job or showing up in areas where they do not belong.

An EAP can be an integral part of the substance abuse program. Employees are referred to EAP counselors by supervisors for either poor job performance, a positive urine screen test, or the employee admitting to a drug or alcohol problem.

Counselors will attempt to assist employees in solving their problems and return a more productive worker to the supervisor. This task may be accomplished through group therapy, one-on-one sessions, or encounters with the employee's immediate family and support groups. Periodic monitoring of an employee's drug habit may be done in conjunction with the company's ongoing urine drug screening program.

The key to a successful drug abuse program is the development of a comprehensive policy. This policy must be detailed and broad enough to address all situations that are likely to be encountered. It is important to define which employees will be covered by the policy.

Consolidate as many elements as possible into one overall comprehensive policy to avoid confusion when it is necessary to discipline employees. The policy also provides excellent documentation in the defense of a legal challenge.

STRESS MANAGEMENT PROGRAMS

The newest, and potentially most costly, of the subjective injury problems in the Workers' Compensation payout as a result of illness is due to psychological stress on the job. While this problem is large enough to deserve much greater treatment than in one chapter, here is an overview of the problem and some suggested control areas.

The problem is large. In 1900, our top killers were not stress related illnesses, but by 1970, four of the top five killers were stress related. These shifts are beginning to cost industry in Workers' Compensation also. In 1980, stress claims were minimal in this country. By 1982, the problem had grown to where 11 percent of the Workers' Compensation health payout had become stress related problems.

By 1985, the problem was the number one occupational health payout under Workers' Compensation. The Center for Disease Control stated:

> There is increasing evidence that an unsatisfactory work environment may contribute to psychological disorders.
> Studies show the factors contributing to unsatisfactory jobs include lack of control over working conditions, non-supportive supervisors or co-workers, limited job opportunities, role ambiguity or conflict, rotating-shift work, and machine-paced work.
> Mental stress can produce such illnesses as neuroses, anxiety, irritability, amnesia, headaches, and gastrointestinal symptoms.
> Average medical costs and indemnity payments in 1981–1982 for these forms of mental stress actually surpassed the average amounts for other occupational diseases.

The trends discussed by the Center for Disease Control indicate clearly the direction for the future. Without some legislative intervention, the problem could be massive for industry. One of the reasons for this is the variety of diseases caused by stress that could end up compensable. For instance, here's a partial list:

— Coronary heart disease	— Diabetes
— Hypertension	— Gout
— Ulcers	— Migraines
— Colitis	— Glaucoma
— Anxiety	— Epilepsy
— Depression	— Hemmorhoids
— Allergies	— Asthma
— Arthritis	— Acne
— Cancer	— Back pain

The reasons for the connection between stress and these illnesses can be better understood when examining the body's responses to stress.

Our natural physiological responses to stress cause physical problems. Those responses (see Exhibit 14.1) were completely appropriate in another age. For the caveman faced with a stressful situation, the physiological responses prepared him for action to deal with the stress either to fight or for flight. Today usually the threat is not real, but holds only symbolic significance. Our lives

Natural Response	Original Benefit	Today's Drawback
Release of cortisone from the adrenal glands	Protection from an allergic reaction to dust from a fight	Cortisone destroys resistance to cancer, other illnesses
Thyroid hormone increases in the bloodstream	Speeds up body's metabolism providing extra energy	Shaky nerves, exhaustion, jumpiness, insomnia
Release of endorphin from the hypothalamus	Provides a potent pain killer so can't feel wounds incurred	Aggravates migraines, backaches, even pains of arthritis
Reduction in sex hormones testoseronec progesterone	Decreased fertility, helps in overcrowding; and loneliness	Anxieties and failures in sex frustration, irritation
Shutdown of the digestive tract	Blood diverted to muscles More power for fight	Dry mouth, bloating, diarrhea, discomfort, cramps
Release of sugar into the blood, with insulin to metabolize it	Quick energy supply to escape danger	Diabetes, hypoglycemia
Increase of cholesterol in blood from liver	Long distance energy to the muscles for fight or flight	Coronary, heart problems
Racing heartbeat	More blood to muscles and lungs for the fight	High blood pressure, strokes
Increased air supply	Extra oxygen to the lungs for the fight or flight	Dangerous for smokers
The blood thickens	More capacity to carry oxygen; fight infection from wounds	Strokes, heart attacks

Exhibit 14.1 Natural responses to stress.

are not in danger, only our egos. Physical action is not warranted and must be subdued, but for the body organs it is too late: what took only minutes to start will take hours to undo. The stress products are flowing through the system will activate various organs until they are reabsorbed back into storage or gradually used by the body. And while this gradual process is taking place, the body organs suffer.

This fight or flight reaction has helped ensure our survival and continues to do so; no amount of relaxation training can ever diminish the intensity of this innate reflex. Stress is physical, intended to enable a physical response to a physical threat; however, *any* threat—physical or symbolic—can bring about this response. Once the stimulation of the event penetrates the psychological defenses, the body prepares for action. Increased hormonal secretion, cardiovascular activity, and energy supply signify a state of stress, a state of extreme readiness to act as soon as the voluntary control centers decide the form of the action, which in our social situation is often no "action".

The damaging effects of stress can be categorized as either (1) changes in the physiological processes that alter resistance to disease, or (2) pathological changes, that is, organ system fatigue or malfunction, that result directly from prolonged overactivity of specific stress organs. There is a gastrointestinal system response, a brain response, a cardiovascular response, and a skin response; thus the connection to so many diseases.

By definition, stress is an arousal reaction to some stimulus—be it an event, object, or person— characterized by heightened arousal of physiological and psychological processes. The stimulus that causes this arousal reaction is the *Stressor* (see Exhibit 14.2).

Is stress a safety problem? As shown in Exhibit 14.2, stress is clearly work caused. All of the psychosocial stressors occur at work. Any one who has ever worked in an organization has experienced and is probably now experiencing the pressure to adapt to someone else's will and ideas; is experiencing some form of frustration (a block between you and your goal); has been overloaded at some point, either physically, physiologically or psychologically; or has felt deprived of filling some basic human needs on the job. Some of the bioecological and personality causes are also job related. Stress related illnesses are clearly job caused and there is little rationale behind saying they should not be compensable.

The problem is that these same stressors occur elsewhere also; at home or in some social situations.

The greater degree to which persons perceive themselves in control of a situation, the less severe their stress reaction. This suggests that feeling helpless and feeling a lack of sufficient power to change one's environment may be a fundamental cause of distress. Thus, anything that adds to the feeling of self-control is likely to reduce the severity of the stress reaction.

Psychosocial	
Adaptation	Overload
Frustration	Deprivation
Bioecological	
Biorhythms	Noise
Nutrition	
Personality	
Self-perception	Anxiety
Behavioral patterns	

Exhibit 14.2 Stressors.

In 1977 NIOSH examined the health records of 22,000 workers in 130 occupations in Tennessee. Some occupations had a higher than expected incidence of stress-related disorders and 12 of those were especially high; 13 occupations had fewer stress-related disorders than expected. The top twelve, in order, are:

— Laborer
— Secretary
— Inspector
— Clinical lab technician
— Office manager
— Foreman
— Manager/administrator
— Waitress/waiter
— Machine operator
— Farm owner
— Mine operator
— Painter (not artist)

The bottom thirteen (not in order) are:

— Sewer worker
— Checker, examiner
— Stockhandler
— Craftsman
— Maid
— Farm laborer
— Heavy equipment operator
— Freight handler
— Child care worker
— Packer, wrapper
— College or university professor
— Personnel/labor relations
— Auctioneer huckster

A quick analysis of these 25 job categories suggests some proof of the control issue. The most stressful occupations (laborer, secretary) were those where there is little self control on the job. For the most part the individual is under the direct control of someone else all day in these occupations. It is the feeling of lack of control that leads to the stressful situation.

It is now generally accepted that stress is a significant part of performing in a job for an employer or even for oneself (as an entrepreneur). Furthermore, empirical evidence points toward the notion that excessive job stress is associated with negative health consequences. Despite the general acceptance of the pervasiveness of stress and the growing empirical evidence about stress's effects on health, only a limited amount of research rigorously evaluates the effectiveness of stress management programs within organizations.

As shown in Exhibit 14.3, the elements of a stress control program in an organization fall into two categories—those things the organization can do and those things the organization can sponsor that the individual can do. It is interesting to note that many of the organizational strategies are the exact same strategies that we discussed earlier in this book for injury control (Chapter 7).

Organizational Strategies

Goal Setting	Performance Feedback Systems
Participative Decision Making	Role Specification
Job Enrichment	Employee Surveys
Work Scheduling	Training Programs
Culture Change	Wellness Programs

Individual Strategies

Cognitive Appraisal	Breathing
Restructuring	Progressive Relaxation
Transcendential Meditation (TM)	Biofeedback
Relaxation Response	Exercise
Social Support Systems	Diet
Employee Assistance Programs	Cranial Electrotherapy Stimulation
Rehearsal	

Exhibit 14.3 Stress control program elements.

They have to do with improvement of the quality of work life goal setting, participation, job enrichment, improving the culture and climate, performance feedback, specification of roles, responsibilities and accountabilities (defuzzying things), surveying employee perceptions, training programs, wellness programs, and employee assistance programs.

The reason for the similarity is obvious. Since stress arises from mature adult human beings, who feel as if they are able to control their own behavior, finding themselves in a work environment where they are totally under the control of someone or something else, then the answer to stress problems is to exert less control over those individuals, or rather to shift that control from organization to self.

OTHER PROGRAMS

There are numerous other specific individual programs, or procedures, that need to be an integral part of a comprehensive safety system. Not all can be dealt with here, but two of them, fleet safety and product safety, are important enough to deal with separately in the next two chapters.

REFERENCES

Kuhlman, R. "The Need for Off-job Safety Programs, *Professional Safety,* March 1986.
Laciviti, M. "On and Off-the-job Safety," *Professional Safety,* October 1983.

Chapter Fifteen

Safety in the Adjunct Fleet

The staff safety specialist usually has no defined responsibility in fleet safety. Often no one in the organization has such responsibility. The safety specialist ignores this part of the operation and fleet management ignores the safety segment of its responsibility. Too often the fleet operation is allowed to go uncontrolled.

Statistics show that in many states (perhaps all states) the motor vehicle is the number one cause of occupational death—the number one industrial killer.

Stanley Abercrombie, writing in *Professional Safety,* offers these statistics:

> Mounting evidence shows that safety professionals should raise motor fleet safety to a position of greater importance on their agendas. Although business/industry losses from motor vehicle accidents have been a matter of record for many years, the full significance of these losses seems to have been muffled or shoved aside while attention focused mostly on the usual array of occupationally related accidents—those injurious events involving machines, equipment, processes, and amblent conditions common to work environments.
>
> Let's look at some examples of the mounting evidence of motor-vehicle-related losses to business and industry:
>
> - In 1971 the American Trucking Associations found that accidents involving over-the-road motor vehicles (including insurance) cost common carrier fleets nearly 5 percent of their gross revenues—*slightly more* than profits before taxes that same year.
> - In private sector establishments with 11 or more employees, according to the U.S. Department of Labor, 29 percent of the 4,950 work-related deaths in 1978 resulted from car/truck accidents. Exhibit 15.1 shows the percentage breakdowns by industry divisions for that year.
> - In 1981 an Oregon Workers' Compensation Department official, expressing pleasure with a sharp downturn in occupational deaths, said "we're still very concerned about the continuing number of deaths caused by traffic accidents."
>
>   These accidents led fatality causes in the logging, manufacturing, and construction industries. A year later Oregon again found motor vehicle accidents to be the leading cause of work-related deaths in the logging, manufacturing, and transportation and public utilities industries.

To ignore this problem is to ignore one of the major causes of injury and death. In short, the safety professional cannot afford to ignore the fleet of vehicles that his or her company runs.

This chapter is not intended for the safety professional who works with the true fleet operator, but rather for one whose responsibility extends to a small fleet of vehicles that is adjunct to the primary operations of the company. This might be the contractor's fleet, the manufacturer's pickup and delivery trucks, the fleet of bakery or dairy trucks, or the sales

Industry Division	Percent
Finance, insurance, and real estate	74
Transportation and public utilities	49
Wholesale and retail trade	33
Agriculture, forestry, and fishing	32
Services	26
Mining—oil and gas extraction only	19
Manufacturing	19
Construction	16
Total (all industry divisions)	29

Exhibit 15.1 Percent of occupational fatalities involving over-the-road motor vehicles, by industry, 1978.

department's private passenger fleet. This chapter gives broad outlines of the accident controls found to be effective with this type of fleet operation.

In dealing with safety in the adjunct fleet, keep in mind the basic principles expressed in Chapter 2. Principle 2 states that certain sets of circumstances can be predicted to produce severe injuries. The adjunct fleet offers such a set of circumstances, and statistics show that severity is a distinct possibility, if not a probability. Why, then, do safety professionals so often tend to ignore this source of high-potential accidents in our operations?

FAILURE TO DEAL WITH FLEET SAFETY

Lack of Communication

One reason the adjunct fleet is ignored is that safety staff are often so positioned in the organization that it is impossible to effectively reach and influence fleet management. Often safety staff report to industrial relations or to production management, whereas the fleet reports to an officer who is not closely connected with either industrial relations or production. The adjunct fleet may be a part of the sales department, or report to a traffic manager, to the head of material handling, or elsewhere. In terms of the principles expressed in Chapter 4, the safety professional who reports directly to the manager in charge of a particular activity probably will not be able to influence other areas, such as the fleet, easily.

Differences in Fleet Safety Work

Industrial safety specialists may not extend their influence to the fleet because they may not feel comfortable in fleet safety work; they may feel that fleet safety is different from plant safety and that the approaches normally used to effect results will not work here. In fact, fleet safety is different from plant safety in two major ways, discussed below.

— *Lack of Supervision.* In the plant or on the construction job all employees are under constant supervision. On the road employees are not supervised; they are on their own. This is perhaps the biggest difference between plant safety and fleet safety. Since supervisory control is so much more limited in fleet safety, we must rely more heavily on some

other controls—in this case, on deciding who will drive initially. Selection of personnel becomes critical in fleet safety.
— *The Environment.* In industrial safety we lean heavily on controlling the environment— on making it as safe as possible. In fleet safety we can control the condition of our own vehicle, but most of the rest of the driving environment is not under our control. The other driver, the condition of the other vehicle, and the road are uncontrollable. Much of this lack of environmental control can be balanced by the quality of the driver. Consequently, the selection and training of that driver are of paramount importance.

The Safety Professional's Role in Fleet Safety

Although fleet safety is different in some aspects from industrial safety, the approach of the safety professional remains unchanged. Principle 5 of Chapter 2 is a case in point. The following are the areas that might be looked at first. Notice the similarity between these areas and those in industrial safety.

— Management's policy
— Driver selection
— Driver training
— Vehicle maintenance and design
— Records

Policy

A statement of management intent is as essential in fleet accident control as it is in plant safety. The policy in fleet accident control may be separate, or it may be an integral part of the total management safety policy, depending on the organization.
In either case the following must be spelled out:

— Management considers safety on the road important.
— The corporate safety program will apply to the driver.
— Employee cooperation is expected.
— Specific responsibilities for safety have been assigned to the various levels of management.
— Accountability will be fixed.

Driver Selection

Selection is the single most important control that management has in fleet safety. Proper selection of drivers requires that management determine the abilities and skills of applicants for the driving job. To determine this, it attempts to obtain information on the driver's experience and performance on previous driving jobs, job knowledge (technical know-how), and attitude toward safety. Management should also consider the driver's job performance during the probationary period, which is often overlooked.
The specific tools of selection are:

— The application form
— The interview

— The reference check
— The license check
— The physical examination
— Written tests
— Road and yard tests

In the screening process, management's first job is to determine how applicants have lived and worked in the past. This is the best single indicator of how they will live and work in the future. Of the above list of selection tools, those concerning past performance are the most important indicators.

One of the best tools is the application form, since it covers the essential facts:

— Driving experience—local, long haul.
— Job performance record. Does the applicant stay with a job?
— Responsibility, maturity, and stability.
— Past safety performance.

The personal interview provides the face-to-face encounter with the applicant. Here the interviewer can appraise the person's knowledge, attitude, character, and maturity. The purpose of the interview is to gather facts. In the interview, these ought to be discussed:

— Driving experience
— Knowledge and education
— Knowledge of the vehicle
— Experience with the vehicle, maaintenance
— Record—arrests, violations

After the application and the interview, some reference checks should be made. Checking prior work references establishes the validity of the information obtained and can be done simply and inexpensively over the telephone. The following should be determined:

— Was the applicant employed by the company as stated?
— What type of work was he or she engaged in?
— What was the applicant's absentee record?
— What was his or her wage record?
— What were the appicant's reasons for leaving?
— Would the company rehire this person?

In addition to making the reference check, management should ensure that the applicant has a valid driver's license. Also, a routine check with the applicable state agency for past violations is well worthwhile.

Every applicant should be given a physical examination to determine whether he or she meets the physical requirements for driving. The driving job, perhaps more than most other jobs in industry, requires this physical check. It seems inconceivable that any manager would put an expensive vehicle, and the possibility of a million-dollar lawsuit, into the hands of a person who cannot see or is subject to blackouts. And yet this happens daily. It happens even in companies with sophisticated safety programs, where those programs do not seem to apply to the salesperson in the company car.

The application, interview, and physical examination are the central part of any selection system. Management would do well to ensure that the potential employee comes through these well before adding some of the "frills."

Various kinds of tests in addition to the above are often valuable. These tests can take many forms and generally include those questions which the company thinks its drivers should be able to answer. Standardized tests are also available. For example, the yard test is used to determine a driver's skill in handling the equipment without going into traffic. Some exercises often included are the parallel park and the alley dock. This test does not predict whether a person will be a good driver—only whether he or she can maneuver the equipment. The road test (in traffic) should be made over a predetermined route which approximates the kind of driving that the applicant will be required to do if hired.

Driver Training

After a driver is finally put on the payroll, he or she must be oriented and then trained. Most of the principles discussed in the section on training as a tool of motivation apply to driver training also. The training given should, as much as possible, aim at defined needs and should accomplish stated training objectives. Objectives are essential to quality training. Basically, the content of a training program is what management wants the workers to know.

Orientation might include:

— Company policies and practices.
— State, county, and local traffic laws.
— Defensive driving.
— Customer or public relations.
— Concepts of safety.

The ongoing training of drivers might include the following:

— Vehicle operation.
— Vehicle condition, including how to check the vehicle daily and what to report.
— Use of company forms.
— Emergency procedures, including roadside warning devices, use of fire extinguishers, proper use of accident report forms, witness cards, conduct at the scene of an accident.
— Demonstrations, in the use of the brake detonator, skill-exercising equipment, and other training devices.
— Psychophysical testing devices. The validity of these lies in their training value, not in their use as measurement devices. They can be used to train drivers on reaction time, night vision, and stopping distance.
— Films and visual aids.

Vehicle Maintenance

Fleets usually evolve their own systems of preventive maintenance, which are devised to fit their special needs. Maintenance of leased fleets varies according to the contract between lessee and lessor. A fixed-cost or full-maintenance arrangement usually includes maintenance, repair, and insurance. A financed lease is one in which the leasing company supplies the vehicles and the lessor conducts the maintenance. Regardless of the setup, a preventive maintenance system must be used. It is the role of the safety staff specialist to ensure that whatever the arrangement for maintenance, the program will be effective.

Generally some kind of maintenance record system is necessary to ensure effectiveness. Records should be designed to:

— Show when maintenance is needed.
— Provide a schedule of what needs to be done.
— Record what has been done.
— Give costs.

Usually, the fleet maintenance record system uses the following forms:

— *Driver's condition report*—a checklist of things to be checked daily by the driver and an order for necessary corrections (to be sent to the shop).
— *Maintenance scheduling form*—a means of ensuring that the shop schedules all vehicles periodically for routine service.
— *Service report*—a record keeping detailed inspection by mechanics when in for routine service and also repairs needed and when made.
— *Vehicle history card or folder*—a complete history of each vehicle.

Probably the best sources for comprehensive record-keeping systems and forms for preventive maintenance are those of the truck manufacturers and oil companies.

Records

Two other types of records are necessary in fleet operations driver records and accident records.

— *Driver Records.*   Every fleet needs to keep some records on each driver. The minimum is a file which contains information on:

1. Hiring
   a. Application form completed for screening
   b. Character references
   c. Results of work reference checks
   d. Record of violations and accidents
   e. Results of interviews
   f. Results of tests given
   g. Physical examination report
   h. Information on previous driver training
2. Job performance
   a. Supervisor's reports
   b. Any commendations
   c. Status in company's safe-driving award program
   d. Performance in maintenance functions
   e. Record of training received
   f. Vehicles driven or assigned

g. Losses of cargo, money, etc.
h. Any road observation reports
i. Any violations of warning notices
j. Accidents
k. Property damage reports
l. Complaints
m. Appraisals

This folder thus contains a brief account of all information pertaining to a driver's performance on a job. A summary record card can be used to coordinate all this information for ready access.

— *Accident Records.* Accident records start with the driver submitting an accident report. In the case of serious accidents, investigations are usually made by the company and the insurance carrier. Management may wish to collect, analyze, and summarize these accident reports to determine trends. Often the information and/or reports are evaluated for type of accident, immediate causes, and driver chargeability. Causes are generally classified in terms of driver, roadway, and vehicle. Of primary importance is proper analysis of driver-related causes. Usually, analyses list improper driver actions as:

— Failure to yield right-of-way
— Following too close
— Failure to signal intentions
— Excessive speed
— Failure to obey traffic signals or signs
— Improper passing
— Improper turn
— Improper backing
— Wrong traffic lane

— *Rate Formulas.* In fleet safety, as in plant safety, accident records are analyzed to determine how the company's record compares with its own past records and with the records of other companies. A number of different rate formulas can be used.

• Annual vehicle-accident frequency per million miles: Multiply the annual number of accidents for a given year by 1 million miles and divide by the actual mileage covered in that year.

$$\text{Accident frequency} = \frac{\text{annual number of accidents} \times 1 \text{ million miles}}{\text{annual mileage driven}}$$

• Annual vehicle-accident loss rate: Divide gross revenue for one year by the total dollar losses of vehicle accidents for that year.

$$\text{Vehicle-accident loss rate} = \frac{\text{annual gross revenue}}{\text{annual dollar losses from vehicle accidents}}$$

• Annual accident rate per driver: Divide the total number of accidents for one year by the total number of drivers.

$$\text{Accident rate per driver} = \frac{\text{total number of accidents}}{\text{number of drivers}}$$

- Annual employee-injury loss rate: Divide the annual gross revenue for one year by employee-injury dollar losses for that year.

$$\text{Employee-injury loss rate} = \frac{\text{annual gross revenue}}{\text{annual dollar losses from employee injury}}$$

Finally, some additional suggestions from Stanley Abercrombie:

What is needed is a whole range of new efforts to improve the quality of performance in motor fleet operations. Here are some starter ideas:

1. Raise qualifications for candidate drivers—age, education, driving experience. Also check the state department of motor vehicles for traffic violations and accidents and require a clean record over a longer period before employing.
2. Consider revising driver pay scales upward to create strong leverage in attracting better educated, more experienced, more reliable drivers. Higher compensation can be a powerful motivator in building a corps of motor fleet drivers who can be expected to perform with greater efficiency and dependability.
3. Set up safety circles among drivers to address problems and devise better operating procedures, such as (a) how to assure driver readiness upon reporting for work (proper rest, no alcohol or marijuana, not emotionally upset), or (b) how to assure habitual use of wheel chocks or other devices to prevent movement of docked vehicles.

   Having a sysem for drivers to discuss and deal with problems will serve to keep solutions fresh in their minds. And the dependable drivers will help raise less competent ones a notch or two on the performance scale.
4. Develop new check-in procedures for drivers reporting for work to weed out those who evidence recent drinking and those who have not had enough rest in the previous 8 hours. Union participation will be necessary to help develop such new procedures.

   Supervisors who achieve genuine rapport with drivers may accomplish these same goals by personally seeing and talking with each driver as he/she reports for work.
5. Employ qualified companion drivers to share the driving on extended trips or when transporting hazardous materials or cargo of very high value.

   Developing new modes of team driving could prove of great value under certain conditions and for selected trips by reducing the risk of strain or pressure build-up sometimes felt by a lone driver.
6. Trucking associations, safety councils, and insurance carriers should obtain copies of pertinent reports from the National Transportation Safety Board, and have committees study the report recommendations with a view to actions that could improve motor fleet operations. This should include actions to urge and/or help others (e.g., state motor vehicle departments, highway departments, various federal agencies) to follow through on NTSB recommendations addressed to them. Even the best recommendations are no good unless they're carried out.
7. Develop new programs of refresher instruction for drivers, emphasizing such factors as: (1) the changing mix of vehicles on the highways (increasing numbers of small cars and motorcyles that are not easily spotted in the traffic stream); (b) defensive driving techniques to counteract the many kinds of "improper driving" encountered; and (c) fuel-efficient driving techniques, which also contribute to safety.

8. Set up means for periodic driver/supervisor reviews of routes and schedules to identify changes in road or traffic conditions and to work out countermeasure strategies including petitioning appropriate authorities for selective enforcement, installation of reflectors on curves, better traffic signal timing, road surface repairs, use of alternate routes.

   Finding ways to relieve undue pressure from tight schedules can help reduce driver errors that lead to accidents.

9. Establish and continue an effective accident review committee, as described in chapter 10 of the *Motor Fleet Safety Manual*. Successful accident review committees involve rotation of members, recognize high standards of safe driving performance, and provide procedures for drivers to appeal committee findings.

10. Consider extending the use of convoy techniques to enhance safety in motor fleets: two or more trucks traveling the same route at the same time; a lead vehicle (pickup truck with amber flashing light) preceding one or more trucks as is common practice in transporting wide loads; lead vehicle and following vehicle (pickups with amber flashers) to escort trucks carrying dangerous cargo; police escort for hazardous segments of given trips.

11. Initiate or extend use of two-way radio and telephonic communication between terminals and vehicles so that drivers and dispatchers/supervisors can check with each other frequently about road, weather, and traffic conditions, trip schedules, vehicle performance, and emergencies.

12. Utilize tachographs and other technology to obtain details of drivers' performance (e.g., speed, stops, trip time) as a means of spotting problems which otherwise may go unnoticed. Union participation and/or driver safety circles can help establish reasonable goals for using "tracking" technology and thereby create positive acceptance of such activity among drivers.

   There is no reason for drivers to be left free to depart from routes and schedules, abuse vehicles, or counteract company policy while carrying out duties for which the company pays them.

18. Contract with qualified consultants to conduct external reviews of operations to uncover aspects of present practice that should be improved. Specialists looking in from the outside can often spot inefficiencies that escape the notice of internal review teams.

19. Computerize fleet data on drivers, vehicle maintenance and parts inventory, accidents, routes, and schedules to make it more readily available for sound management decision-making.

   Fleets too small to set up their own computerized data banks can join together and have a single firm provide this kind of service, much as companies contract for computerized financial services.

Conclusion

The foregoing "starter" ideas are intended as just that—to stimulate serious consideration of novel ways to increase motor fleet safety. No single suggestion is necessarily appropriate for every fleet. Some call for joint action by groups of fleets; others, by fleets working closely with concerned associations and government agencies. The "trying on for size" of these ideas may well spark effective alternatives or entirely new methods of getting at stubborn problems.

The continuation of motor fleet accidents, with their attendant economic and other costs, attests to the need for substantially upgrading present approaches and controls relating to basic practices in motor fleet operation. With 25 percent or more of all occupationally related fatalities occurring in motor fleet operations,

these accidents deserve more attention than safety professionals are now giving
them.

REFERENCES

National Safety Council: *Motor Fleet Safety Manual,* Chicago, 1966.
Abercrombie, S. A. "Enlarging the Focus on Motor Fleet Safety," *Professional Safety,* July
1983.

Chapter Sixteen

Product Safety

In product safety, as in fleet safety, often no one in the company, including the safety professional, has responsibility for controlling potential losses. This is so even though almost everyone in the organization has an interest in product safety: design, quality control, manufacturing, sales, advertising, safety, field service, and legal. This problem has become so acute that something must be done about it.

Ralph Dudley and John Heldack, writing in *Professional Safety*, describe the problem:

> During the past decade, corporate and individual U.S. consumers have become increasingly concerned over, and aware of, the safety of the products and services they use. Indicative of this concern and awareness is that in 1961 there was the first $1 million product liability award. (Product Liability Awards statistics furnished by Judy Bello of Jury Verdict Research, Solon, Ohio. There may be more awards in the U.S.A., but these statistics represent information in their files.) Between 1962 and 1975 there were 205 awards in excess of a million dollars. In the past few years the number of these awards has increased dramatically. In 1977 there were 49, in 1978 there were 58 and in 1979 there were 67 awards in excess of a million dollars. All these figures are for the U.S.A. only. Manufacturers are also concerned about the safety of their products as well as the drastic increase in the products liability premiums they must pay. Nobody knows the exact amount of these premiums. The insurance services office indicates that they have reflected the costs of products liability awards over the past decade, as well as a very substantial increase in the types of situations in which a manufacturer or other seller will be held liable. Manufacturers and other sellers now have broad exposure for damage and/or injury which may arise with respect to their products and services. In the Kollsman Instrument case for example, decided in 1963, a crash of an American Airlines Electra was attributed to a defective altimeter, exposing Kollsman to liability for all deaths and damage which had occurred.
>
> The number of parties with payment is made up of 60% Bodily Injury only, (Insurance Serves Office Product Liability Closed Claim Survey: A Technical Analysis of Survey Results, New York 1977), 37% Property Damage only, and 2.4% both Bodily Injury and Property Damage. However, when consideration is given to dollar of payment the perspective is different. The overwhelming dominance of Bodily Injury only (83%) is seen relative to Property Damage only (13.4%) and both Bodily Injury and Property Damage (3.6%). For this reason I will constrain the remaining statistics to Bodily Injury only.
>
> Seventy-three percent of all bodily injury cases are closed without a suit being filed, which involves only 6.6% of total dollar payment. Only 22.1% of cases are resolved after suits are filed, but before trial. This sounds relatively inexpensive, but it isn't. It involves a whopping 73.5% of the total payment dollar. While settlements reached during trial before court verdict, and as a result of a verdict are only 4.8%

of cases, they account for 20% of payment dollar. These are really big expenses to the manufacturer even if he wins!

For 75% of the dollar value litigations, the time from product manufacture to first report of an incident is 61 months. Similarly for 75% of the dollar value litigations from first report of an incident is 58 months. This implies that only 75% of the dollar value litigations are resolved within 10 years of manufacture. This illustrates the significant duration of the trial.

Exhibit 16.1 shows the increase in product liability claims. Exhibit 16.2 shows the type of product most involved.

This chapter is intended to present only a broad outline of the product safety problem. Before we begin discussing product loss-control principles, however, we should look at and define product liability. A knowledge of product liability and its current trends will help us to see more clearly the significance of product loss controls.

PRODUCT LIABILITY

Product liability concerns the legal responsibility of a manufacturer or seller of a product to compensate a consumer who has been harmed by the product. Under present court rulings, product liability falls into one of two basic categories: (1) negligence and (2) breach of warranty.

Negligence

A manufacturer of a product may be liable for harm caused by the product because of a failure to exercise the care of a reasonable person of ordinary prudence to see that the product would do no harm to the buyer: That is negligence.

A manufacturer's affirmative duty to exercise due care includes the making of reasonable tests and inspections to discover latent (hidden or concealed) hazards. Failure to detect what would have been noticed if reasonable inspections and tests had been performed can constitute a basis for liability should a loss occur. That the manufacturer conforms to standard practices of the industry is not necessarily a valid excuse. The injury determines whether such practices are consistent with due care.

	1963	1966	1971	1976	1980
Number of Patents Filed	85,700	88,525	104,430	102,350	104,380
Number of Products Liability Claims	50,000	100,000	375,000	600,000	895,000
Average Bodily Injury Product Claim Settlement	— —	$1,500	$5,000	$26,000	$41,100
Consumer Protection Laws in Effect (Fed. Gov't.)	12	17	32	41	44

Exhibit 16.1 Product liability claims.

Consumer Product Safety Council Estimated* National
Injuries By Product
(October 1, 1978—September 1, 1979)

Item or Activity	Thousands of Injuries	Item or Activity	Thousands of Injuries
1. Stairs	662	26. Lawn Mowers	74
2. Bicycles	526	27. Power Saws (home)	71
3. Baseball	432	28. Wrestling	65
4. Football	425	29. Tableware	64
5. Basketball	384	30. Snow Skiing	56
6. Nails, Tacks	249	31. Guns	55
7. Tables	219	32. Porches	53
8. Glass Doors	196	33. Fishing	53
9. Chairs	196	34. Poles, Exterior Walls	52
10. Playground	165	35. Doors	51
11. Beds	163	36. Sleds	51
12. Skating	162	37. Chain Saws	51
13. Bottles	130	38. Gymnastics	50
14. Cutlery	126	39. Skateboards	48
15. Desks	109	40. Hammers	47
16. Ladders	98	41. Hockey	46
17. Lumber	92	42. Razors	43
18. Drinking Glasses	91	43. Motor Scooters	43
19. Swimming	89	44. Pencils, Pens	41
20. Cans	89	45. Outerwear	40
21. Fences	87	46. Exercise Equipment	39
22. Bathtubs, Showers	80	47. Interior Walls	39
23. Soccer	79	48. Bricks	38
24. Tennis, Badminton	77	49. Boats, Motors (Recreational)	38
25. Volleyball	76	50. Cleaning Agents	38

*NOTE: These estimates are developed from National Electronic Injury Surveillance System data from 119 emergency rooms throughout the country.

Exhibit 16.2 Annual injuries by product.

In addition, the manufacturer is required to exercise care in planning and designing a product so that it is reasonably safe not only for the purpose for which it was intended but often also for possible uses to which it might be put that were unintended by the manufacturer. The manufacturer is also obligated to provide adequate warning of possible and latent dangers that may be present when the product is put to its proper and intended use.

In the past, only those who bought the product directly from the manufacturer could recover damages if product failure caused them to sustain a loss. Now anyone who suffers a loss from a product which is defective as a result of negligence by the manufacturer can recover.

Breach of Warranty

Failure of the product to perform as represented by the manufacturer constitutes breach of warranty. There are two kinds of warranty: (1) expressed or published, and (2) implied or unwritten. Until recently, courts held that there exists an implied warranty that all food products sold for human consumption are wholesome and fit to eat. The courts ruled that not only the direct purchaser but also the ultimate consumer can recover damages resulting from impure food. Now this concept has been extended to the point where courts have held that manufacturers are liable to anyone—not just direct purchasers—for breach of implied warranty involving any kind of product.

Express warranty is a written statement made by the manufacturer to the purchaser, guaranteeing that the product will or will not perform in certain ways and meet certain standards. Recently, some courts have rules that advertising and sales literature amount to warranties guaranteeing the product to the ultimate consumer and that any remote consumer can recover damages for breach of these warranties.

It is apparent that the laws dealing with product liability are changing rapidly and that the changes have tended to increase the liability of the manufacturer. Because of these changes and because of greater public knowledge concerning lawsuits, both the number of product liability cases and the average size in cost to the manufacturer are increasing rapidly. Jury Verdict Research, Incorporated, has reported that during a recent three-month period there were more of such cases in the courts than in the two preceding years combined. The average penalty in product liability cases is more than twice as high as it is in personal general liability cases. Clearly, manufacturers must evaluate what they are doing in the product loss-control area. They must determine the adequacy of their programs and formulate a plan of action for the future.

In general, the following guidelines (Montgomery & Owen, 1976; Fischer, 1974) have been held by Courts to be considered by juries when deliberating whether to impose penalties on manufacturers for unsafe products:

— The degree of risk acceptance by the consumer.
— The state of the art in this particular product line.
— The technical and economic feasibility of eliminating or of minimizing the risk of injury.
— The extent to which the manufacturer warned the consumer of the hazard. This includes graphic design, color, language, conspicuousness, and label permanency.
— Likelihood and seriousness of injury.
— Usefulness, and availability of a safer replacement product.
— Consumer's expectation and general knowledge of the product in reasonable and foreseeable use.
— Bargaining power of the manufacturer as compared with that of the consumer.

THE ROLE OF PRODUCT SAFETY IN AN ORGANIZATION

Dudley and Heldack define the product safety role this way:

> Hazards which must be guarded against include those which may be encountered at any point during fabrication, assembly, inspection, test, use, foreseeable misuse, installation, maintenance, and disposal. An example of foreseeable misuse, a manufacturer of a chair must anticipate that it will be used as a footstool, and design it accordingly. Since a seller is responsible for hazards in products obtained from others, products purchases for resale must be examined for hazards and appropriate corrective action taken. The contract of purchase should also require evidence of specified insurance, preferably with a vendor's endorsement, or provide that your company be designated as a named insured under the vendor's liability policies. A contractual indemnification agreement may be an acceptable alternative, where the seller is larger and solvent. Exhibit 16.3 shows that safety requirements should be defined at the systems levels through adherence to various codes. In

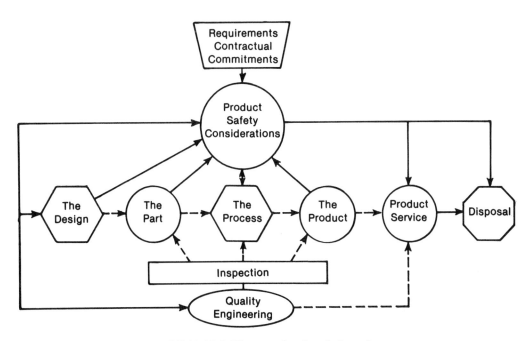

Exhibit 16.3 The product safety role.

general, recognized standards such as Underwriters Laboratories, National Electric Code, and others will be viewed by the courts as the *minimum* required, and noncompliance which causes injury will normally result in liability. Courts feel free to judge the adequacy of such standards, however, as well as the adequacy of customary industry practice, and to judge them inadequate where sufficient safety is not provided.

These requirements should be applied to all levels of hardware assembly and software by designers, and safety, reliability, maintainability people. This should be done on a continuing basis throughout the design process. It is very important that communication be maintained among all the participants.

Qualitative analysis top down (deductive), or synthesis bottom up (inductive) studies are performed to identify failure modes which would lead to hazards. Having identified the hazards, then define candidate countermeasures. In evaluating the candidate countermeasures the following hierarchy should be used to satisfy the fundamental duties imposed upon manufacturers.

a. Duty to design out hazards wherever possible.
b. Duty to provide guards and interlocks for hazards which cannot be designed out.
c. Duty to warn with respect to hazards which cannot be designed out and which remain after utilizing guards and interlocks.
d. Duty to provide procedures and instructions on proper use, servicing, etc., so as to minimize the risk of hazards.

LOSS CONTROL

The section looks briefly at the following areas and suggests how the safety professional might attack loss control problems in each:

- Management policy.
- Design and engineering.
- Purchasing.
- Manufacturing.
- Sales and advertising.
- Field service.

Management Policy

Stated policy is essential for articulating management's desires, assigning responsibility, and fixing accountability for the control function. Product safety policy may be issued separately or it may be part of management's general policy statement.

A stated policy illustrates to all personnel the need for continued emphasis on product loss control. It expresses management' s interest and concern and it commits the organization to goals for improvement. The written policy serves as the foundation for all the activities and efforts that are conducted to reduce losses. Stated policy must be followed up by stated procedure, that is, specific ways of handling product claims. This is mentioned here because in cases of claims against products, there is usually no past routine which ensures the immediate availability of accurate information needed for defense against the claim, as there is in cases of employee injuries.

Management's policy also assigns basic responsibilities, not only to those in the organization who have an essential part in ensuring that losses are controlled, but also to those who have a part in minimizing the likelihood of a successful lawsuit against the company. This means that the policy assigns responsibility to the line for manufacturing control and to the staff for their part in product safety. For instance, purchasing will have defined responsibility for incoming quality; advertising will have responsibility to ensure that copy is not legally damaging; and legal may have defined responsibility to recheck all advertising copy and service manuals.

Management may wish to center responsibility under one staff specialist or loss-control coordinator. This coordinator serves initially to determine policy and procedures for management and then to coordinate all departments under his or her direction.

Design and Engineering

Loss control is limited in design and engineering because the controls are primarily a function of the qualifications and performance of the design staff. Although the safety professional is not competent to make any judgment on such things, certain questions can be posed:

- Have written job descriptions and standards of performance been established for engineering and design personnel? Are they regularly reviewed?
- Does the head of the engineering department know whether each employee is practicing the things necessary to produce a good design?
- Is all work checked by a superior before release to manufacturing?

- Do all product failures receive a full analysis so that design and manufacturing can make any necessary changes?
- Do design and engineering personnel receive training? If so, what is the purpose of such training?
- Are rigid specifications written for each part? Are drawings and specifications tied together so that purchasing, manufacturing, and quality control follow the desires of the original design?
- Are all changes in materials and manufacturing approved by design?

Dudley & Helhack offer the following comments regarding loss control in design and engineering:

Before any prospective product is permitted to enter the development phase, it is necessary to establish that it is not unreasonably hazardous. Liability may be imposed when death, injury or damage occurs if a product is found to be unreasonably hazardous, notwithstanding the lack of any other fault or basis of liability. The following criteria are used to determine whether a product is unreasonably hazardous.

- The usefulness and desirability of the product.
- The availability of other, safer products to meet the same needs.
- The likelihood of injury and its probable severity.
- The obviousness of the danger.
- Common knowledge and normal public expectation of the danger (particularly for established products).
- The avoidability of injury by care in use of the product, including the effect of instructions and warnings.
- The ability to eliminate or reduce the danger without seriously impairing the usefulness of the product or making it unduly expensive.
- The type of user: i.e., child, consumer, sophisticated professional, etc.

An aluminized kite would be a good example of an unreasonably hazardous product, with a significant hazard from power line electrocutions, to a protected class (children), trivial benefit (appearance only), and safer alternatives.

Knives are good examples of hazardous products which are not unreasonably hazardous. Knives unavoidably need to be sharp to cut, and have significant widespread utility. At least until a safer alternative became generally available, the Pasteur treatment for rabies was another good example. There was no alternative, it was painful and dangerous, but it was lifesaving.

Exhibit 16.4 shows that plans are needed to define the direction of the safety program for each product and against which progress can be measured. Establish and implement the design safety procedure. Get the proper people to talk to each other and establish appropriate controls, acquire data, and identify problems. This process should alert management—and the customer—to potential problems and questions which require resolution. Identify candidate alternatives for corrective action; establish criteria and use these criteria to evaluate the candidate alternatives for corrective action. The preferred course of action should then be selected based on benefit cost methodology. These corrective actions are fed back into the system. Develop a set of "lessons learned" through which to develop even better, safer, products.

The design review process. Exhibits 16.1 and 16.2 emphasized product safety throughout the entire life cycle of a product. However, the upper limit of safety of hardware, maintaining reliability, quality, and cost if truly limited by the design.

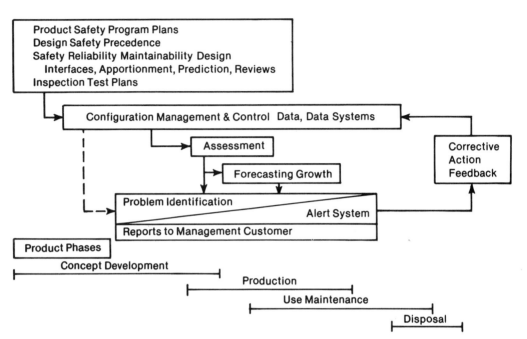

Exhibit 16.4 Product safety program.

Not only that, the initial phases of the program are the points at which the most cost effective improvements can be made in that product. So, this article will discuss the design, and especially the design review, process. It is truly a process. As with most processes, it can be described by the input, implementation, and output as shown in Exhibit 16.5.

The input to the process must be either safety analysis, working from the top down by deductive logic, or alternatively, safety synthesis, working from the bottom (or the piece parts) up by a process of inductive reasoning.

The design review is the process implementation.

The output of the process is: *specifications* at the concept formulation/early development phase; the *design* to go into production; *test* and *inspection plans* to produce good hardware; *data systems* to enable appraisal of what is going on; and *controls* to keep the system working properly.

PRODUCT HAZARD ANALYSIS

Joseph P. Ryan suggests the use of a product hazard analysis in the design phase:

In the following, critical factors relating to product safe use are presented. By considering these factors in the design stages, reasonable safety can be achieved.

  (1) Product Design Objectives
      A. Consumer Use: safety, functional, efficient, reliable, attractive, utility, life.
      B. Constraints: cost, market pricing, economic life, sales, competition.
  (2) Product Hazards
      A. Energy level: pressure, force, temperature, explosive, flammable, mass, stability.

Input
- Product Safety and Reliability Program Plans
- Analysis
    Apportionment Tradeoff Studies
    Functional and Interface Analysis
    Design Safety
    Hazards
    FMECA (Top Down)
- Synthesis
    Design Including Redundancy
    Prediction
    FMECA (Bottom up)
- Documentation Control/Review Designs, Design Changes, Test Plans and Procedures, Data, Data Systems

---

Implementation
- Product Safety Representation
- Design Review Participation (Chairman or Secretary)

Output
- Design Specifications
- Design
- Integrated Test and Inspection Plan
- Data System
- Process Controls

Exhibit 16.5 The design review process.

    B. Consumer Limbs: cutting, crushing, pinching, burning, abrasive, sharp, tearing.
    C. Consumer Health: hygiene, toxic, radiation.
    D. Surface: rough, irregular, sharp, hard, soft, slippery, floppy, hot, cold, unstable.
    E. Geometry: edges, bulky, grabbing, holding, small, too large, unwieldy.
    F. Motion: static, dynamic, rotation, reciprocating, intermittent, continuous.
    G. Safeguards: guards, presence-sensing devices, emergency-stop, fail-safe, personal clothing.
    H. Warnings and Instructions: labels, signs, instructions, operator manuals, spare parts, assembly and disassembly, clarity.
(3) Consumer Identification
    A. Skill required: minimum, mechanical, chemical, electrical, prior experience, education.
    B. Mental: age, analytical, alert, impairment, mature.
    C. Physical: strength, male, female, dexterity, right- or left-handed, tall, balanced, agile, stamina.
    D. Cognitive Processes: perceptual, long- and short-term memory, risk-taking, comprehension, decision.
    E. Psychological: behavior, emotional, calm, panic, impulsive, deliberate, cautious.
(4) Alternate Design
    A. Trade-offs: product revision, competition, consumer safety
    B. Safeguards: consumer, costs, practical, advanced technology

        C. Constraints: competition, benefit-cost analysis, pricing.
        D. Discontinue, or Revise.
  (5) Operating Manual
        A. Safety: prominent first page location, graphics, pictograms.
        B. Instructions: safe use, pictograms, simple assembly steps, common tools, strength requirements, assistance requirements, cleaning, maintenance, storage.
        C. Warnings: cautionary instructions, labels, signs, conspicuousness, durability, attention/alert emphasis, pictograms.
        D. Spare Parts: illustrations, identification, ordering.
        E. User Post-Purchase Safety Notification Mailing List.

Purchasing

In purchasing, the staff specialist can do a good job of analyzing present controls and spotting weakness. Some of the things to observe are:

— What specifications are there for purchasing?
— Are the purchasing specifications set by engineering? Are they based on product tests and known standards?
— How are incoming parts or materials tested?
— Are proper records kept of these tests?
— Are materials or incoming parts identified so that they can be traced to the supplier?
— What procedures ensure that design changes are communicated to purchasing?
— What is the warranty setup with suppliers?

Manufacturing

Most manufacturers have some product quality controls in effect. The product safety specialist need only ensure that such controls are adequate and are properly in effect. The following might be determined:

— What kinds of inspections are made in manufacturing?

— Trial run	— Percentage
— First piece	— Preassembly
— Pilot piece	— Functional
— Working	— Efficiency
— Key operation	— Endurance
— Sampling	— Destructive
— Product	

— What inspection systems are used?

— Patrolling	— Centralized
— Before	— After
— During	

— What methods of inspection are in use?

— Visual	— Measuring
— Gauging	— Testing

- Are major parts and assemblies identified with parts numbers? Are records kept of major parts and assemblies?
- Does the finished product have a key identification number? Where is it located? Will the number stay on the product despite rough usage?
- What records are kept for production?
- What type of rejects result from production? Do they have any salvage value?
- What happens to factory rejects? Are they destroyed? Sold as scrap? Sold to the public at reduced prices?
- What procedure is followed when a change is made during a manufacturing run? Are dealers or field service notified of changes which may affect service of the product?
- Are the products certified or inspected by Underwriters Laboratories, American Gas Association, or a similar agency?

Sales and Advertising

As previously said, either an expressed or an implied warranty may go with a product. Occasionally, manufacturers also make use of disclaimers, which are written statements that disclaim or limit the liability for a product. Such disclaimers are of little use with consumer products but can be of value with contracts in which the parties are of equal strength.

When a product has inherent hazards, the laws require that the customer be warned of these hazards. Even when laws are not specific on this point, the company may be liable if warning is *not* given.

The product safety specialist in sales and advertising should not attempt to become a legal expert. Rather, his or her role should be that of a liaison person—one who ensures that legal personnel do, in fact, review all labels, literature, manuals. Some things that should be watched for are:

- Invalid claims concerning product performance.
- Incorrect specifications.
- Statements to the effect that a product is "perfect," "foolproof," or safe for use "under all conditions."

Labelling

In general, acceptable warning hazard labels are based on the following criteria:

- The label should use the generic name of the material that is hazardous. For example, print the words "sodium hydroxide" rather than "cleaner."
- The language should be simple, direct, and state the message briefly. Further, the message conveying the hazard should be separate from the description of the product constituents.
- The label should be colorful, using graphics to assist in cognition. Further, the label should be placed in a conspicuous location and not hidden from full view.
- The label should instruct the consumer what to do if an injury takes place. For example: Wash skin quickly with water, or Admit to hospital immediately.
- The label should caution and instruct the consumer to provide personal protection, if necessary. For example: Wear safety glasses when using this saw.

Field Service

It is important that a communications network be established between the manufacturer and the field salespeople, agents, and retail outlets which handle the product. Often statements made by the salespeople can be damaging to a product claim.

Also, when a loss occurs, the field representative of the manufacturer is the person closest to the customer and hence the one most likely to get accurate information. A system must be set up, however, to ensure that this information is obtained. Some questions that a product specialist might ask when investigating controls in this area are:

- Does the manufacturer clearly state what maintenance and service are required for example, in service manuals and owner manuals)?
- Does the dealer have preparation and installation responsibilities?
- Is the dealer assisted through factory training and bulletin services?
- Who does installation or assembly?
- Are installation or assembly instructions clearly written and, if necessary, properly illustrated?
- Are adequate warnings given on key aspects of installation or assembly?
- Is the manufacturer the only source of replacement parts?
- Are parts lists provided? What is the location of parts depots?
- Are parts available for obsolete products? What is the source of such parts? Is there a need to replace entire assemblies?
- What procedure is followed when manufacturing changes require parts to be replaced with dissimilar parts?
- Are major hazards identified? Are there warnings concerning load limits, electrical grounding, overspeeding, and excessive temperature and pressure?
- Are operating instructions clearly written? Are major controls identified? Are starting directions, etc., included?
- Does the manufacturer provide operator training programs?
- Can recall or field repair be accomplished?
- What procedures are set up for accomplishing recall or field repair?
- Is there a standard procedure for dealing with customer complaints?

Loss Procedures

A major role of the product specialist in loss procedures is to ensure that the procedures instituted allow for the best possible handling of product claims. Such procedures are for the following purposes:

- To get prompt reports from the field of incidents which may lead to potential product liability cases.
- To get complete information regarding the incident.
- To promptly report the incident, with as much information as possible, to the insurance carrier.
- To assist the insurance carrier in the investigation of the incident.
- To inform company management of the incident and of possible ramifications in regard to sales, design, and manufacturing.
- To assist in the examination of the defective product and to preserve evidence.
- To improve customer relations through prompt action on customer complaints.

REFERENCES

Duckworth, W. "Determining the Degree of Hazard for Product Loss Control." *Professional Safety*, August 1982.

Dudley, R. and J. Heldack. Product Liability Planning in the Corporation, *Professional Safety*, October 1981.

Getzoff, B. 'The New Theory of Product Liability," *National Safety News*, February 1969.

———. "Product Liability: Guides for the Corporate Manufacturing Executive," *National Safety News*, February 1967.

Mazel, J. "Zero Defects," *Factory*, July 1965.

Philo, H. "Negligent Design," *American Trial Lawyers*, June 1966.

Robb, D. "Safety Is Not Just Common Sense," *Journal of the ASSE*, December 1965.

Ryan, J. Hazard Analysis Guidelines in Product Design," *Professional Safety*, March 1988.

Schroeder, R. "An Insurance Engineer Looks at Product Liability," *Journal of the ASSE*, January 1968.

Zinch, W. C. "The Foreman and Quality," *Supervision*, April 1966.

———. "A Guide to Zero Defects," Office of the Secretary of Defense, 1965.

Part VI

Appendixes

Appendix A
How to Analyze Your Company

This appendix contains two outlines that can be used to obtain information on a company and evaluate its progress and performance in safety. They are:

1. *A loss-control analysis guide.* This outline, typical of those used by many insurance companies, can be used to spot-check areas that should be examined and analyzed in order to determine the effectiveness of the current safety program. It is intended to trigger the safety engineer's thoughts—to remind the engineer to look closely at the most pertinent areas of management as they apply to safety.
2. *A safety program appraisal.* This outline was developed for the purpose of appraising each plant in a multiplant operation in a similar manner. In a multiplant operation with a corporate safety department, it could be used to audit plant safety operations. The outline was compiled from several systems used by a number of different companies.

A LOSS-CONTROL ANALYSIS GUIDE

To understand an organization's safety performance, first obtain facts about how accidents are presently being controlled. Next, analyze those facts to determine where weaknesses exist. Then, in the course of everyday work or in a special report to management, we offer suggestions for minimizing those weaknesses. Some of the areas in which to obtain facts are:

I. Management Organization

A. Does the company have a written policy on safety?
B. Draw an organizational chart and determine the line and staff relationships.
C. To what extent does executive management accept its responsibility for safety?
    1. To what extent does it participate in the effort?
    2. To what extent does it assist in administering?
D. To what extent does executive management delegate safety responsibility? How is this accepted by:
    1. The superintendent or top production people?
    2. The foremen or supervisors?
    3. The staff safety people?
    4. The employees?
E. How is the company organized?
    1. Are there staff safety personnel? If so, are their duties clear? Are responsibilities and authorities clear? Where is staff safety located? What can it reach? What influence does it have? To whom does it report?

2. Are there safety committees?
   a. What is the makeup of the committees?
   b. Are their duties clearly defined?
   c. Do they seem to be effective?
3. What type of responsibility is delegated to the employees?

F. Does the company have written operating rules or procedures?
   1. Is safety covered in these rules?
      a. Is it built into each rule, or are there separate safety rules?

II. Accountability for Safety

A. Does management hold line personnel accountable for accident prevention?
B. What techniques are used to fix accountability?
   1. Accountability for results:
      a. Are accidents charged against departments?
      b. Are claim costs charged?
      c. Are premiums prorated by losses?
      d. Does supervisory appraisal of supervisors include looking at their accident records? Are bonuses influenced by accident records?
   2. Accountability for activities:
      a. How does management ensure that supervisors conduct toolbox meetings, inspections, accident investigations, regular safety supervision, and coaching?
      b. Other?
C. Are any special systems set up?
   1. SCRAPE
   2. Other

III. Systems to Identify Problems—Hazards

A. Are routine inspections performed?
   1. Who is responsible for inspection functions?
   2. Who makes inspections?
   3. How often are inspections made?
   4. What types of inspections are made?
   5. To whom are the results reported?
   6. What type of follow-up action is taken?
   7. By whom?
B. Are any special inspections made?
   1. Boilers, elevators, hoists, overhead cranes, chains and slings, ropes, hooks, electrical insulation and grounding, special machinery such as punch presses, x-ray equipment, emery wheels, ladders, scaffolding and planks, lighting, ventilation, plant trucks and vehicles, materials handling equipment, fire and other catastrophe hazards, noise and toxic controls.
C. Are any special systems set up?
   1. Job safety analysis
   2. Critical incident technique
   3. High-potential accident analysis
   4. Fault-tree analysis
   5. Safety sampling

    D. What procedure is followed to ensure the safety of new equipment, materials, processes, or operations?

    E. Is safety considered by the purchasing department in its transactions?

    F. When corrective action is needed, how is it initiated and followed up?

    G. When faced with special or unusual jobs, how does the company ensure safe accomplishment?
        1. Is there adequate job and equipment planning?
        2. Is safety a part of the overall consideration?

    H. What are the normal exposures for which protective equipment is needed?
        1. What are the special or unusual exposures for which personal protective equipment is needed?
        2. What personal protective equipment is provided?
        3. How is personal protective equipment initially fitted?
        4. What type of care maintenance program is instituted for personal protective equipment?
        5. Who enforces the wearing of such equipment?

IV. Selection and Placement of Employees

    A. Is an application blank filled out by prospective employees?
        1. Does it ask the right questions?

    B. What type of interview and screening process is the prospective employee subjected to before being hired?

    C. How are the prospective employee's references and past history checked?

    D. Who actually does the final hiring?

    E. Is the physical condition of the employee checked before hiring?
        1. If a physical exam is given, how complete is it?
        2. How is the information used?

    F. Are any skill, knowledge, or psychological tests given?

    G. Are job physical requirements specified from job analysis?
        1. Are these requirements considered in new hires?
        2. Are they considered in job transfers?

V. Training and Supervision

    A. Is there safety indoctrination for new employees?
        1. Who conducts it?
        2. Of what does it consist?

    B. What is the usual procedure followed in training a new employee for a job?
        1. Who does the training?
        2. How is it done?
        3. Are written job instructions based on the job analysis used?
        4. Do they include safety?

    C. What training is given to an older employee who has transferred to a new job?

    D. What methods are used for training the supervisory staff?
        1. How are new supervisors trained?
        2. Is there continuous training for the entire supervisory force?
        3. Who does the training?
        4. Is safety a part of it?

E. After employees have completed the training phases of their job, what is their status?
    1. What is the quality of the supervision?
    2. What use is made of the probation period?

VI. Motivation

A. What ongoing activities are aimed at motivation?
    1. Group meetings, literature distribution, contests, film showings, posters, bulletin boards, letters from management, incentives, house organs, accident facts on plant operations, other gimmicks and gadgets, and activities in off-the-job safety.
B. What special-emphasis campaigns have been used?

VII. Accident Records and Analysis

A. What injury records are kept? By whom?
B. Who sees and uses the records?
C. Are standard methods used?
D. What type of analysis is applied to the records?
    1. Daily analysis
    2. Weekly analysis
    3. Monthly analysis
    4. Annual analysis
    5. Analysis by department
    6. Cost analysis
    7. Other
E. What is the accident investigation procedure?
    1. What circumstances and conditions determine which accidents will be investigated?
    2. Who does the investigating?
    3. When is it done?
    4. What type of reports are submitted?
    5. To whom do they go?
    6. What follow-up action is taken?
    7. By whom?
F. Are any special techniques used?
    1. Estimated costs
    2. Safe-T-Scores
    3. Statistical control charts

VIII. Medical Program

A. What first-aid facilities, equipment, supplies, and pesonnel are available to all shifts?
B. What are the qualifications of the people responsible for the first-aid program?
C. Is there medical direction of the first-aid program?
D. What is the procedure followed in obtaining first-aid assistance?
E. What emergency first-aid training and facilities are provided when normal first-aid people are not available?
F. Are there any catastrophe or disaster plans?
G. What facilities are available for transportation of the injured to a hospital?
H. Is a directory of qualified physicians, hospitals, ambulances, available?
I. Does the company have any special preventive medicine program?
J. Does the company engage in any activities in the health education field?

A SAFETY PROGRAM APPRAISAL FOR MULTIPLANT OPERATIONS

Appraisal Factors

I. Goals

    A. What safety objectives were set for this period?

    B. What progress was made toward achieving those objectives?

II. Growth of Personnel

    A. How are we improving the safety knowledge of our personnel?

	Subject	Time	Effectiveness
Line supervisor Intermediate manager Plant engineer Product engineer Industrial engineer Plant manager			

III. Inspections

    A. Are inspections effective?

    B. General plant inspections made:

Item	When inspected	By whom	Items corrected		Comment
			Last period	This period	

    C. Below are some items that might be checked in making general safety inspections:

General layout	Noise
Flow of material	Radiation
Traffic flow	Fire protection
Aisles	Guarding
Machine controls	Rest rooms
Fumes	Washrooms
Illumination	Access ladders
Temperature	Stairs
Floor loads	Elevated platforms
Pits and excavations	Protective equipment
Areas under construction	Electrical equipment
Flammable liquids	Others

D. Preventive maintenance inspections:

Item	Does standard procedure exist?	When inspection	By whom	Effectiveness

E. The following are typical pieces of equipment for which preventive maintenance inspections will probably be necessary:

Elevators	Lift trucks
Hitching equipment	Other trucks
Ladders	Boilers
Scaffolds	Ovens and furnaces
Hoists and cranes	Power presses
Chains and slings	Portable grinders
Crane runways	Stand grinders
Exhaust systems	Power conveyors

IV. Procedures

A. Are regular standard operating procedures effective in these areas?

	Does standard method exist?	Effectiveness
Equipment lockout Labeling of containers and pipes Welding or burning permits Glove issuance Approval of equipment overloads Entering enclosed spaces, tanks Others		

V. Experience

A. On-the-job

	Last period	This period	Significance*

Number of industrial first-aid injuries
Number of disabling injuries
Number of permanent potential
    disabilities and permanent total
    disabilities			

	Last period	This period	Significance*
Number of fatalities			
Time lost because of first-aid injuries (estimate in days)			
Days lost or charged because of disabling injuries			
Disabling injury frequency rate			
Disabling injury severity rate			
Combined frequency-severity indicator			
Estimated cost of first-aid cases			
Medical and compensation costs paid out			
Medical and compensation costs incurred			
Cost of property damage and public liability damages for motor vehicle accidents			
Cost of other public liability damages			
Cost of fire losses			
Total industrial injury and accident costs			
Insurance loss ratio			

B.   Off-the-job

	Last Period	This period	Significance
Number of off-the-job injuries (first-aid cases)			
Number of injuries reported to health and accident insurance carrier			
Number of disabling nonindustrial injuries			
Frequency rate for off-the-job injuries			
Time lost from work because of off-the-job first-aid cases (estimated in days)			
Time lost from work because of disabling off-the-job injuries			
Cost of treating nonindustrial first-aid cases (estimate)			
Cost of off-the-job injuries reported to health and accident insurance carrier			
Total cost of first-aid and insurance for off-the-job injuries			

C.   Injury summary

	Last period	This period	Significance
Total disabling injuries—industrial and off-the-job			
Total time lost—industrial and off-the-job			
Total industrial and off-the-job injury and accident costs			
Industrial and off-the-job cost factor (per 1,000 worker-hours)			

VI. Investigations

A. What operational factors may be indicators of accident trends?

	Last period	This period	Significance

Accident causes and corrective action:
   Number of incidents reported
   Number where no effective action
     was possible
Where action was taken, give number
   where:
     Guarding was changed
     Work method was changed
     Equipment or facility was modified
     Employee was reinstructed in method
     Protective equipment was changed
     Employee was cautioned
     Employee personal factor was given
       added consideration
What causes and agencies were most
   predominant?			

VII. Expenses

A. What has been the total safety program expense?

	Last period	This period	Significance
Total staff expense Total cost of protective equipment Total safety expense Safety expense factor (per 1,000 worker-hours)			

VIII. Outside Activities

A. Are we participating in community safety affairs?

List projects	No. of people involved	Time allotted	Evaluation of project

Appendix B
Sample Management Safety Policies

This appendix discusses actual management safety policies which have been notably successful in industry. Each policy is included here for a reason. The policies are for the following:

1. Columbia River Paper Company, a division of Boise Cascade Corporation, Boise, Idaho. This is a perfect example of a success story starting with policy. This policy was written by a young man who had just been given the job of mill safety director. He was inexperienced in safety, but he learned fast. The first thing he did after accepting the position was to write a policy and submit it to management for signature.
2. Richardson Paint Company, Austin, Texas. Policy does not have to be elaborate. The letter in this example from the president of the company actually was safety policy—and very effective policy—as it was written by the president himself and spelled out his desires.
3. An unidentified contractor. The construction industry is thought of as different. Policy is essential in construction, as it is in any operation. Management is management, regardless of the type of operation. Policy is nothing more than management in action, expressing its desire and intent. This policy worked for one contractor.

example I
COLUMBIA RIVER PAPER COMPANY,
BOISE CASCADE CORPORATION

The following memo was sent from the newly appointed safety director of this mill to the mill manager. The policy was approved by the manager, starting this mill on a remarkable safety record.

TO: _____ *Mill Manager*
FROM: _____ *Safety Director*
SUBJECT: Proposed Safety Policy

Since the job of safety director is a newly created position in this company and since no written policy is available, I felt that my first duty should be to develop a basic outline of company safety policy, for two reasons:

1. Nearly every company with an effective safety program has a written safety policy.
2. This is the best way I know to acquaint you with my basic plan for our safety program.

When I was assigned the job of developing a safety program for this mill, paper mill safety was brand new to me. So I have used the following sources of information in making up this proposal:

351

The safety seminar put on by our insurance carrier
The Portland staff of our carrier
The Pacific Coast Association of Pulp & Paper Manufacturers
The Camus mill safety staff of Crown Zellerbach Corporation
The Oregon Stage Industrial Accident Commission
The National Safety Council

From these sources I have been able to study the essential parts of the safety policies of probably a hundred mills. This proposed is not patterned after that of any particular company, but is a combination developed from many different programs along with some original ideas of my own.

I am not optimistic enough to believe that my thinking will dovetail exactly with yours. Therefore, I am submitting this as a proposal only. I would like you to read it over and make whatever deletions, corrections, or additions you feel are necessary. If, in this way, it can be made into an acceptable statement of safety policy for our mill, then it can be used as a guide for future safety activities of everyone concerned. If you think it is workable for us or not at all suitable, or if you should like to discuss any part of it with me, let me know.

Fire prevention and protection are not mentioned because I have not yet had an opportunity to look into that field; and for now I think it would be better to keep them separate.

SAFETY POLICY
Columbia River Paper Company

Introduction

A. In order to clarify the safety activities of Columbia River Paper Company, division of Boise Cascade Corporation, the following is set forth as a basic program to clearly establish its existence. It consists essentially of an outline of the relationships of management and employee responsibilities which are necessary for an effective safety program.
B. We wish it to be known that this program will become a basic part of our management policy and will govern our judgment on matters of operation equally with considerations of quality, quantity, personnel relations, and other phases of our management policy.
C. The program is developed and administered, upon approval, by the safety director. In doing so, the safety director is performing a function of the resident manager. His primary purpose is to assist management by encouraging safety consciousness and employee participation. This is accomplished by the development of promotional material, program planning, motivation incentives, safety meetings, inspections, etc.

Purpose

The management of Columbia River Paper Company holds in high regard the safety, welfare, and health of its employees. We believe that "Production is not so urgent that we cannot take time to do our work safety." In recognition of this and in the interest of modern management practice, we will constantly work toward:

A. The maintenance of safe and healthful working conditions.
B. Consistent adherence to proper operating practices and procedures designed to prevent injury and illness.

C. Conscientious observance of all federal, state, and company safety regulations.

Responsibility

A. The resident manager has taken the responsibility to develop an effective program of accident prevention.
B. Plant superintendents are responsible for maintaining safe working conditions and practices in the areas under the jurisdiction.
C. Department heads and supervisors are responsible for the prevention of accidents in their departments.

1. They are directly responsible for maintaining safe working conditions and practices and for the safety of all men under their supervision.
2. Nowhere is the quality of supervision more apparent than in housekeeping. Good housekeeping is not only essential for safety but also indicative of an efficient department.
3. Each supervisor is responsible for the proper training of the employees reporting to him. Job hazards and safe procedures should be fully explained to each employee before he begins work.
4. It is also the supervisor's responsibility to see that required personal protective equipment is used in accordance with safety rules and practices.
5. Supervisors should encourage employee safety suggestions and given them immediate consideration.
6. Supervisors will schedule departmental safety meetings as often as necessary to effect safe practices and work methods.

D. Foremen are responsible for the prevention of accidents in their crews.

1. They will enforce all general and departmental safety rules and regulations.
2. They must see that all accidents are reported and that first aid is rendered in case of injury.
3. They will investigate all accidents and near misses and prepare foreman's reports of accident.
4. Each foreman must have a current first aid card in his possession.

E. The safety director is delegated by the resident manager and has the responsibility to provide advice, guidance, and any such aid as may be needed by supervisors in preventing accidents, including:

1. Safety indoctrination of new employees.
2. Protective-equipment emphasis (safety shoes, ear protectors, etc.).
3. Safety meeting planning and assistance.
4. Supplying information and educational material for meetings.
5. Providing forms for safety meeting minutes, hazard analysis, etc.
6. Accident investigation follow-up.
7. Suitable job placement for employees able to return after injury.
8. Statistical reporting and study.
9. General publicity—handouts, management reports, memos, letters, notices, etc.
10. Arranging periodic safety inspection of the department.
11. Scheduling and conducting special safety meetings for all employees.
12. Arranging first aid training and other special instruction.

F. The safety director will also:

1. Prepare for approval an annual program designed to encourage a broad safety awareness in the plant.
2. Prepare and keep adequate records of all accidents, and from these prepare such charts as will best show the way to eliminating these accidents.
3. Keep in touch with new developments in the field of accident prevention, personal protective equipment, and first aid equipment and procedures, so that he will be an effective guide to everyone connected with the basic safety program.
4. Coordinate the joint efforts of management and labor, and direct these toward totally safe operations.
5. Coordinate development of safety educational material. Prepare and schedule sessions for safety training of supervisors, foremen, committeemen, and other employees.
6. Promote the public relations aspect of the safety program including preparing for approval releases to news media outside the company.
7. Conduct plant safety committee meetings, providing the material necessary for departmental safety meetings.

G. Employees are responsible for the exercising of maximum care and good judgment in preventing accidents.

1. No job shall be considered efficiently completed unless the worker himself has followed every precaution and safety rule to protect himself and his fellow employees from bodily injury throughout the operation.
2. Employees should report to their foreman and seek first aid for all injuries, however minor these may be.
3. Unsafe conditions, equipment, or practices should be reported as soon as possible.
4. Employees are to read and abide by Columbia River Paper Company safety regulations and all departmental safety rules.
5. Employees will be provided with whatever personal protective equipment is necessary—they are expected to use it.
6. Each employee should consider safety meetings as part of his regular job. Reasons to be excused must be just as important as those for missing any of his regular shift.

Following this basic policy and list of responsibilities, there were, of course, lists of other rules and regulations of the company, as well as descriptions of the operation of several safety committees and of several inspection systems.

This policy was accepted by mill management, signed and published by the mill manager, and put into effect. The results were remarkable. Before the existence of this policy, the mill operated with a frequency and severity rate slightly worse than national average for paper mills. Beginning with the signing of this policy, the company operated for 2½ million worker-hours without a disabling injury. This policy was the starting point for all the activities that went into making this safety record. The author of this policy later became the corporate safety director of Boise Cascade Company, one of the nation's largest corporations.

example II
RICHARDSON PAINT COMPANY, AUSTIN, TEXAS

The following memo from the president was sent to all supervisors of this company. It constitutes policy.

TO: *All Foremen*
FROM: *S. R. Richardson, President*
SUBJECT: SAFETY LETTER

After looking over the large number of reported accidents in October and November and the year-to-date summary, I think it is high time that we pause and do some serious thinking and planning. If our current accident rate keeps up, it is inevitable that someone is going to get seriously hurt or killed.

All of us, whether we are driving a car, walking on the sidewalk or street, working around home, or working on the job, are exposed to hazards of one kind or another. Whether we stay healthy is entirely within our own control, whether we use the small amount of time it takes to recognize a hazard when we see one, whether we are living by the five safety rules, or whether we are only mounting them.

We have talked thousands of words and written reams of paper to all of you, but the evidence is at hand to show that the necessity of safety is not soaking through. It is high time to quit yakking by mouth or paper and get down to some solid action.

I am making five suggestions. No, suggestion is the wrong word—and will be changed to orders.

1. Immediately upon receipt of this letter, get your crew together and read it to them. You and your men are each to sign and return the attached form immediately. If any man has not studied the safety book and does not know how it is to be used, he is to be laid off without pay until he learns it. Only when he feels that he knows the book from cover to cover, may he return to work—whether it takes one, two, three, four hours or a full week.
2. Until we get this accident situation under control, all safety meetings will be held on the men's own time. We will not pay for them nor are they to be charged to the customer.
3. Even though 99 percent of all accidents are caused by one man's carelessness, nevertheless, unsafe acts that are the basic cause of all accidents are a matter of crew discipline. Each man has a responsibility toward his fellow worker and toward his crew to watch for unsafe acts on the part of a co-worker and call them to his attention.

In order that each of you realize your individual responsibility, I am giving you two orders:

a. Whenever a reportable accident occurs where no time is lost, the entire crew will lose 15 minutes time that day, the time to be used to discuss the reportable accident and how it could have been prevented.
b. Whenever a lost-time accident occurs, the entire crew will lose 1 hour for that day, the time to be used in the same manner as in *a* above.

4. Each week, each crew is to discuss safety suggestions, select the best one, and send that one to me signed by every member of the crew. Forms are attached for that purpose.

5. Heretofore, whenever an accident occurred, the foreman and the men injured were to write me a personal letter. That is now being changed. Whenever an accident occurs, the entire crew will use the 15 minutes for only reportable accidents and the 1 hour for lost-time accidents to discuss it with the foreman and the men injured, and the foreman will write up a full report signed by all men and send the report to me.

I know this is tough, but it is not half as tough as we can get unless these accidents come to a screaming halt. They are *not* necessary, and they *are* controllable.

For the past year, we have left accident control and accident prevention pretty much up to the individual, but it is evident that the individual has not assumed that responsibility; and so managemant must of necessity step back into the picture—and not with padded gloves.

Your life and money are at stake. Your life is much more important to us than our money, and it certainly ought to be of paramount importance to you and your family. Anything that we at the management level can do or that you at the crew level can do is of maximum importance if it will save the life of one man.

I am willing to listen to squawks on these orders, and I am willing to listen to alternative suggestions, but unless the crews can come up with workable plans to take their place, the order will stick.

It is in your lap, boys. Either come up with some good, workable safety suggestions, or accept the orders as they are written.

We will not put these orders into effect until we have heard from each foreman. If some really good alternative workable suggestion for accident prevention comes in, we will adopt it in lieu of the orders. We will inform you of our decision well in advance and will, if necessary, be willing to discuss the decision with your own appointed committee. But something must be done—and soon.

This memo was sent out from the office of Mr. S. R. Richardson, then president. Attached to it was the return memo. Each supervisor was required to fill it out and return it:

TO: *Mr. Richarson*
FROM:_____ *Foreman*

We have had a meeting and have read your letter.

Each of us has the company safety book, and we know the safety rules.

Each of us has not only read but also studied the entire book, and we know the responsibility of each man toward the safety program.

Answer one of the following:

I. We agree that your orders are needed and will follow them:

Yes   No

II. We agree with some and disagree with others checked as follows:

1. Yes   No
2. Yes   No
3. Yes   No
4. Yes   No
5. Yes   No

We have the following alternative suggestions to take the place of the ones we have checked "No."

(Signed) _____ *Foreman*

Date:
Mechanics:

example III
A CONTRACTOR

This policy represents direction in safety by the management of a construction company. It consists of a broad management safety policy, procedures to be followed by superintendents and supervisors, and a set of employee work rules.

MANAGEMENT SAFETY POLICY

It is the policy of this company that every employee is entitled to work under the safest possible conditions for the construction industry. To this end, every reasonable effort will be made in the interest of accident prevention, fire protection, and health preservation.

The company will endeavor to maintain a safe and healthful workplace. It will provide safe working equipment and necessary personal protection, and, in the case of injury, the best first aid and medical service available.

It is our belief that accidents which injure people, damage machinery, and destroy materials cause needless personal suffering, inconvenience, and expense.

We believe that practically all accidents can be prevented by taking common-sense precautions.

Because of the large number of jobs in progress at one time, the varied nature of the work, and the widespread location of the jobs, we must "formalize" our safety program, utilizing written reports and records, to achieve the maximum use and effectiveness of accident prevention information.

To coordinate the safety program, Mr._____ , Vice-president, will continue to act as safety director. The overall effectiveness of the safety program is his responsibility. His duties include the review and analysis of accident information, safety meeting reports, etc., and the communication of pertinent information to all the jobs and shops.

The responsibility for safety on each individual job and at each shop remains with the superintendent. His duties include the reviewing of all accident investigation and safety inspection reports for the job or shop. He is also responsible for passing safety information along to all his foremen, and for maintaining an accident log to help in identifying accident trends and problem areas so that additional safety effort can be directed as needed. When necessary, he will advise subcontractors, etc., of physical changes or new safety regulations. The superintendent will also conduct regularly scheduled foremen's meetings at which job or shop work progress, hazards, accidents, and other work and safety items will be discussed. A written record of these meetings will be maintained.

The National Safety Council states that "The workmen's attitude toward safety depends absolutely on the attitude of the foreman." The foreman is responsible not only for the quality and quantity of the work produced by the men under him, but for their safety as well. The foreman has the greatest opportunity to create safe working conditions and safe attitudes.

The foreman's daily safety inspections will be supplemented by periodic checklist typewritten safety inspections, as specified by the superintendent.

All work-related accidents requiring professional treatment, and all "near" accidents and accidents requiring first aid treatment where conditions were such that a more serious injury could have resulted, shall be investigated by the foreman; and a written report, in duplicate, shall be submitted, by the foreman, to the superintendent. The foreman will then make sure that the necessary action to prevent similar-type accidents has been taken and/or that others working under his direction are reinstructed or cautioned as needed.

The foreman is responsible for the holding of toolbox meetings with his workers at which accident and other information concerning the safety of his workers and their work is discussed.

The foreman is also responsible for the employees' performance in adhering to the work rules set forth in the employees' "A Safety Policy."

Foremen must advise_____ , Vice-president, two days in advance of any blasting to be done on the job by anyone.

It is wrong to believe that accidents are unavoidable and will always happen. If all of us do our part, including acting and talking safety at all times, a healthy attitude toward accident prevention and an improved safety record can be achieved.

Reducing accident and related insurance costs will permit us to be more competitive in our industry, thus helping to safeguard our jobs.

(Signed) _____ *President*
_____ *Vice-president*
_____ *Safety Director*

RESPONSIBILITIES

Superintendents

ACCIDENT INVESTIGATION

1. All accident investigation reports will be reviewed by the superintendent to make sure that the proper action has been taken by the foreman to prevent a recurrence of a similar type of accident.
2. Additional action will be directed by the superintendent, as needed.
3. The superintendent will inform the other foremen under his direction, whose operations are such that similar conditions exist, of the accident and of the necessary actions to be taken to prevent similar accidents.
4. The superintendent, when necessary, will advise subcontractors, etc., of physical changes and new safety regulations.
5. The original accident investigation report will be forwarded to Mr._____ _____, Vice-president, for review. The carbon copy will be retained at the job or shop.

ACCIDENT RECORDS

1. The superintendent is responsible for maintaining a record of all work-related accidents and injuries, using the construction-work monthly accident record form.
2. This form will be made out in duplicate.
3. At monthly intervals the duplicate copies will be forwarded to the main office. The originals will be retained at the job or shop.
4. The superintendent will analyze the reports in order to identify any accident trends or problems so that safety effort can be directed as needed.
5. Accidents requiring only on-the-job first aid should be listed and indicated by use of the letters FA in the left-hand margin.

SAFETY INSPECTION

1. The safety inspector's report forms will be reviewed for completeness and accuracy.
2. At monthly intervals these reports will be forwarded to the main office.

SAFETY MEETINGS

1. The superintendent will schedule monthly foremen's meetings at the shops and at the large job sites, or where conditions warrant.
2. Job or shopwork progress, hazards, accidents, and other work and safety items should be discussed at these meetings.
3. The superintendent will make sure that a written report of the meeting is maintained. The report is to be forwarded to the office of Mr._____ , Vice-president.
4. When conditions warrant, additional meetings should be held as needed.
5. The superintendent is expected to maintain frequent contact with his foremen to discuss safety and work progress.

FOREMEN

ACCIDENT INVESTIGATION

All work-related accidents requiring professional treatment, and all "near" accidents and accidents requiring first aid treatment where conditions were such that a more serious injury could have resulted will be investigated by the foreman.

1. Every accident should be investigated as soon as possible to determine the cause and what action should be taken to prevent a recurrence of similar-type accidents.
2. The accident should be discussed with the injured worker after medical treatment has been given, and with others who saw the accident and/or are familiar with the conditions involved.
3. The supervisor's accident investigation report form will be filled out in duplicate. Sufficient information will be included to allow the job superintendent and others to reconstruct the accident so that they can determine whether additional action is needed.
4. Corrective action to prevent similar-type accidents should be taken as soon as possible. It may be necessary to check back to see that this has been accomplished satisfactorily.
5. Both copies of the supervisor's accident investigation report form will be turned in to the job superintendent for review by him and the main office.

SAFETY INSPECTION

1. Each foreman will make a daily safety inspection; and unsafe conditions, including unsafe worker actions, will be corrected promptly. Corrective action should be taken in all areas where a hazard to the worker, physical property, or the public exists.
2. In addition to the above, a weekly written safety inspection will be made, using the safety inspector's report form. As noted above, corrective action will be taken in all areas where a hazard to the worker, physical property, or the public exists. The completed safety inspector's report will be given to the job superintendent for review.

SAFETY MEETINGS

1. Attendance at safety meetings scheduled by the job superintendent is compulsory.
2. Each foreman will conduct a weekly "toolbox" safety meeting with his workers where accident and other information can be discussed. The well-being of physical property and the safety of the public should also be discussed.

Safety Policy for Employees

All employees:

1. MUST wear hard hats on all jobs where there is danger of being struck by falling or moving objects, and on all roadwork.
2. MUST wear safety glasses—when chipping, grinding, operating a jackhammer, drilling above chest height, or whenever an eye injury hazard exists.
3. MUST wear sturdy shoes which are in good repair. Safety shoes are recommended for all persons working on the ground and on decks.

4. MUST use grounded electrical equipment and hand tools which are in good condition.
5. MUST observe all safety precautions and report unsafe conditions to the foreman.
6. MUST understand the foreman's instructions. If you do not know how to do the job safely, ask your foreman.
7. MUST report all injuries immediately.
8. MUST keep their mind on the job; this is the best way to prevent accidents.
9. TRUCK and EQUIPMENT operators MUST be absolutely certain that the path of movement is clear, particularly when they are backing up. The horn or other audible signal must be sounded while backing up.

THIS POLICY SHALL BE OBSERVED BY ALL EMPLOYEES

(Signed)_____ *President*
_____ *Vice-president*
_____ *Superintendent*

Appendix C
A Special Emphasis Program

This appendix presents an example of a special-emphasis program: the NO STRAIN campaign of Industrial Indemnity Company, conceived and written by Ray Campbell and me, former safety analyst of that company. Following is a complete description of the campaign as outlined in the "Manager's Guide," which is an integral part of the campaign packet.

MANAGER'S GUIDE TO A NO STRAIN CAMPAIGN

THE PROBLEMS OF BACK INJURIES

Back strains and lifting injuries are among the most frequent and troublesome on-the-job injuries. Every material handling job involves a risk, and every employee is potentially likely to attempt an unsafe lift at any time. Strains may be caused by falls, slipping, reaching too far, twisting, lifting too heavy a load, or even repeatedly lifting an acceptable load for a long time.

Whatever the cause, these injuries hurt backs and groins. Such injuries can cause real misery. They also can create feelings of distrust, suspicion, and hopelessness. The net result may be impaired morale, lessened operating efficiency, and higher costs, including raised insurance rates.

In California alone in one year the costs from back and lifting injuries totaled $50 million. This represented about 50,000 injuries—or an average cost of $1,000 per back injury.

WHAT CAN BE DONE?

In many companies chronic frustration over this seemingly insoluble problem creates wrong attitudes that further aggravate the problems. Something can be done. By implementing a NO STRAIN campaign you can take positive steps to eliminate the problem of back and lifting injuries. But the key to the success of the campaign is to begin with the attitude that something *can* be done.

The key to the NO STRAIN campaign is to make your employees realize that pain and possible disability are connected with back injuries. Demonstrate by your own attitude how concerned you and your supervisors are with the problem. Connect this idea with the NO STRAIN symbol in your people's minds, and keep them thinking about it through repetition of the NO STRAIN symbol.

KEY ELEMENTS OF THE CAMPAIGN

1. A NO STRAIN folder for employees detailing material handling and lifting techniques (Exhibit C-1)
2. A NO STRAIN folder for supervisors with facts on material handling, job analysis, and employee motivation (Exhibit C-2)

363

Exhibit C.1 NO STRAIN employee folder.

Exhibit C.2 NO STRAIN supervisor folder.

3. A NO STRAIN multiple-image wallet card, with rules on correct lifting (Exhibit C-3)
4. A NO STRAIN lifting instruction chart for use by supervisors in department training meetings (Exhibit C-4)
5. Two NO STRAIN posters, each 11 by 16½ inches, in red, blue, and black colors (Exhibit C-5)
6. Three NO STRAIN stickers, each 2 by 4 inches, in red, blue and black colors, for use at points of exposure (Exhibit C-6)
7. Two NO STRAIN banners, each 8 by 60 inches, in red and blue colors, to be hung in each department during the campaign following the kickoff meeting (Exhibit C-7)

All these materials carry the NO STRAIN symbol—four basically vertical red lines with a sharp angle jog on a blue background. This abstract symbol conveys no specific lifting instruction, but is intended to serve as a reminder to use care when lifting is involved. Display these materials prominently. See that your people learn what the NO STRAIN symbol means.

STARTING THE CAMPAIGN

Do your part. Read the plan of action in this guide. When you understand how the NO STRAIN program works, draw up a plan of attack that meets your needs. Familiarize your supervisors with the campaign, since they will be responsible for implementing it. Study your work methods and the work layout. Make sure that the right equipment is provided for each job. Improved methods increase profits.

Study your training needs. Help your supervisors to encourage proper lifting techniques among all employees.

ENLIST YOUR SUPERVISORS

Enlist your supervisors in the NO STRAIN campaign. Give them the facts they will need to train their men in safe lifting in their jobs. Tell your supervisors what the

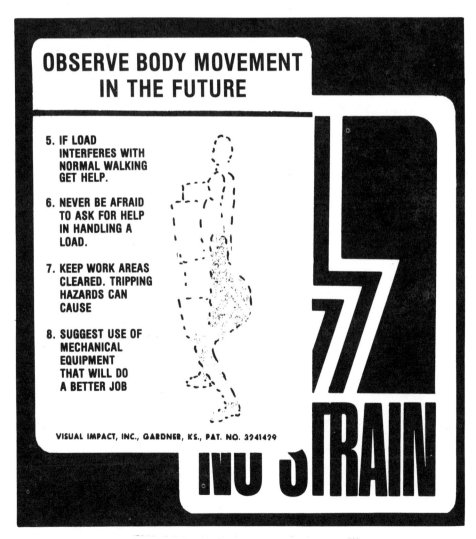

Exhibit C.3 NO STRAIN wallet card.

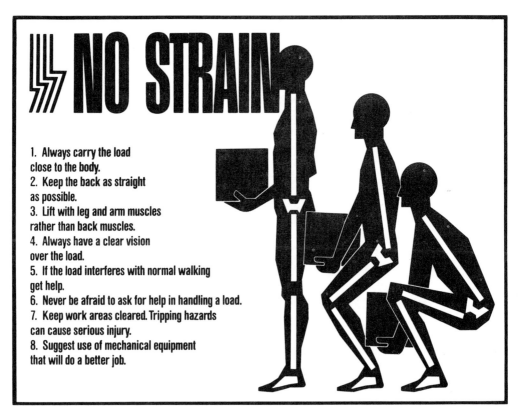

Exhibit C.4 NO STRAIN instruction chart.

problem is, what you hope to accomplish, how this will help them in their own departments, and what their roles will be. Stress the importance of their giving the campaign their wholehearted support.

Establish an accounting system which will allow you to allocate work injury and material damage costs to each department. Tell the supervisors that you are doing this.

It is essential that the supervisor convey to each employee both his and management's real concern with the problem of back injuries. Each employee must be made to understand that you really care.

Tell your supervisors that you, the manager, will provide them with new employees who are able to work with NO STRAIN, any equipment they need to correct material handling problems, and all the help they need to train and motivate employees in safe lifting. Tell the supervisors that any strains that occur to their employees in the future will be their responsibility and will affect their own performance evaluations. NO STRAIN can mean higher productivity.

Exhibit C.5 NO STRAIN posters.

Exhibit C.6 NO STRAIN stickers.

Exhibit C.7 NO STRAIN banners.

FUTURE HIRES

After the campaign is rolling smoothly and employees have been instructed in the correct methods of lifting, it is important to direct attention to your hiring program. Set up safeguards to ensure that you hire the right man for the job. You will know how much and what kind of strength each job requires if you have analyzed it properly.

Establish a system to find out about prospective employees' problems, if any. Here are the important points to include:

1. A frank discussion of job requirements during the employment interview. Ask the prospective hire whether he has had any back trouble and whether he thinks he can handle the job you are considering him for.
2. A check of previous employers. Find out whether he has had claims for back injuries.
3. A check with the state Workers' Compensation board to round out the picture if you are suspicious about his compensation claim history (where this is possible).

By asking the right questions, you should be able to determine which persons are obviously unfit for heavy work. Because some people have less obvious limits on their ability, which only a doctor can discover, you may want to invest in a pre-employment or preplacement medical examination. The more completely you describe the job conditions to the doctor, the better he will be able to tell you how capable the person is for the job. In some cases, low-back X-rays may be needed.

When a new employee is hired, start him out right. See that he is properly oriented— told what he can and cannot do, and trained to meet the requirements of the job. This is the place where your supervisors can help.

A SUGGESTED NO STRAIN SCHEDULE

The coordinated NO STRAIN campaign outlined here is intended as a guide, and not as a hard and fast program that must be adhered to without variation. Select a convenient interval between each two of the various phases of the campaign. This plan has been set up on a weekly basis, but you can schedule the individual steps at shorter or longer intervals if it better suits your requirements. Regularity, however, is important, so draw up a schedule and stick to it.

BEFORE THE CAMPAIGN

Order supplies of all forms, folders, posters, stickers, and banners that you will need for the program. Explain the employment procedure, medical examination program, and use of the NO STRAIN interview guide to your employment interviewers. Establish an accident cost-by-department system.

Write your supervisors a letter telling them about the problem of back injuries, describing the need for the NO STRAIN campaign, and enlisting their support. Announce the NO STRAIN kickoff meeting. This kickoff meeting should be scheduled for week 1 just after the supervisors receive the letter.

WEEK 1

Mail the NO STRAIN letter to employees so that it will reach them before the department kickoff meeting. Hold the supervisors' meeting. Issue the materials to supervisors. Tell them to make individual contacts with employees. Instruct the supervisors to put up the posters and banners in their departments. Let each supervisor know that you will be checking with him. Tell him when to give you his first checklist report. Explain that the NO STRAIN campaign will include these activities:

- A NO STRAIN folder will be given to each employee. As the supervisor hands out the folders he must impress each employee with the harsh reality of pain and costly work time lost if the employee injuries his back.

- NO STRAIN posters will be displayed on bulletin boards, and banners posters above aisles.

- Each supervisor will be issued enough NO STRAIN point-of-exposure stickers for all his employees—to be given out at the meeting held in week 2. Each employee should be instructed to place the sticker where he thinks it will be most effective as a reminder on safe lifting.

- Each supervisor will receive a NO STRAIN checklist with space for employee names and space for him to list each campaign activity. The supervisor should fill in the date when each activity is completed for each employee and submit the finished form to management.

- Each supervisor will receive an instruction chart showing the correct way to lift. He will hold a meeting with his crew to announce the purpose of the campaign and to describe and demonstrate proper lifting techniques.

- Supervisors should be instructed to make any changes in work method, equipment, or employee placement that are necessary to eliminate a dangerous condition.

WEEK 2

Make sure that your supervisors hold department meetings and that they know how to instruct their crews in proper lifting techniques.

WEEK 3

Either telephone or see each supervisor. Discuss his first NO STRAIN work analysis. Ask him about equipment needed and decide how to get it. Continue to emphasize the company's sincere desire to substitute mechanical equipment for manual handling wherever possible; to take decisive action in the reporting, treatment, and investigation of every back strain; and to hire only people who are able to work with NO STRAIN.

WEEK 4

Review job strength inventories with your supervisors. Remind them to make individual contact with each employee and to issue a second NO STRAIN sticker.

WEEK 5

Begin the first of a series of monthly reviews with each supervisor. Discuss the first month's checklist with him to see how well he has followed the intent of the NO STRAIN campaign.

Hang the second banner (Again with NO STRAIN) and the second poster. Repeat all activities. Give special recognition to supervisors who have done a good job the first month.

TO SUM UP

As a manager you can do something about strains if you:

- Begin with the attitude that something can be done.

- Instill a sense of responsibility in your supervisors for carrying out the program.

- Plan work methods.

- Plan work layout.

- Hire the right people.

- Train your supervisors.

- Get your people to think.

PERMANENT NO STRAIN ACTIVITIES

Supervisors should continue training employees and promoting safe lifting after the campaign. Analysis of work methods, careful job placement, and individual employee contacts should become good habits with them. You should devise your own program for continually following up your supervisors to see that they carry out their responsibilities.

Although the intensive period of the NO STRAIN campaign lasts for only a short time, obviously the concepts of lifting safely are always valid. Refresher meetings should be scheduled at regular intervals in the postcampaign period to ensure that your employees continue to use correct material handling methods. The practice of careful interviewing to determine "back histories" of prospective employees should have become a basic part of your hiring process.

Continue to involve your supervisors in analyzing jobs for accident traps within their departments. Encourage your supervisors to solicit suggestions from their crews on how to correct problems in their work environment. Assign individual workmen to act as safety investigators to bring to light problems that employees may be reluctant to discuss with their supervisors. During the follow-up period every effort should be made to strengthen the safe attitudes developed in the NO STRAIN campaign.

Good luck with your NO STRAIN campaign. Let us work to ensure that back and lifting injuries soon become a thing of the past.

SAMPLE FORMS AND LETTERS

Following are samples of the forms, checklists, guides, and other materials that are available as a part of the NO STRAIN campaign, should you wish to make use of them:

1. A NO STRAIN campaign checklist for charting the progress of the program (Exhibit C-8)
2. A letter for your supervisors telling them what they must do in the campaign (Exhibit C-9)
3. A letter for employees telling them about the campaign and what will be expected of them to make the program a success (Exhibit C-10)
4. A job strength inventory to help you analyze the special requirements of each job (Exhibit C-11)
5. A work-analysis data sheet to help you find and eliminate work-area hazards (Exhibit C-12)
6. A NO STRAIN interview guide for hiring new employees (Exhibit C-13)

The NO STRAIN campaign is a fairly complete special-emphasis program, packaged and ready for use by management. Special-emphasis programs have been most effective. Not only will a well-devised and well-run campaign reduce numerically the type of injury being attacked, but it will also measurably reduce all other types of injuries at the same time. This has been true in industry in most cases where these campaigns have been conducted.

KEY ELEMENTS OF A SPECIAL-EMPHASIS PROGRAM

The following are the key elements of a true special-emphasis program:

1. *A defined program.* The program should be aimed at a particular defined problem, usually the one that is causing the most trouble. Examples are falls, motor-vehicle accidents

No Strain Checklist

This checklist is to assist you, the supervisor, by reminding you of your role in administering the **NO STRAIN** campaign. The campaign depends heavily on the supervisor. You must do certain things with your team in order for the campaign to succeed. When these things are accomplished, you should so state on this checklist. Send it to management following the fifth week of the campaign and as often after as management desires.

Department (or craft) _____ Supervisor _____

I. Supervisors' kickoff meeting _____ NO STRAIN banner up _____

NO STRAIN poster up _____ (Fill in completion dates)

Individual contacts (folder to each employee)

Employee		Date	Employee		Date

II. Meeting of employees (card to each) _____ (Date held) _____

Follow-up contacts

Employee		Date	Employee		Date

III. Second banner and poster up _____ (Date) _____

Work analyses completed   IV. Job strength analyses completed

Material or part		Date	Job		Date

V. Return this checklist to management.   1E202-55-NS

Exhibit C.8 NO STRAIN checklist.

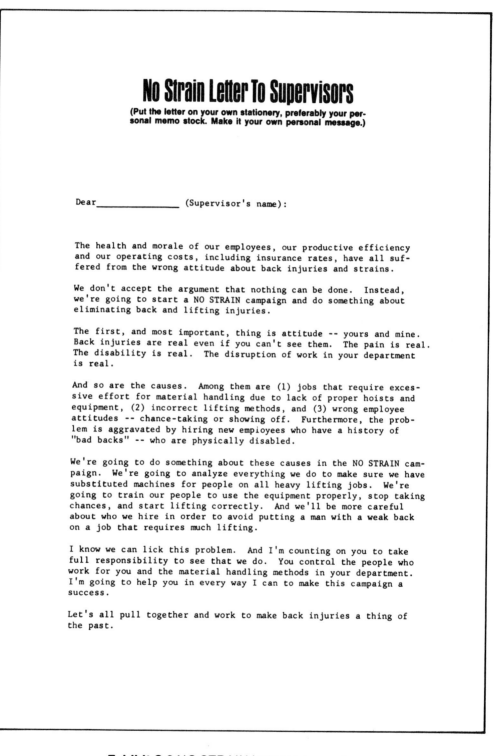

No Strain Letter To Supervisors

(Put the letter on your own stationery, preferably your personal memo stock. Make it your own personal message.)

Dear_____ (Supervisor's name):

The health and morale of our employees, our productive efficiency and our operating costs, including insurance rates, have all suffered from the wrong attitude about back injuries and strains.

We don't accept the argument that nothing can be done. Instead, we're going to start a NO STRAIN campaign and do something about eliminating back and lifting injuries.

The first, and most important, thing is attitude -- yours and mine. Back injuries are real even if you can't see them. The pain is real. The disability is real. The disruption of work in your department is real.

And so are the causes. Among them are (1) jobs that require excessive effort for material handling due to lack of proper hoists and equipment, (2) incorrect lifting methods, and (3) wrong employee attitudes -- chance-taking or showing off. Furthermore, the problem is aggravated by hiring new employees who have a history of "bad backs" -- who are physically disabled.

We're going to do something about these causes in the NO STRAIN campaign. We're going to analyze everything we do to make sure we have substituted machines for people on all heavy lifting jobs. We're going to train our people to use the equipment properly, stop taking chances, and start lifting correctly. And we'll be more careful about who we hire in order to avoid putting a man with a weak back on a job that requires much lifting.

I know we can lick this problem. And I'm counting on you to take full responsibility to see that we do. You control the people who work for you and the material handling methods in your department. I'm going to help you in every way I can to make this campaign a success.

Let's all pull together and work to make back injuries a thing of the past.

Exhibit C.9 NO STRAIN letter to supervisors.

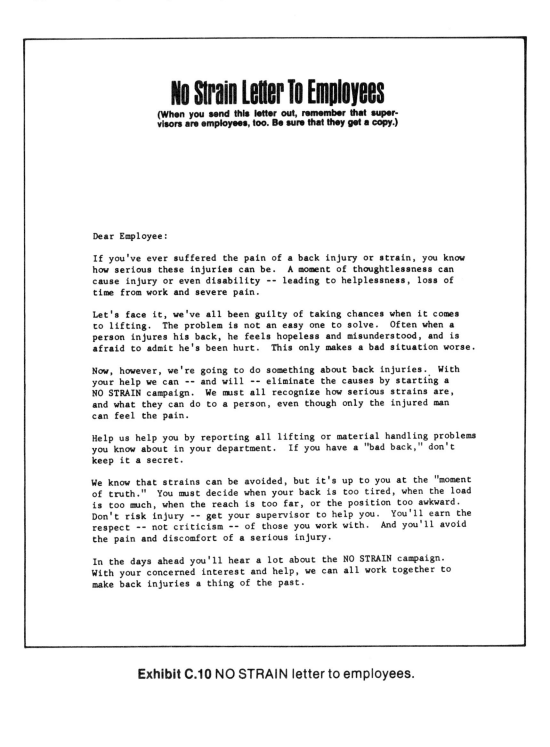

No Strain Letter To Employees
(When you send this letter out, remember that supervisors are employees, too. Be sure that they get a copy.)

Dear Employee:

If you've ever suffered the pain of a back injury or strain, you know how serious these injuries can be. A moment of thoughtlessness can cause injury or even disability -- leading to helplessness, loss of time from work and severe pain.

Let's face it, we've all been guilty of taking chances when it comes to lifting. The problem is not an easy one to solve. Often when a person injures his back, he feels hopeless and misunderstood, and is afraid to admit he's been hurt. This only makes a bad situation worse.

Now, however, we're going to do something about back injuries. With your help we can -- and will -- eliminate the causes by starting a NO STRAIN campaign. We must all recognize how serious strains are, and what they can do to a person, even though only the injured man can feel the pain.

Help us help you by reporting all lifting or material handling problems you know about in your department. If you have a "bad back," don't keep it a secret.

We know that strains can be avoided, but it's up to you at the "moment of truth." You must decide when your back is too tired, when the load is too much, when the reach is too far, or the position too awkward. Don't risk injury -- get your supervisor to help you. You'll earn the respect -- not criticism -- of those you work with. And you'll avoid the pain and discomfort of a serious injury.

In the days ahead you'll hear a lot about the NO STRAIN campaign. With your concerned interest and help, we can all work together to make back injuries a thing of the past.

Exhibit C.10 NO STRAIN letter to employees.

Job Strength Inventory

(To be completed for each lifting job and sent to doctor who examines new employees.)

Physical Requirements

What part of average work period does employee sit? Give percentage. _____

What part of average work period must employee stand? Give percentage. _____

How much lifting does the job require? Give details. _____

_____

What is the maximum weight the employee must lift? _____ Percentage of total _____

What is the weight the employee must lift most often? _____

How much lifting is:
(Give percentages)   At floor level? _____   Waist high? _____   Shoulder high? _____

How much lifting is overhead? Give percentage. _____

How much carrying does the job require? _____

How much turning or twisting is required? _____

How much reaching is required? _____

Special Requirements

Give details of any special requirements of the job such as high work, work in enclosed spaces, crawling, climbing or kneeling.

Recommended Job Improvements

Evaluate the requirements of the job. Is there an easier way to do the job? Consider all possible new layouts and mechanical equipment which could reduce manual material handling operations. List your recommendations:

Exhibit C.11 Job strength inventory.

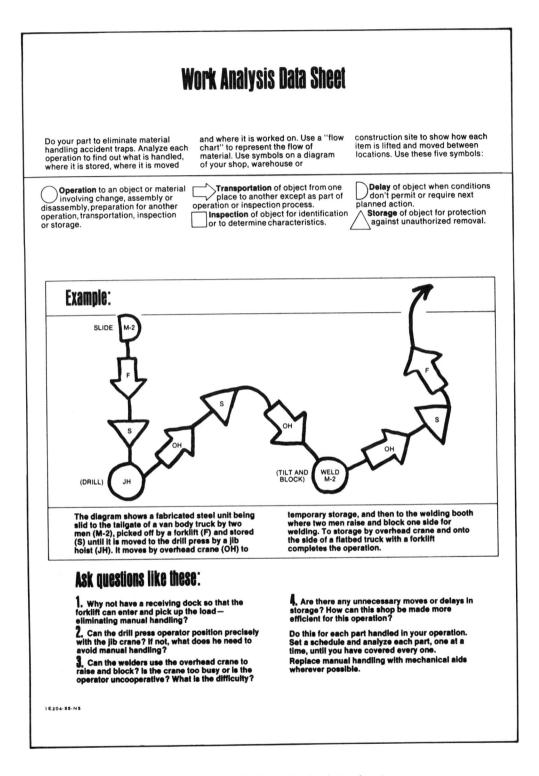

Work Analysis Data Sheet

Do your part to eliminate material handling accident traps. Analyze each operation to find out what is handled, where it is stored, where it is moved and where it is worked on. Use a "flow chart" to represent the flow of material. Use symbols on a diagram of your shop, warehouse or construction site to show how each item is lifted and moved between locations. Use these five symbols:

○ **Operation** to an object or material involving change, assembly or disassembly, preparation for another operation, transportation, inspection or storage.

⇨ **Transportation** of object from one place to another except as part of operation or inspection process.

☐ **Inspection** of object for identification or to determine characteristics.

⟩ **Delay** of object when conditions don't permit or require next planned action.

△ **Storage** of object for protection against unauthorized removal.

Example:

The diagram shows a fabricated steel unit being slid to the tailgate of a van body truck by two men (M-2), picked off by a forklift (F) and stored (S) until it is moved to the drill press by a jib hoist (JH). It moves by overhead crane (OH) to temporary storage, and then to the welding booth where two men raise and block one side for welding. To storage by overhead crane and onto the side of a flatbed truck with a forklift completes the operation.

Ask questions like these:

1. Why not have a receiving dock so that the forklift can enter and pick up the load—eliminating manual handling?

2. Can the drill press operator position precisely with the jib crane? If not, what does he need to avoid manual handling?

3. Can the welders use the overhead crane to raise and block? Is the crane too busy or is the operator uncooperative? What is the difficulty?

4. Are there any unnecessary moves or delays in storage? How can this shop be made more efficient for this operation?

Do this for each part handled in your operation. Set a schedule and analyze each part, one at a time, until you have covered every one.
Replace manual handling with mechanical aids wherever possible.

1E204-55-NS

Exhibit C.12 Work analysis data sheet.

No Strain Interview Guide

Instructions: Have a frank discussion with the prospective employee. Explain the lifting requirements of the new job from the Job Strength Inventory. Try to get candid answers from the applicant on the lifting requirements of his past jobs. Find out if he has had any back injuries or strains which could be a problem on the new job. Phone the previous employer(s) to verify the lifting requirements and injury experience. Record the replies. Ask the applicant to sign the medical exam approval (bottom of page) before completing the rest of this form.

Last name	First	Middle	Age	Sex	Height	Weight

Health History (Include dates and other pertinent details.)

Have you had any serious falls? _____   Any auto accidents? _____

Back or spinal operations? _____   Back muscle spasms? _____

Chronic backache? _____   Other back problems? _____

Any time on light work or off work during the last two years? _____

Have you received any compensation awards? Give amounts. _____

Work History (Include addresses and phone numbers from employment application. Start with the most recent job and work back at least two years.)

Last employer _____   Maximum weights handled _____

Location _____   How often handled? (Hours daily) _____

Phone _____   How high lifted? (Waist, shoulder) _____

Type of job _____   Date phoned to verify _____

Comments _____

Next-to-last employer _____   Maximum weights handled _____

Location _____   How often handled? (Hours daily) _____

Phone _____   How high lifted? (Waist, shoulder) _____

Type of job _____   Date phoned to verify _____

Comments _____

(Use other side if necessary)

I have no objection to a medical examination or having X-rays taken.

_____ Applicant's signature   _____ Date

Exhibit C.13 NO STRAIN interview guide.

of a certain type, backing accidents, burns, and hand injuries. The problem should be defined on the basis of the company's previous experience.

2. *A measurable base period.* A past record which indicates the problem will give a base against which to compare results.

3. *Measurable results.* Records should be kept to make sure you know the results of your efforts.

4. *Defined responsibilities.* Responsibility for certain activities must be fixed—usually to the line organization.

5. *Procedures to fix accountability.* Management must know whether the line is carrying out those activities.

6. *Planned steps.* Decisive steps should be planned before the start of the program.

7. *A recall symbol.* Have some symbol that can be seen everywhere and will remind employees over and over again of the program.

Appendix D
Ergonomic Information

In this appendix background information is presented on some basic ergonomic information that can be used for design and ergonomic analysis purposes. Most of the information comes from *Human Engineering Guide to Equipment Design*, by Van Cott and Kinkade, published by the U.S. Government Printing Office.

ANTHROPOMETRIC CONSIDERATIONS

Anthropometric considerations involve the relationships between body dimensions, strength, dexterity, and mobility of the product user. It is easy to understand that because of differences between large people and small people, strong people and weak people, it may be difficult to design and fit everyone, and in some cases economically unfeasible.

Generally speaking, one should consider large people where clearance is concerned, small people where reach is concerned, weak people when force is concerned. Mobility and dexterity are not so much a matter of size. However, there is an interaction which must be considered. For example, if a product has to be used in a position which is selected as a compromise for both ends of the size population, one or the other (i.e., large or small) may be forced to assume positions which interfere with mobility and dexterity.

Basic Body Dimensions

— Stature—This dimension should be considered wherever head clearance may be a problem. Since an overhead which clears the tallest person also accommodates the shortest person, this value is the determining criteria.
— Eye height—The position of the operator's eyes are important in determining proper placement of visual displays and for providing adequate clearance for seeing over intervening objects.
— Shoulder height—Generally, it is recommended that controls be placed somewhere between waist and shoulder height for most convenient operation.
— Arm reach—The person with the shortest arm reach generally establishes the primary constraint in deciding where to locate controls.

NORMAL HABIT PATTERNS

A number of useful and operationally meaningful conclusions can be drawn about human behavior. These observations represent simplifications and generalizations about some very complex, normal human behavior.

The following table contains some of the recognized observations about typical human behavior characteristics which may lead to injury or to unsafe acts. Each statement is followed by a design-oriented precaution or injunction, the summarized results of which are subsequently incorporated in human engineering safe-design criteria listings.

381

List of Typical Human Behaviors That May Lead To an Unsafe Act or to Injury

Behavior Description	Design Consideration
Many people do not consider the effects of surface friction on their ability to grasp and hold an article.	Design surface texture to provide friction characteristics commensurate with functional requirements of task or device.
Most people cannot estimate distances, clearances, or velocities very well; people tend to over-estimate short distances and under-estimate long distances.	Design products so that users need not make estimates of critical distances, clearances, or speeds. Provide indicators of these quantities where there is a functional requirement.
Most people do not look where they put their hands and feet, especially in familiar surroundings.	If hand or foot placement is a critical aspect of the user-product interface, design so that careless, inadvertent placement of hands or feet will not result in injury. Provide guards, restraints, and warning labels.
People often utilize the first thing handy as an aid in getting to where they want to go or to manipulate something.	Either design the product so that the "first thing handy" simply cannot be functionally useful, or so that it can!
Many people do not think about such things as: high temperatures; becoming an electrical ground when they grasp an article which is "hot"; picking up an object that is wet; or walking carefully on a slick floor.	This lack of awareness may result from ignorance, inattenion, carelessness or sheer disregard for safety. Where possible, observe "Murphy's Law"—design so the person simply cannot make the wrong motion, or touch the "hot" spot.
People seldom anticipate the possibility of contact with sharp edges of corners.	Except to meet functional requirements, eliminate sharp edges on surfaces or units where inadvertent human contact is even remotely possible.
People rarely think about the possibility of catching their clothing on a handle or other protruding object.	Same as item 6, above; also provide proper warnings and labels. Conduct user task (operator function) analysis and always mount handles in most functionally advantageous places.
Very few people think about the possibility of fire or explosion from overheated objects or cooking oil suddenly exposed to the air.	Unless it is a functional requirement, eliminate configurations which will permit such possibilities, even with product misuse.
People seldom give serious thought to the innate curiosity of children in exploring the	Obviously, no simple injunction approaches adequacy; follow all available human engi-

List of Typical Human Behaviors That May Lead To an Unsafe Act or to Injury (Cont.)

Behavior Description	Design Consideration
wonderful mysteries of appliances, pots, and pans.	neering specifications and safety precautions. Eliminate sharp edges, i.e., heavy spring-loaded closures. Use glass and ceramic material with special fracture characteristics.
Many people do not take the time to read labels or instructions, or to observe safety precautions.	Make labels *brief, bold, simple, clear.* Repeat (place) same labels on various parts of a product. Make use of color coding, fail-safe innovations, and other attention-demanding devices.
Many designers do not recognize the existence of response stereotypes, i.e., that the average user "expects" something to operate in a certain fashion.	Become aware of the more common stereotype behaviors (see material following this table). Also, do not depart from common design on objects or items, which have demonstrated utility and user acceptance just for sake of change, i.e., standard positions for ON in wall electric switch is *UP*; don't change to *DOWN*.
Some housewives have little mechanical aptitude, and, therefore, do not recognize mechanical relationships which would suggest a mode of operation.	For most home-use products, routine operation and handling *can* be made very simple. For example, even in such complex items as cassette-type sound recorders and play-back units there is a tremendous range in operational use-complexity, all for accomplishing essentially the same function. Keep it simple.
Some housewives have a penchant for operating an appliance while their hands are full.	If possible, design so that they can do just this! Conversely, design so that the hand/product interface is a full challenge to the operator.
Many people perform most tasks while thinking about something else.	Accept this as a way of life; design so they can continue this practice. No really satisfactory design-out on this one, except—**KEEP IT SIMPLE.**
Most people perform in a perfunctory manner, utilizing previous habit patterns. Under stress, they almost always revert to these habit patterns.	Don't change an established design (if satisfactory) just for the sake of change. Base "design innovations" on functional requirements changes.

List of Typical Human Behaviors That May Lead
To an Unsafe Act or to Injury (Cont.)

Behavior Description	Design Consideration
Most people will use their hands to test, examine, or explore.	Design in such a way that exploring hands and feet won't get hurt.
Many people will continue to use a faulty article even though they suspect that it may be dangerous.	Two general alternatives: (1) do not build in "graceful degradation" characteristics; or (2) provide clearly marked, evident, fault or failure warning indicators.
Very few people recognize the fact that they cannot see well enough either because of poor eyesight or because of lack of illumination.	Make all critical labels such that they can be clearly and easily read by individuals with no better than 20–40 vision (uncorrected) in one eye. Provide task, item, or area illumination within product where functional requirements justify same.
Children act on impulse even though they have been told *to* do or *not* to do something.	Any designer who has children is aware of this behavior tendency. Task or operator requirement analysis may lead to specific solutions.
Most people are reluctant to (and seldom do) recheck their operational or maintenance procedures for errors or omissions.	Keep it simple; provide concise, brief instructions; design for step-by-step procedures with a minimum of procedural interactions and interdependencies. For example, during recent product survey (riding lawn mowers) two very similar units were compared for simplicity of operation; both performed identical functions and outward appearances were strikingly similar. However, one machine had such complicated instructions and such a confusion of levers and controls that it almost defied ordinary usage or understanding.
In emergency situations, people very often respond irrationally and with seemingly random behavior patterns.	Same as item 20 and many others. Keep operation simple; provide for fail-safe operation; follow stereotypes and standard configuration precepts where possible.
People often are unwilling to admit errors or mistakes of judgment or perception, and thus will continue a behavior or action originally initiated in error.	Keep your design simple; provide for fail-safe operation. Where functionally justifiable, design for sequence-checking. Design for automatic product shut-off in the event appropriate sequence is not followed.

List of Typical Human Behaviors That May Lead To an Unsafe Act or to Injury (Cont.)

Behavior Description	Design Consideration
Foolish attitudes and emotional biases often force people into apparently irrational behaviors and improper use of products.	Be aware of this. No matter how you design it, there is one end of a gun you can't make safe *and* simultaneously meet the functional requirements of guns.
A physically handicapped person will often undertake tasks and operations of which he or she is incapable, largely out of false pride.	Study the limitations of the physically handicapped. Design for one-hand, one-eye, one-leg operators.
People often misread or fail to see labels, instructions, and scale markers on various items, thereby improperly setting or adjusting them and thus creating a hazard.	Follow detailed human engineering practice and techniques for placement of labels, design of scales, displays and markers.
Many people become complacent after long-term successful use of (or exposure to) generally hazardous products.	Design in such a way that the fail-safe characteristics of a device cannot be avoided by simple means. Provide attention-demanding but simple, brief warning displays. Consider using very dramatic warning devices to indicate potential failure such as flashing lights or loud sounds. In some cases provide for fairly complicated operating procedures associated with automatic system shutdown where sequence of steps is violated.
The span of visual (sensory) comprehension is limited by certain innate human characteristics and possibly further constrained by training and experience. However, the average adult tends to "fill in" according to some previously experienced pattern or relationship, even though the stimulus or object-condition of interest does not contain all of the elements upon which a decision or act should be based. For example, in reading a new set of instructions, even though incomplete (or containing a departure from the ordinary), the average adult is prone to "fill in" or go on the assumption that the item or element was or was not present according to an "expectation set." Another example is the ability or tendency of a person to fill in the details of a geometric pattern, even though its rendition is imperfect or incomplete.	Although the related behavior tendency seems to be very complex, a fairly simple, straightforward design practice will provide the solution to many of the safety aspects of these issues. The general precaution is directly applicable—KEEP IT SIMPLE. Make labels large. Keep instructions brief, concise and clear, avoid long sentences. Do not make changes in configuration just for the sake of change. Sometimes, providing alternate wording to give the same instructions or operating procedure is very useful. Be sure not to place such labels or instructions side by side or where both can be seen simultaneously by the same person. Here, as in many other instances, an operator task analysis will assist in providing the safe-design information or requirement specifications.

POPULATION STEREOTYPES

The term population stereotypes has more than one meaning. The most common meaning relates to what people do to or with things and the extent to which this behavior becomes relatively universal across both geography and time. For example, a very common behavioral stereotype is for persons to *push* outward on exit doors and to *pull* towards themselves on entry doors to public buildings.

Common practice in the USA is for electric switches to be pushed up or to the right for ON; whereas for fluid control, the value is turned counter-clockwise for increase or ON and clockwise for decrease or OFF.

In Europe, and much of the rest of the world, these conventions do not apply; thus the specific population stereotype may be the opposite of that which prevails in the United States.

VISUAL DISPLAYS DESIGN

Although engineering-oriented persons learn from experience to interpret rather complex displays, the majority of product users do not have this experience. Therefore, displays should be selected which are simple to interpret. Several factors should be considered in selecting or designing a visual display. For example, the display should not contain any more information than absolutely necessary to convey the intended message.

In addition, display formats should follow certain rules based upon the typical expectations of the average operator. For example, people interpret or readout displayed information according to certain stereotypes (e.g., they read from left to right, from top to bottom, they expect values to increase from bottom to top, left to right or clockwise).

Other display formatting rules include sizing and spacing of display. For example, the position of a switch provides system condition information, thus the switch position and/or direction of movement should be compatible with the same rules noted above.

INSTRUCTIONS & LABELING

It should be assumed that all products may at some time be used or operated by someone who is unfamiliar with the product and inexperienced in its use. It should not be assumed that instructions provided by means of an instruction book, pamphlet, or sheet will be used or retained for later reference. Clear, visible, legible instruction labels should be placed on the product in a conspicuous place. All controls and displays, as well as critical parts, also should be labeled. Do not assume that the user will recognize a component or understand what it is for, and how it is to be used.

CONTROL MODE AND DESIGN

- To provide an increase in function, control motion should be up, forward, to the right, or clockwise.
- Go/No-Go or ON-OFF controls should operate so that the GO or ON position is up, forward, to the right.
- Push-pull handles should pull out of a control panel for ON.
- Handles used for braking action should "pull" ON.

— Finger-actuated "trigger-type" controls should be compatible with a squeeze action of the operator's hands and fingers; i.e., contraction of finger actuates function.
— Control lever path should be designed so that it is normal to the pivot point of the limb (i.e., radiates out from the point rather than perpendicular to it).
— All foot-operated controls should be pushed for ON or increase in function.
— All controls shall be located within maximum reach limits of the operator when he is in the normal operating position. Controls, preferably, should be located at less than 75 percent of the maximum reach. Controls should be located where they can be reached by a direct or straight path. Panel controls should never be located behind other controls (behind a steering wheel, or example).
— Controls should not be located where they cannot be seen.
— The most frequently used controls should be located in the most convenient place, near the hand or foot normally used to operate them.
— When a control may be used from one or more operating position, it should be located for the best compromise between the two positions.
— Avoid locating controls below waist or above shoulder height. If short, fast reaction time is important, locate controls directly in front of the operator, not on the side.
— Provide sufficient spacing between controls to preclude inadvertent actuation. When there is not enough space, provide separation guards or use a dual-mode safety design.

Ergonomic Tables

Following are some tables helpful for design or for making the ergonomic analysis.

VISUAL RECOMMENDATIONS

Task condition	Type of task or area	Illumi- nance level (Ft.-c)	Type of illumination
Small detail, low contrast, prolonged periods, high speed, extreme accuracy.	Sewing, inspecting dark materials, etc.	100	General plus supplementary, e.g., desk lamp.
Small detail, fair contrast, speed not essential.	Machining, detail drafting, watch repairing, inspecting medium materials, etc.	50-100	General plus supplementary.
Normal detail, prolonged periods.	Reading, parts assembly; general office and laboratory work	20-50	General, e.g., overhead ceiling fixture.
Normal detail, no prolonged periods.	Washrooms, power plants, waiting rooms, kitchens	10-20	General, e.g., random natural or artificial light
Good contrast, fairly large objects.	Recreational facilities..	5-10	General.
Large objects..........	Restaurants, stairways, bulk-supply warehouses.	2-5	General.

Exhibit D.1 General illumination levels and types of illumination for different task conditions and types of tasks.

Color	Reflectance	Color	Reflectance
White 85			
Light:		Dark:	
Cream 75		Gray 30	
Gray 75		Red 13	
Yellow 75		Brown 10	
Buff 70		Blue 8	
Green 65		Green 7	
Blue 55			
Medium:		Wood Finish:	
Yellow 65		Maple 42	
Buff 63		Satinwood 34	
Gray 55		English Oak 17	
Green 52		Walnut 16	
Blue 35		Mahogany 12	

Exhibit D. 2 Approximate reflectance factors for various surface colors.

	Recommendations		
Condition of use	Lighting technique	Luminance of markings (ft.-1)	Brightness adjustment
Indicator reading, dark adaption necessary.	Red flood, integral or both, with operator choice.	0.02-0.1	Continuous throughout range.
Indicator reading, dark adaptation not necessary but desirable.	Red or low-color-temperature white flood, integral, or both, with operator choice.	0.02-1.0	Continuous throughout range.
Indicator reading, dark adaptation not necessary.	White flood...	1-20...	Fixed or continuous.
Reading of legends on control consoles, dark adaptation necessary.	Red integral lighting red flood, or both, with operator choice.	0.02-0.1..	Continuous throughout range.
Reading of legends on control consoles, dark adaptation not necessary.	White flood...	1-20..	Fixed or continuous.
Possible exposure to bright flashes.	White flood...	10-20	Fixed.
Very high altitude, daylight restricted by cockpit design.	White flood...	10-20..	Fixed.
Chart reading, dark adaptation necessary.	Red or white flood with operator choice.	0.1-1.0 (on white portions of chart).	Continuous throughout range.
Chart reading, dark adaptation not necessary.	White flood...	5-20...	Fixed or continuous.

Exhibit D.3 Recommendations for indicator, panel, and chart lighting.

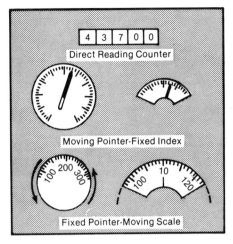

Basic types of symbolic indicators.

For—	Counter is—	Moving pointer is—	Moving scale is—
Quantitative reading.	Good (requires minimum reading time with minimum reading error).	Fair.....	Fair.
Qualitative and check reading.	Poor (position changes not easily detected).	Good (location of pointer and change in position is easily detected).	Poor (difficult to judge direction and magnitude of pointer deviation).
Setting....	Good (most accurate method of monitoring numerical settings, but relation between pointer motion and motion of setting knob is less direct).	Good (has simple and direct relation between pointer motion and motion of setting knob, and pointer-position change aids monitoring).	Fair (has somewhat ambiguous relation between pointer motion and motion of setting knob).
Tracking...	Poor (not readily monitored, and has ambiguous relationship to manual-control motion).	Good (pointer position is readily monitored and controlled, provides simple relationship to manual-control motion, and provides some information about rate).	Fair (not readily monitored and has somewhat ambiguous relationship to manual-control motion)
Orientation...	Poor....	Good (generally moving pointer should represent vehicle, or moving component of system).	Good (generally moving scale should represent outside world, or other stable frame of reference).
General...	Fair (most economical in use of space and illuminated area, scale length limited only by number of counter drums, but is difficult to illuminate properly).	Good (but requires greatest exposed and illuminated area on panel, and scale length is limited).	Fair (offers saving in panel space because only small section of scale need be exposed and illuminated, and long scale is possible).

Exhibit D.4 Relative evaluation of basic symbolic indicator types.

| Nature of markings | Height (in.)* | |
	Low luminance†	High luminance‡
Critical markings, position variable (numerals on counters and settable or moving scales)	0.20-0.30	0.12-0.20
Critical markings, position fixed (numerals on fixed scales, control and switch markings, emergency instructions)	0.15-0.30	0.10-0.20
Noncritical markings (identification labels, routine instructions, any markings required only for familiarization	0.05-0.20	0.05-0.20

*For 28-in viewing distance. For other viewing distances, increase or decrease values proportionately.
† Between 0.03 and 1.0 ft.-L.
‡ Above 1.0 ft.-L.

Exhibit D.5 Recommended numeral and letter heights.

Factors	Optimum	Preferred limits	Acceptable limits
Ratio of $\frac{\text{viewing distance}}{\text{image width}}$	4	3-6	2-8
Angle off center line—degrees	0	20	30
* Image luminance-ft.-Lamberts (no film in projector).	10	8-14	5-20
Luminance variation across screen-ratio of maximum to minimum luminance.	1	1.5	3.0
Luminance variation as a function of seat position-ratio of maximum to minimum luminance.	1	2	4.0
Ratio of $\frac{\text{ambient light}}{\text{highest part of image}}$	0	0.002-0.01	0.2†

* For still projections higher values may be used.
† For line drawings, tables, not involving gray scale or color.

Exhibit D.6 General recommendations for group viewing for slides and motion pictures.

AUDITORY RECOMMENDATIONS

Alarm	Intensity	Frequency	Attention-getting ability	Noise-penetration ability	Special features
Diaphone	Very high...	Very low...	Good...	Poor in low-frequency noise. Good in high-frequency noise.	
Horn....	High.....	Low to high.	Good.......	Good.......	Can be designed to beam sound directionally. Can be rotated to get wide coverage.
Whistle.....	High....	Low to high.	Good if inter-mittent.	Good if frequency is properly chosen.	Can be made directional by reflectors.
Siren.....	High....	Low to high.	Very good if pitch rises and falls.	Very good with rising and falling frequency.	Can be coupled to horn for directional transmission.
Bell.....	Medium....	Medium to high.	Good...	Good in low-frequency noise.	Can be provided with manual shutoff to insure alarm until action is taken.
Buzzer....	Low to medium.	Low to medium.	Good....	Fair if spectrum is suited to background noise.	Can be provided with manual shutoff to insure alarm until action is taken.
Chimes and gong.	Low to medium.	Low to medium.	Fair....	Fair if spectrum is suited to background noise.	
Oscillator...	Low to high.	Medium to high.	Good if inter-mittent.	Good if frequency is properly chosen.	Can be presented over inter-com system.

Exhibit D.7 Types of alarms, their characteristics and special features.

Conditions	Design recommendations
1. If distance to listener is great—	1. Use high intensities and avoid high frequencies.
2. If sound must bend around obstacles and pass through partitions—	2. Use low frequencies.
3. If background noise is present—	3. Select alarm frequency in region where noise masking is minimal.
4. To demand attention—	4. Modulate signal to give intermittent "beeps" or modulate frequency to make pitch rise and fall at rate of about 1-3 cps.
5. To acknowledge warning—	5. Provide signal with manual shutoff so that it sounds continuously until action is taken.

Exhibit D.8 Summary of design recommendations for auditory alarm and warning devices.

Distance between talker and listener (ft)	Speech-interference level (dB)*			
	Normal†	Raised†	Very loud†	Shouting†
0.3	71	77	83	89
1.0	65	71	77	83
2.0	59	65	71	77
3.0	55	61	67	73
4.0	53	59	65	71
5.0	51	57	63	69
6.0	49	55	61	67
12.0	43	49	55	61

Exhibit D.9 Speech-interference levels that barely permit reliable conversation.

CONTROL RECOMMENDATIONS

System Response		Acceptable Controls	
Type	Examples	Type	Examples
Stationary		Linear or rotary	
Rotary through an arc less than 180°		Linear or rotary	
Rotary through an arc more than 180°		Rotary	
Linear in one dimension		Linear or rotary	
Linear in two dimensions		Linear or Two Rotary	

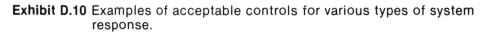

Exhibit D.10 Examples of acceptable controls for various types of system response.

Class A

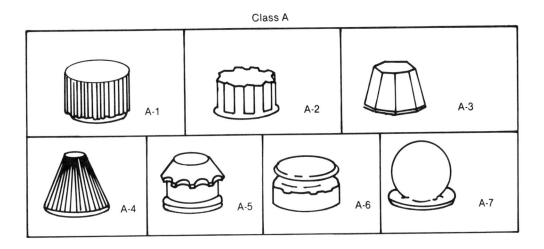

Class B

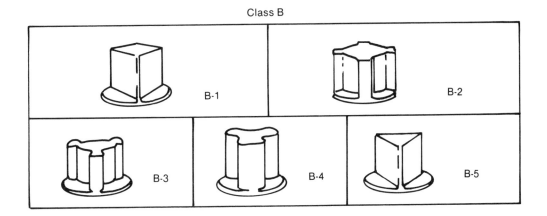

Class C

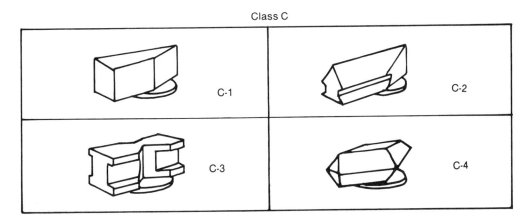

Examples of three classes of knobs: (A) those for twirling or spinning; (B) those to be used where less than a full turn is required and position is not so important; and (C) those where less than a full turn is required and position is important.

Exhibit D.11 Knob examples.

Appendix E
Sources of Help

This section looks at sources of help available in analysis and improvement of the behavioral system, safety system, and physical environment. Probably the first point to be made is that the task of analysis and improvement cannot be delegated to the outside. There are people and agencies to help, but the responsibility cannot be delegated. Secondly, the analysis is better if made internally because you know your organization better than any outsider. There are, however, times and circumstances when an outsider's help can be invaluable.

WHEN TO SEEK OUTSIDE HELP

Kenneth Albert suggests that there are six specific times when outside help should be obtained:

1. When special expertise is essential.
2. For a politically sensitive issue.
3. When impartiality is necessary.
4. If time is critical.
5. If anonymity must be maintained.
6. When the prestige of an outsider would be helpful.

While Alpert is speaking in a nonsafety vein, some of the above points seem valid in determining the need for an outside safety consultant. If there is no one in the organization familiar with OSHA, it might be helpful to call in an OSHA expert to assist. Perhaps, more realistically, if there is no one in the organizational structure able to analyze the behavioral system, it might serve you well to locate such a person.

Often a safety problem is intertwined with managerial personalities and is extremely difficult to solve internally. Here an outsider might well offer a solution acceptable to top management. To obtain an analysis in a hurry with some suggested approaches to a solution, it often can be done faster by an outside consultant, and often an outsider's recommendation will seem to carry more weight than the insider's.

In the safety field, it appears as though most organizations need outside help on the behavioral system analysis, many on the safety system analysis, and few on the physical condition analysis. However, in looking at what is available in safety consultants, the exact opposite is true. Physical condition consultants abound everywhere, safety system analysts are quite rare, and behavioral analyst experts are almost nonexistent.

WHAT EXTERNAL HELP IS AVAILABLE

External help might be classified into these categories:

1. Insurance company field safety engineers or consultants.
2. Government safety consultants under OSHA direction.

3. Private consulting firms.
4. Part-time private consultants.

Government Consultants

It is only recently that Congress allowed consulting in addition to compliance checking. It is now a requirement in all states. Most states are now gearing up to provide consulting to businesses to comply with this change. Who provides the consulting service varies from state to state. In most cases the state contracts with Federal OSHA to do the consulting. In others, private organizations get the contract, and in some the state university will provide the consulting service to the state's employers who wish it. Utilizing government consultants is completely voluntary. They will not visit any place of business unless invited. Before choosing to invite them, however, any organization ought to find out a little about the program.

The stated goal of the on-site consultation program by OSHA is to obtain "safety and healthful workplace for employees." Thus by definition the consulting will pertain only to physical condition, not the safety system or the behavioral system.

The defined on-site responsibilities of the OSHA consultant are:

1. To identify and properly classify hazards.
2. To recommend corrective measures (short of engineering assistance).
3. To arrange abatement dates for serious hazards.
4. To report to the OSHA supervisor any employer-unabated serious hazards.
5. To follow up on employer actions.

There are some aspects of receiving consulting service from OSHA that are unusual. Their purpose is to help improve physical conditions, but in two instances, the consultants have additional duties:

1. In the case of serious violations of the OSHA standards, they must set abatement dates and follow-up.
2. In the case of imminent violations of the OSHA standards, they must refer them to either their supervisor and on up the Bureau of Labor hierarchy or to the compliance staff for your immediate compliance.

In other words, OSHA consulting is true consulting only when nothing seriously wrong is found. If anything serious or imminent is found, the "customer" has lost control of the decision process of how and when to correct.

Insurance Consultants

Most of the consultants in this country are insurance consultants. They comprise a large percentage of the safety engineers in this country. Many others (if not most) get their safety start in insurance companies. Almost all companies, except the very large and self-insured, are helped routinely by insurance loss-control representatives.

The safety services department of any large insurance company is charged with three specific functions:

1. A sales assist function.
2. An underwriting assist function.
3. A customer service function.

Only the third of these is of value to the customer. Historically, most companies started with a primary emphasis on function 2, the underwriting assist function, where the field representative is the "eyes and ears" of the underwriting department. This field job is one of observing and reporting what is going on at the policyholder's place of business back to the desk-bound underwriter, who can then, based upon the information, determine desirability and pricing of the insurance. The third function, that of customer service, seems to have evolved a little later and never in some organizations. This function consists of doing something for the customer that improves the safety program and reduces the likelihood of that customer having accidents and financial loss. What that something is varies considerably from company to company. The first function, that of helping to sell insurance to more customers, is a by-product of the service function, convincing the potential customer that the insurer can provide some help.

As a result of the past, different philosophies have emerged which dictate the value of the service the insurance company is able to provide. In some companies the safety services department is still very much a part of the underwriting function. Their duties are to observe and report. In some companies the engineering department reports to the underwriting department. In other, more modern insurance companies the loss-control department is independent, existing primarily to serve the customer and secondarily to sell and to assist the underwriter.

If the loss-control department is an arm of underwriting, it is almost impossible to get any real safety service from them. They simply will not be geared up to provide that service. Their people will not be trained in consulting; they will not have the tools to help.

If the loss-control department is not an arm of underwriting, it has the chance of being able to provide good service to a customer. It also may be quite ineffective, since there are all kinds of ways to make sure the service provided is not effective.

When the service has as its primary mission to assist sales, the customer service will suffer. When the service is an inspection-only service, as is very prevalent, the service suffers, since the safety system and the behavioral system will be totally overlooked. When the service consists of the delivery of safety aids and materials and nothing else, it is a meaningless service. When the service consists primarily or totally of merely holding safety meetings for a customer or merely of ensuring that physical conditions are up to code, it is a weak service. Perhaps the least effective service (although fairly common) consists of the carrier's representative bringing out the canned safety program that the carrier's home office has devised for use at all insured companies.

Depending upon what philosophies are behind the service of the carrier, additional services may have been developed beyond the service of the representative that calls on the customer.

The bottom line on the value of the insurance carrier service is the individual that calls on you, who can be helpful or a waste of your time.

Private Consulting Firms

A third source of external help is the private consultant (full-time) or the private consulting firms. They can be called in to consult in any area—behavioral, safety system, or physical—and with none of the strings mentioned above using OSHA consultants. The only difficulty is in choosing wisely and obtaining a consultant with the necessary skills and knowledge to do what you want done.

Part-Time Private Consultants

The fourth place to locate a private consultant is among those individuals that consult on a part-time basis to supplement their incomes. This usually is either a retired safety professional who would like to keep active or a college or university professor who needs to supplement his income and hopes to keep knowledgeable about the real world.

Here again the problem is locating these poeple and hiring a person with the appropriate competencies. Just because a man has 40 years in safety, or because he teaches safety, does not mean that these competencies are ensured.

HELP FROM ASSOCIATIONS

Listed below are a number of associations that provide assistance in safety.

Standards and Specification Groups

American National Standards Institute, 1430 Broadway, New York, NY 10018. Coordinates and administers the federated voluntary standardization system in the United States.

American Society for Testing and Material, 1916 Race Street, Philadelphia, PA 19103. World's largest source of voluntary consensus standards for materials, products, systems and services.

Fire Protection Organizations

Factory Insurance Association, 85 Woodland Street, Hartford, CT 06015. Composed of a group of capital stock insurance companies to provide engineering, inspections, and loss adjustment service to industry.

Factory Mutual System, 1151 Boston-Providence Turnpike, Norwood, MA 02062. An industrial fire protection, engineering and inspection bureau established and maintained by mutual fire insurance companies.

National Fire Protection Association, 470 Atlantic Ave., Boston, MA 02210. The clearing house for information on fire protection and fire prevention. Nonprofit technical and educational organization.

Underwriter Laboratories Inc., 207 East Ohio Street, Chicago, IL 60611. Not-for-profit organization whose laboratories publish annual lists of manufacturers whose products proved acceptable under appropriate Standards.

Insurance Associations

The American Association of State Compensation Insurance Funds, P.O. 5922, San Francisco, CA 94101. Produces technical information available to safety engineers.

American Mutual Insurance Alliance, 20 North Wacker Dr., Chicago, IL 60606. Dissemination of information on safety subjects, conducts specialized courses for member company personnel. Prints several publications.

American Insurance Association Engineering and Safety Service, 85 John Street, New York, NY 10038. Issues many publications on Accident Prevention.

Trade Associations

Air Pollution Control Association, 4450 Fifth Avenue, Pittsburgh, PA 15213.

American Chemical Society, 1155 Sixteenth Street, NW, Washington, DC 20036.

American Conference of Governmental Industrial Hygienists, P.O. Box 1937, Cincinnati, OH 45201.

American Foundrymen's Society, Golf and Wolf Roads, Des Plaines, IL 60016. Serves as a clearinghouses for information in this industry.

American Gas Association, 1515 Wilson Boulevard, Arlington, VA 22209. Serves as a clearinghouse and advisor to its member companies.

American Industrial Hygiene Association, c/o William E. McCormick, Managing Director, 66 S. Miller Road, Akron, OH 44313.

American Iron and Steel Institute, 1000 16th Street, NW, Washington, DC 20036.

American Meat Institute, 1600 N. Wilson Boulevard, Arlington, VA 20007.

American Medical Association, Committee on Occupational Toxicology and Council on Occupational Health, 535 North Dearborn Street, Chicago, IL 60610.

American Mining Congress, 1200 18th Street, NW, Washington, DC 20036.

American National Standards Institute, 1430 Broadway, New York, NY 10018.

American Paper Institute, 260 Madison Avenue, New York, NY 10016.

American Petroleum Institute Safety and Fire Protection Service, 1801 D Street, NW, Washington, DC 20006.

American Public Health Association, 1015 Eighteenth Street, NW, Washington, DC 20036.

American Pulpwood Association, 605 Bird Avenue, New York, NY 10016.

American Road Builders Association, 525 School Street, SW, Washington, DC 20024.

American Society for Testing and Materials, 1916 Race Street, Philadelphia, PA 19103.

American Society of Tropical Medicine and Hygiene, PO Box 15208, Emory University Branch, Atlanta, GA 30333.

American Trucking Association, Inc., 1616 P Street, NW, Washington, DC 20036. Membership available to any person concerned with truck safety.

The American Waterways Operators, Inc., 1250 Connecticut Avenue, NW, Suite 502, Washington, DC 20036.

American Water Works Association, 2 Park Avenue, New York, NY 10016.

American Welding Society, 2501 NW 7th Street, Miami, FL 33125. Devoted to the proper and safe use of welding by industry. Booklets and other publications available.

Associated General Contractors of America, Inc., 1957 E Street, NW, Washington, DC 20006.

Association of American Railroads, 1920 L Street, NW, Washington, DC 20006.

Can Manufacturers Institute, Inc., 1625 Massachusetts Avenue, NW, Washington, DC 20036. Acts as a clearinghouse of information on the metal can industry in the United States.

The Chlorine Institute, 342 Madison Avenue, New York, NY 10017. Provides a means for chlorine producers and firms with related interests to deal with problems.

Compressed Gas Association, Inc., 500 Fifth Avenue, New York, NY 10036. Provides information on safe handling and storage of gases.

Edison Electric Institute, 90 Park Avenue, New York, NY 10016.

Graphic Arts Technical Foundation, 4615 Forbes Avenue, Pittsburgh, PA 15213.

Gray and Ductile Iron Founder's Society, Inc., Cast Metals Federation Building, Rocky River, OH 44126.

Industrial Health Foundation, Inc., 5231 Centre Avenue, Pittsburgh, PA 15232.

Industrial Medical Association, 150 North Wacker Drive, Chicago, IL 60606.

Industrial Safety Equipment Association, Inc., 2425 Wilson Boulevard, Arlington, VA 22201. Devoted to the promotion of public interest in safety.

Institute of Makers of Explosives, 420 Lexington Avenue, New York, NY 10017. Booklets on the safe transportation, handling and use of explosives.

International Association of Drilling Contractors, 211 N. Ervay, Suite 505, Dallas, TX 75201.

International Association of Refrigerated Warehouses, 7315 Wisconsin Avenue, NW, Washington, DC 20014.

Linen Supply Association of America, 975 Arthur Godfrey Road, Miami Beach, FL 33140.

Manufacturing Chemists' Association, Inc., 1825 Connecticut Avenue, NW, Washington, DC 20009. Information available on safe handling, transportation and use of chemicals.

National Association of Manufacturers, 1776 F Street, NW, Washington, DC 20006.

National LP-Gas Association, 79 W. Monroe Street, Chicago, IL 60603. Programs and information to train the public in safe handling and uses of LP gas.

National Safety Council, 444 N. Michigan Avenue, Chicago, IL 60611.

Printing Industries of America, Inc., 1730 N. Lynn Street, Arlington, VA 22209.

Scaffolding and Shoring Institute, 2130 Keith Building, Cleveland, OH 44115. Booklets on shoring available.

The Society of the Plastics Industry, Inc., 250 Park Avenue, New York, NY 10017. Information available on plastics and safety.

Society of Toxicology, Robert A. Scala, Secretary, Medical Rsearch Division, Research and Engineering Company, PO Box 45, Linden, NJ 07036.

The American Academy of Occupational Medicine, The American College of Preventive Medicine, Ward Bentley, Executive Director, 801 Old Lancaster Road, Bryn Mawr, PA 19010.

The Industrial Medical Committees on State and County Medical Societies.

UNITED STATES FEDERAL AGENCIES

Atomic Energy Commission, Division of Biomedical and Environmental Research, Washington, DC 20545.

Department of Commerce

National Bureau of Standards, Washington, DC 20234.

National Technical Information Service (NTIS), 5285 Port Royal Road, Springfield, VA 22151.

Department of Labor

Bureau of Labor Statistics, Fourteenth Street and Constitution Avenue, NW, Washington, DC 20210.

Occupational Safety and Health Administration, Fourteenth Street and Constitution Avenue, NW, Washington, DC 20210.

Department of Interior

Bureau of Mines, C Street between Eighteenth and Nineteenth Streets, NW, Washington, DC 20240.

Mining Enforcement and Safety Administration, C Street between Eighteenth and Nineteenth Streets, NW, Washington, DC 20240.

Environmental Protection Agency

Air and Water Programs Office, 401 M Street, SW, Washington, DC 20460.

Pesticides Program Office, Chamblee, GA 30341.

Radiation Program Office, Rockville, MD 20852.

Water Program Operations Office, 401 M Street, SW, Washington, DC 20460.

Air Pollution Technical Information Center (APTIC), Office of Technical Information and Publications (OTIP), Air Pollution Control Office (APCO), P.O. Box 12055, Research Triangle Park, NC 22709.

National Environmental Research Center, Research Triangle Park, NC 27711 (also located in Cincinnati, OH 45628 and Corvallis, OR 97330).

Western Environmental Research Laboratory, P.O. Box 15207, Las Vegas, NV 89114.

Public Health Service

Center for Disease Control, Atlanta, GA 30333.

National Institute for Occupational Safety and Health (NIOSH), Parklawn Building, 5600 Fishers Lane, Rockville, MD 20852 (also located in the U.S. Post Office Building, Fifth and Walnut Streets, Cincinnati, OH 45202).

Health Resources Administration

National Center for Health Statistics, 330 Independence Avenue, SW, Washington, DC 20201.

National Institutes of Health

National Cancer Institute, Bethesda, MD 20014.

National Clearinghouse for Mental Health Information (NCMHI), National Institute of Mental Health, 5660 Fishers Lane, Rockville, MD 20852.

National Heart, Lung and Blood Institute, Bethesda, MD 20014.

National Institute of Environmental Health Sciences (NIEHS), P.O. Box 12333, Research Triangle Park, NC 22079.

EDUCATIONAL RESOURCE CENTERS

Alabama Educational Resource Center
University of Alabama at Birmingham
School of Public Health
Medical Center, University Station
Birmingham, AL 35294
205/934-6080

Arizona Educational Resource Center
University of Arizona
Arizona Health Sciences Center
1145 N. Warren-ACOSH
Tuscon, AZ 85724
602/626-6835

California Educational Resource Center
– Northern
  University of California, Berkeley
  206 Earl Warren Hall
  Berkeley, CA 94720
  415/642-0761

California Educational Resource Center
– Southern
  University of California, Irvine
  Dept. of Community & Environmental
  Medicine
  Irvine, CA 91717
  714/833-6269

Cincinnati Educational Resource Center
  University of Cincinnati
  Institute of Environmental Health
  3223 Eden Ave.
  Cincinnati, OH 45267
  513/872-5701

Harvard Educational Resource Center
  Dept. of Environmental Health Sciences
  Harvard School of Public Health
  665 Huntington Ave.
  Boston, MA 02115
  617/732-1260

Illinois Educational Resource Center
  University of Illinois
  School of Public Health
  P.O. Box 6998
  Chicago, IL 60680
  312/996-2591

John Hopkins Educational Resource Center
  Johns Hopkins University
  School of Hygiene & Public Health
  615 N. Wolfe St.
  Baltimore, MD 21205
  301/955-3900 or 3720

Michigan Educational Resource Center
  University of Michigan
  Dept. of Industrial & Operations
  Engineering
  2260 S.G. Brown Laboratory
  Ann Arbor, MI 48109
  313/763-2245

Minneasota Educational Resource Center
  University of Minnesota
  School of Public Health
  1162 Mayo Memorial
  420 Delaware St., S.E.
  Minneapolis, MN 55455
  612/373-8080

New York/New Jersey Educational
Resource Center
  Mt. Sinai School of Medicine
  1 Gustave Levy Place
  New York, NY 10029
  212/650-6174

North Carolina Educational Resource Center
  109 Conner Dr.
  Professional Village
  346 A, Suite 1101
  Chapel Hill, NC 27514
  919/962-2101

Texas Educational Resource Center
  The University of Texas Health
  Science Center at Houston
  School of Public Health
  P.O. Box 20186
  Houston, TX 77025
  713/792-7450

Utal Educational Resource Center
  Rocky Mountain Center for
  Occupational & Environmental Health
  University of Utah Medical Center
  DFCM Room BC 106
  Salt Lake City, UT 84132
  801/581-8719

Washington Educational Resource Center
  University of Washington
  Dept. of Environmental Health SC-34
  Seattle, WA 98195
  206/543-4383

BIBLIOGRAPHY OF SAFETY MANAGEMENT BOOKS

American Engineering Council. *Safety and Production.* New York: Harper & Brothers, 1928.

American National Standards Institute. *Safety Codes and Standards.* New York: ANSI.

American Society of Safety Engineers. *A Selected Bibliography of Reference Materials in Safety Engineering.* Chicago: ASSE, 1967.

Anderson, R. *OSHA and Accident Control Through Training.* New York: Industrial Press, 1975.

Anton, T. *Occupational Safety and Health Management.* New York: McGraw-Hill, 1979.

Associated General Contractors. *Manual of Accident Prevention in Construction,* 6th ed. Washington, DC: AGC, 1971.

Berman, H. H., and H. W. McCrone. *Applied Safety Engineering.* New York: McGraw-Hill, 1943.

Binford, C., C. Fleming, and Z. Prust. *Loss Control in the OSHA Era.* New York: McGraw-Hill, 1975.

Bird, F. *Management Guide to Loss Control.* Atlanta, GA: Institute Press, 1974.

Bird, F., and G. Germain. *Damage Control.* Macon, GA: Academy Press, 1966.

Bird, F. and R. Loftus. *Loss Control Management.* Loganville, GA: Institute Press, 1976.

Blake, R. *Industrial Safety.* 3rd ed. Englewood Cliffs, NJ: Prentice-Hall, 1963.

Brown, D. *Systems Analysis and Design for Safety.* Englewood Cliffs, NJ: Prentice-Hall, 1976.

Bush, V. *Safety in the Construction Industry: OSHA.* Reston, VA: Reston Publishing, 1975.

Calabresi, G. *The Costs of Accidents.* New Haven: Yale Press, 1970.

De Reamer, R. *Modern Safety Practices.* New York: John Wiley, 1958. (being revised).

Douglas, H. and J. Crowe. *Effective Loss Prevention.* Toronto: Industrial Accident Prevention Association, 1976.

Equipment Guide Book Co. *OSHA Requirements Guide.* Palo Alto, CA: Equipment Guide Book Co., 1976.

Ferry, T. *Elements of Accident Investigation.* Springfield, IL: Charles C. Thomas, 1978.

Ferry, T. and D. Weaver. *Directions and Safety.* Springfield, IL: Charles C. Thomas, 1976.

Firenze, R. *Guide to Occupational Safety and Health Management.* Dubuque: Kendall/Hunt Publishing Co., 1973.

Fletcher, J. *Total Loss Control.* Toronto: National Profile Ltd., 1974.

Fletcher, J., and H. Douglas. *Total Environmental Control.* Toronto: National Profile Ltd., 1972.

Gardner, J. *Safety Training for the Supervisor.* Reading, MA: Addison-Wesley, 1969.

Gilmore, C. *Accident Prevention and Loss Control.* New York: AMA, 1970.

Griffiths, E. *Injury and Incapacity—With Special Reference to Industrial Insurance.* Baltimore: William Wood & Company, 1935.

Grimaldi, J. and R. Simonds. *Safety Management.* Homewood, IL: Irwin, 1975.

Grose, V., Managing Risk, Englewood Cliffs, NJ: Prentice-Hall, 1988.

Gunderson, E., and R. Rahe. *Life Stress and Illness.* Springfield, IL: Charles C. Thomas, 1974.

Haddon, W., E. Suchman, and E. Klein. *Accident Research: Methods and Approaches.* New York: Harper, 1964.

Hammer, W. *Handbook of Systems and Product Safety.* Englewood Cliffs, NJ: Prentice-Hall, 1972.

Hammer, W. *Occupational Safety Management and Engineering.* Englewood Cliffs, NJ: Prentice-Hall, 1976.

Handley, W. *Industrial Safety Handbook.* New York: McGraw-Hill, 1969.

Heinrich, H. W. *Basics of Supervision, and Supervisor's Safety Manual.* New York: Alfred Best, 1944; also *Formula for Supervision.* New London, CT: National Foremen's Institute, 1949.

Heinrich, H., N. Roos, and D. Petersen. *Industrial Accident Prevention,* 5th ed. New York: McGraw-Hill, 1979.

International Labor Office. *Encyclopedia of Occupational Safety and Health.* New York: McGraw-Hill, 1970.

Johnson, W. *The Management Oversight and Risk Tree—MORT.* Washington, DC: U.S. Government Printing Office, 1973.

Judson, H. H., and J. M. Brown. *Occupational Accident Prevention.* New York: John Wiley & Sons, 1944.

Larson, J. *The Human Element in Industrial Accident Prevention.* New York: New York University, 1955.

Lowrance, W. *Of Acceptable Risk.* Los Altos, CA: Kaufman, 1976.

Lykes, N. *A Psychological Approach to Accidents.* New York: Vantage Press, 1954.

Malasky, S. *Systems Safety.* New York: Spartan Books, 1974.

Marcum, E. *Modern Safety Management Practice.* Morgantown, WV: Worldwide Safety Institute, 1978.

Margolis, B., and W. Kroes. *The Human Side of Accident Prevention.* Springfield, IL: Charles C. Thomas, 1975.

Matwes, G. *Loss Control: Safety Guidebook.* New York: Van Nostrand Reinhold, 1973.

McCall, B. *Safety First—At Last,* New York: Vantage Press, 1975.

McClean, A. *Occupational Stress.* Springfield, IL: Charles C. Thomas, 1974.

McGlade, F. *Adjustive Behavior and Safe Performance.* Springfield, IL: Charles C. Thomas, 1970.

National Safety Council. *Accident Prevention Manual for Industrial Operations.* 7th ed. Chicago: NSC, 1974.

National Safety Council. *Supervisors Safety Manual,* 4th ed. Chicago: NSC, 1973.

National Safety Council. *Motor Fleet Safety Manual,* 2nd ed. Chicago: NSC, 1972.

National Safety Council. *Guide to Occupational Safety Literature.* Chicago: NSC, 1975.

National Safety Council. *Safety Communications.* Chicago: NSC, 1975.

Peck, T. *Occupational Safety and Health: A Guide to Information Sources.* Detroit: Gale Research Co., 1974.

Peters, G. *Product Liability and Safety.* Washington, DC: Coiner Publications, 1962.

Petersen, D. *Human Error Reducation and Safety Management.* New York: McGraw-Hill, 1979.

Petersen, D. *SBO—Safety By Objectives.* Goshen, NY: Aloray, 1979.

Petersen, D. *Safety Management—A Human Approach.* Second Edition. Goshen, NY: Aloray, 1975.

Petersen, D. *Safety Supervision.* New York: Amacom, 1976.

Petersen, D. *The OSHA Compliance Manual.* 2d ed., New York: McGraw-Hill, 1979.

Pope, W. *Systems Safety Management.* Alexandria, VA: Safety Management Information Systems, Inc., 1968.

Resnick, L. *Eye Hazards in Industry.* New York: Columbia University Press, 1941.

Robens, J. *OSHA Compliance Manual.* Reston, VA: Reston Publishers, 1976.

Rodgers, W. *Introduction to System Safety Engineering.* New York: John Wiley & Sons, 1971.

Schenkelbach, L. *The Safety Management Primer.* Homewood, IL: Dow-Jones-Irwin, 1975.

Schulzinger, M. *Accident Syndrome,* Springfield, IL: Charles C. Thomas, 1956.

Shaw, L., and H. Sichel. *Accident Proneness.* Oxford, Pergamon Press, 1971.

Stack, H. J., E. B. Siebrecht, and J. D. Elkow. *Education for Safe Living.* Englewood Cliffs, NJ: Prentice-Hall, 1949.

Thygerson, A. *Safety Principles Instruction and Readings.* Englewood Cliffs, NJ: Prentice-Hall, 1972.

Tye, J. *Management Introduction to Total Loss Control.* London: British Safety Council, 1970.

U.S. Dept. of Labor. *Occupational Safety and Health—A Bibliography.* Washington, DC: DOL, 1974.

Vernon, H. *Accidents and Their Prevention.* New York: Macmillan, 1937.

Widener, J. *Selected Readings in Safety.* Macon, GA: Academy Press, 1973.

OCCUPATIONAL HEALTH REFERENCES

Industrial Hygiene and Chemical Engineering, AIHA-ACGTH Committee on Respirators. *Respiratory Protective Devices Manual.* 2nd ed. The Committee, Lansing, MI, 1976.

Air Sampling Instruments for Evaluation of Atmospheric Contaminants. 4th ed. American Conference of Governmental Industrial Hygienists, Cincinnati, 1972. Gives instrument description and source of supply for U.S. distributors.

Beranek, L. L. (Ed.). *Noise and Vibration Control.* Academic Press, New York, 1970. The practical treatment of noise control design and construction.

Drinker, P. and T. Hatch. *Industrial Dust: Hygienic Significance, Measurement and Control.* 2nd ed. New York: McGraw-Hill, 1954.

Environmental Health Monitoring Manual. United States Steel Corporation, 1973. A loose-leaf manual designed to train plant personnel to conduct the monitoring of the exposure of workers to potential toxic hazards and harmful physical conditions.

Hanson, N.W., D. A. Reilly, H. E. Staff (Eds.). *The Determination of Toxic Substances in Air: A Manual of ICI Practice.* W. Heffer & Sons, Cambridge, England, 1965. Includes some procedures not given elsewhere.

The Industrial Environment, Its Evaluation and Control. 3rd ed. National Institute for Occupational Safety and Health, Rockville, 1973. An industrial hygiene textbook, rather than a syllabus, covering a broad range of subjects from mathematics to medicine.

Industrial Ventilation, a Manual of Recommended Practice. 13th ed. Committee on Industrial Ventilation, American Conference of Governmental Industrial Hygienists, Lansing, MI, 1974. Provides new developments and standards for industrial ventilating systems.

Industrial Safety Data Sheets. National Safety Council, 444 North Michigan Avenue, Chicago, IL 60611. Information and recommendations regarding the safe handling of chemicals and safe practices in the work environment.

Safety Guides. Manufacutring Chemists' Association, 1825 Connecticut Avenue, NW, Washington, DC 20009. Information and recommendations pertaining to safe practices in the work environment.

TLVs. Threshold Limit Values for Chemical Substances and Physical Agents in the Workroom Environment . . . . American Conference of Governmental Industrial Hygienists, P.O. Box 1937, Cincinnati, OH 45201. Annual: threshold limits based on information from industrial experience, experimental human and animal studies, intended for use in the practice of industrial hygiene.

DATA SHEETS, GUIDES, MANUALS

AIHA Hygienic Guide Series. American Industrial Hygiene Association, 66 South Miller Road, Akron, OH 44313. Separate data sheets on specific substances giving hygienic standards, properties, industrial hygiene practice, specific procedures and references.

ANSI Standards. Z37 Series, Acceptable Concentrations of Toxic Dusts and Gases. These Guides represent a consensus of interested parties concerning minimum safety requirements for the storage, transportation and handling of toxic substances and are intended to aid the manufacturer, the consumer and the general public. American National Standards Institute, 1430 Broadway, New York, NY 10018.

ASTM Standards, With Related Material. American Society for Testing and Materials, 1916 Race Street, Philadelphia, PA 19103. Annual.

Data Sheets. Manufacturing Chemists' Association, 1825 Connecticut Avenue, NW, Washington, DC 20009. Includes information on properties, hazards, handling, storage, hazard control, employee safety, medical management, etc. of specific chemicals. Thus far about 100 data sheets have been compiled.

ENCYCLOPEDIAS

Encyclopaedia of Occupational Health and Safety. Second Edition, formerly called Occupational Health International Labour Office, Geneva, 1971–1972, 2 vols. A concise, single access (A–Z) illustrated reference giving fully referenced information on all facets of worker safety and health.

GOVERNMENT REPORTS

Annual Survey of Manufacturers. U.S. Bureau of the Census, Industry Division, Washington, DC. Includes statistics on wholesale and retail trade.

Injury Rates by Industry. U.S. Bureau of Labor Statistics, Fourteenth Street and Constitution Avenue, NW, Washington, DC 20210. An annual report on industrial injuries.

DIRECTORIES

American Industrial Hygiene Association Membership Book. AIHA, 66 South Miller Road, Akron, OH 44313. Annual.

Chemical Guides to the United States. Seventh Edition. Noyes Data Corporation, Park Ridge, NJ, 1973. Describes over 400 of the largest U.S. chemical firms with index to companies, no subject index.

Directory of Chemical Producers. Stanford Research Institute, 855 Oak Grove, Menlo Park, CA 94025. Four volumes published continuously on a quarterly installment basis listing a total of 1,600 chemical producers and 10,000 individual commercial chemicals, arranged alphabetically by company, product, and region.

Directory of Federally Supported Information Analysis Centers, Third Edition, Library of Congress, Science and Technology Division, National Referral Center, Washington, DC 20540, 1974.

Directory of Governmental Occupational Safety and Health Personnel. National Institute for Occupational Safety and Health, Rockville, MD. Annual.

A Directory of Information Resources in the United States. Library of Congress, Science and Technology Division, National Referral Center, Washington, DC 20540. Biological Sciences, 1972. Federal Government, 1967. General Toxicology, 1969. Physical Sciences Engineering, 1971. Informative subject-indexed directories to information resources in science. Distributed by the Superintendent of Documents.

Directory of State Control Officials, in *Environment Index.* Environment Information Center, Inc., 124 East 39th Street, New York, NY 10016. Annual.

International Directory of Occupational Safety and Health Services and Institutions. International Labour Office, Geneva, 1969.

NIH Public Advisory Groups, DHEW Publication No. (NIH) 72-11. U.S. Public Health Services, National Institutes of Health. 1972.

Occupational Safety and Health Consultants. National Institute for Occupational Safety and Health, Denver, 1974. 1974 marks the last printing of this list in order to avoid duplication of similar lists published by the American Industrial Hygiene Association and the American Society of Safety Engineers (included here).

Peters, G. A. (ed.): *National Directory of Safety Consultants.* Third Edition. American Society of Safety Engineers, 850 Busse Highway, Park Ridge, IL 60068, 1974.

Schildhauer, C.: *Environmental Information Sources—Engineering and Industrial Applications. A Selected Annotated Bibliography.* Special Libraries Association, 235 Park Avenue South, New York, NY 10003, 1972. Well annotated bibliography of all types of publications in the field of environmental health.

State and Local Environmental Libraries, A Directory. U.S. Environmental Protection Agency, Washington, DC 1973.

OSHA

Binford, C., C. Fleming, and Z. Prust. *Loss Control in the OSHA Era,* McGraw-Hill, New York, 1975.

Occupational Safety and Health Administration, various publications on all aspects.

OSHA Requirements Guide, Equipment Guide Book Co., Palo Alto, Calif., 1976.

Petersen, D. *The OSHA Compliance Manual,* 2d ed., McGraw-Hill, New York, 1979.

Roberts, J. *OSHA Compliance Manual,* Reston, Reston, VA 1976.

Bibliographies

Guide to Occupational Safety Literature, *National Safety Council, Chicago, 1975.*

A Selected Bibliography of Reference Materials in Safety Engineering, *American Society of Safety Engineers,* Chicago, 1967.

U.S. Department of Labor, *Occupational Safety and Health, A Bibliography,* Washington, 1974.

Supervision and Pyschology

Argyris, C. *Interpersonal Competence and Organizational Effectiveness,* Irwin-Dorsey, Hollywood, IL, 1962.

Argyris, C. *Personality and Organization,* Harper, New York, 1957.

Bass, B. M. and V. A. Vaughn. *Training in Industry: The Management of Learning,* Wadsworth, Belmont, CA 1966.

Beaumont, Henry. *Psychology Applied to Personnel,* Longmans, New York, 1946.

Behavioral Science, National Industrial Conference Board, 1969.

Berne, E. *Games People Play,* Grove Press, New York, 1964.

Blake, R. and J. Mouton. *The Managerial Grid,* Gulf, Houston, TX, 1964.

Carlson, S. *Executive Behavior,* Strombergs, Stockholm, 1951.

Dalton, M. *Men Who Manage,* Wiley, New York, 1959.

Drucker, P. *The Practice of Management,* Harper, 1954.

Ferster, C. and M. Perrott. *Behavior Principles,* Appleton-Century-Crofts, New York, 1968.

Gausch, J. P. "Balanced Involvement," Monograph 3, American Society of Safety Engineers, Chicago, 1973.

Gellerman, S. *Motivation and Productivity,* American Management Association, New York, 1963.

Ghiselli, E. and C. Brown. *Personnel and Industrial Psychology,* McGraw-Hill,

Gulton, R. *Personnel Testing,* McGraw-Hill, New York, 1965.

Harris, T. *I'm Ok–You're Ok,* Harper, New York, 1971.

Heinrich, H. W. *Basics of Supervision,* and *Supervisor's Safety Manual,* Alfred Best and Company, New York, 1944; also *Fomula for Supervision,* National Foremen's Institute, New London, CT 1949.

Herzberg, F. *Work and the Nature of Man,* World Publishing, Cleveland, 1966.

Herzberg, F., B. Mausner, R. Peterson, and D. Capwell. *Job Attitudes: Review of Research and Opinion,* Psychological Service of Pittsburgh, Pittsburgh, 1957.

Hovland, C., L. Janis, and H. Kelley. *Communication and Persuasion,* Yale University Press, New Haven, CT 1953.

Humble, J. *Management by Objectives in Action,* McGraw-Hill, New York, 1970.

Kalsem, Palmer J. *Practical Supervision,* McGraw-Hill, New York, 1945.

Katz, D., N. Maccoby, G. Gurin, and L. Floor. *Productivity, Supervision and Morale among Railroad Workers,* Institute for Social Research, Ann Arbor, MI, 1951.

Katz, D., N., Maccoby, and N. Morse. *Productivity Supervision and Morale in an Office Situation,* Institute for Social Research, Ann Arbor, MI, 1950.

Kelly, J. *Organizational Behavior,* Irwin-Dorsey, Homewood, IL, 1969.

Larson, J. *The Human Element in Industrial Accident Prevention,* New York, University, New York, 1955.

Levinson, H. *The Great Jackass Fallacy,* Harvard Business School, Cambridge, MA, 1973.

Lewin, K. *Field Theory in Social Science,* Harper, New York, 1951.

Likert, R. *The Human Organization,* McGraw-Hill, New York, 1967.

Likert, R. *New Patterns of Management,* McGraw-Hill, New York, 1961.

Lindzay, G., and E. Aronsen. *The Handbook of Social Psychology,* Addison-Wesley, Cambridge, MA, 1968.

Mager, R. *Developing an Attitude toward Learning,* Fearon, Belmont, CA, 1968.

Mager, R. and P. Pipe. *Analyzing Performance Problems or You Really Oughta Wanna,* Fearon, Belmont, CA, 1970.

Maier, N. *Psychology in Industry,* Houghton Mifflin, Boston, 1965.

McClelland, D., J. Atkinson, R. Clark, and R. Lowell. *The Achievement Motive,* Appleton-Century-Crofts, New York, 1953.

McGehee, W. and P. Thayer. *Training in Business and Industry.* Wiley, New York, 1961.

McGregor, D. *The Human Side of Enterprise,* McGraw-Hill, New York, 1960.

Odiorne, G., *Management by Objectives,* Pitman, New York, 1965.

Petersen, D. *Safety Supervision,* Amacom, New York, 1976.

Porter, L., and E. Lawler. *Managerial Attitudes and Performance,* Irwin-Dorsey, Homewood, IL, 1968.

Safety Communications, National Safety Council, Chicago, 1975.

Sayles, L. *Managerial Behavior,* McGraw-Hill, New York, 1964.

Schulzinger, M. S. *Accident Syndrome,* Charles C. Thomas, Springfield, IL, 1956.

Skinner, B. *Science and Human Behavior,* Macmillan, New York, 1963.

Supervisors Safety Manual, National Safety Council, Chicago, 1967.

Vernon, H. *Accidents and Their Prevention,* Macmillan, New York, 1937.

Viteles, Morris S. *Industrial Psychology,* Norton, New York, 1932.

Weaver, D. A. *Strengthening Supervising Skills,* Employers Insurance of Wausau, Wausau, WI, 1964.

Index